Next-Generation Big Data

A Practical Guide to Apache Kudu, Impala, and Spark

Butch Quinto

Next-Generation Big Data: A Practical Guide to Apache Kudu, Impala, and Spark

Butch Quinto
Plumpton, Victoria, Australia

ISBN-13 (pbk): 978-1-4842-3146-3
https://doi.org/10.1007/978-1-4842-3147-0

ISBN-13 (electronic): 978-1-4842-3147-0

Library of Congress Control Number: 2018947173

Copyright © 2018 by Butch Quinto

This work is subject to copyright. All rights are reserved by the Publisher, whether the whole or part of the material is concerned, specifically the rights of translation, reprinting, reuse of illustrations, recitation, broadcasting, reproduction on microfilms or in any other physical way, and transmission or information storage and retrieval, electronic adaptation, computer software, or by similar or dissimilar methodology now known or hereafter developed.

Trademarked names, logos, and images may appear in this book. Rather than use a trademark symbol with every occurrence of a trademarked name, logo, or image we use the names, logos, and images only in an editorial fashion and to the benefit of the trademark owner, with no intention of infringement of the trademark.

The use in this publication of trade names, trademarks, service marks, and similar terms, even if they are not identified as such, is not to be taken as an expression of opinion as to whether or not they are subject to proprietary rights.

While the advice and information in this book are believed to be true and accurate at the date of publication, neither the authors nor the editors nor the publisher can accept any legal responsibility for any errors or omissions that may be made. The publisher makes no warranty, express or implied, with respect to the material contained herein.

Managing Director, Apress Media LLC: Welmoed Spahr
Acquisitions Editor: Susan McDermott
Development Editor: Laura Berendson
Coordinating Editor: Rita Fernando

Cover designed by eStudioCalamar

Cover image designed by Freepik (www.freepik.com)

Distributed to the book trade worldwide by Springer Science+Business Media New York, 233 Spring Street, 6th Floor, New York, NY 10013. Phone 1-800-SPRINGER, fax (201) 348-4505, e-mail orders-ny@springer-sbm.com, or visit www.springeronline.com. Apress Media, LLC is a California LLC and the sole member (owner) is Springer Science + Business Media Finance Inc (SSBM Finance Inc). SSBM Finance Inc is a **Delaware** corporation.

For information on translations, please e-mail rights@apress.com, or visit http://www.apress.com/rights-permissions.

Apress titles may be purchased in bulk for academic, corporate, or promotional use. eBook versions and licenses are also available for most titles. For more information, reference our Print and eBook Bulk Sales web page at http://www.apress.com/bulk-sales.

Any source code or other supplementary material referenced by the author in this book is available to readers on GitHub via the book's product page, located at www.apress.com/9781484231463. For more detailed information, please visit http://www.apress.com/source-code.

Printed on acid-free paper

This book is dedicated to my wife, Aileen; and my children, Matthew, Timothy, and Olivia.

Table of Contents

About the Author xvii

About the Technical Reviewer xix

Acknowledgments xxi

Introduction xxiii

Chapter 1: Next-Generation Big Data 1
    About This Book 2
    Apache Spark 2
    Apache Impala 3
    Apache Kudu 3
    Navigating This Book 3
    Summary 5

Chapter 2: Introduction to Kudu 7
    Kudu Is for Structured Data 9
    Use Cases 9
        Relational Data Management and Analytics 10
        Internet of Things (IoT) and Time Series 11
        Feature Store for Machine Learning Platforms 12
    Key Concepts 12
    Architecture 13
    Multi-Version Concurrency Control (MVCC) 14
    Impala and Kudu 15
        Primary Key 15
        Data Types 16
        Partitioning 17

TABLE OF CONTENTS

Spark and Kudu 19
    Kudu Context 19

Kudu C++, Java, and Python Client APIs 24
    Kudu Java Client API 24
    Kudu Python Client API 27
    Kudu C++ Client API 29

Backup and Recovery 34
    Backup via CTAS 34

Copy the Parquet Files to Another Cluster or S3 35
    Export Results via impala-shell to Local Directory, NFS, or SAN Volume 36
    Export Results Using the Kudu Client API 36
    Export Results with Spark 38
    Replication with Spark and Kudu Data Source API 38
    Real-Time Replication with StreamSets 40
    Replicating Data Using ETL Tools Such as Talend, Pentaho, and CDAP 41

Python and Impala 43
    Impyla 43
    pyodbc 44
    SQLAlchemy 44

High Availability Options 44
    Active-Active Dual Ingest with Kafka and Spark Streaming 45
    Active-Active Kafka Replication with MirrorMaker 45
    Active-Active Dual Ingest with Kafka and StreamSets 46
    Active-Active Dual Ingest with StreamSets 47

Administration and Monitoring 47
    Cloudera Manager Kudu Service 47
    Kudu Master Web UI 47
    Kudu Tablet Server Web UI 48
    Kudu Metrics 48
    Kudu Command-Line Tools 48

Known Issues and Limitations 51

Security	52
Summary	53
References	53

Chapter 3: Introduction to Impala 57

Architecture	57
Impala Server Components	58
Impala SQL	63
Data Types	63
SQL Statements	64
SET Statements	71
SHOW Statements	72
Built-In Functions	74
User-Defined Functions	76
Complex Types in Impala	76
Querying Struct Fields	77
Querying Deeply Nested Collections	78
Querying Using ANSI-92 SQL Joins with Nested Collections	79
Impala Shell	79
Performance Tuning and Monitoring	84
Explain	85
Summary	85
Profile	86
Cloudera Manager	87
Impala Performance Recommendations	93
Workload and Resource Management	95
Admission Control	95
Hadoop User Experience	96
Impala in the Enterprise	98
Summary	98
References	98

TABLE OF CONTENTS

Chapter 4: High Performance Data Analysis with Impala and Kudu 101
Primary Key 101
Data Types 102
Internal and External Impala Tables 103
    Internal Tables 103
    External Tables 104
Changing Data 104
    Inserting Rows 104
    Updating Rows 105
    Upserting Rows 105
    Deleting Rows 105
Changing Schema 106
Partitioning 106
    Hash Partitioning 106
    Range Partitioning 106
    Hash-Range Partitioning 107
    Hash-Hash Partitioning 108
    List Partitioning 108
Using JDBC with Apache Impala and Kudu 109
Federation with SQL Server Linked Server and Oracle Gateway 110
Summary 111
References 111

Chapter 5: Introduction to Spark 113
Overview 113
    Cluster Managers 114
Architecture 115
Executing Spark Applications 116
Spark on YARN 116
    Cluster Mode 116
    Client Mode 117

TABLE OF CONTENTS

Introduction to the Spark-Shell 117

    SparkSession 118

    Accumulator 119

    Broadcast Variables 119

    RDD 119

Spark SQL, Dataset, and DataFrames API 127

Spark Data Sources 129

    CSV 129

    XML 130

    JSON 131

    Relational Databases Using JDBC 132

    Parquet 136

    HBase 136

    Amazon S3 142

    Solr 142

    Microsoft Excel 143

    Secure FTP 144

Spark MLlib (DataFrame-Based API) 145

    Pipeline 146

    Transformer 146

    Estimator 146

    ParamGridBuilder 147

    CrossValidator 147

    Evaluator 147

Example 147

GraphX 152

Spark Streaming 152

Hive on Spark 152

Spark 1.x vs Spark 2.x 152

TABLE OF CONTENTS

Monitoring and Configuration 153
    Cloudera Manager 153
    Spark Web UI 154
Summary 157
References 157

Chapter 6: High Performance Data Processing with Spark and Kudu 159

Spark and Kudu 159
    Spark 1.6.x 159
    Spark 2.x 160
Kudu Context 160
    Inserting Data 161
    Updating a Kudu Table 162
    Upserting Data 163
    Deleting Data 164
    Selecting Data 165
    Creating a Kudu Table 165
    Inserting CSV into Kudu 166
    Inserting CSV into Kudu Using the spark-csv Package 166
    Insert CSV into Kudu by Programmatically Specifying the Schema 167
    Inserting XML into Kudu Using the spark-xml Package 168
    Inserting JSON into Kudu 171
    Inserting from MySQL into Kudu 173
    Inserting from SQL Server into Kudu 178
    Inserting from HBase into Kudu 188
    Inserting from Solr into Kudu 194
    Insert from Amazon S3 into Kudu 195
    Inserting from Kudu into MySQL 196
    Inserting from Kudu into SQL Server 198
    Inserting from Kudu into Oracle 201
    Inserting from Kudu to HBase 205

Inserting Rows from Kudu to Parquet	208
Insert SQL Server and Oracle DataFrames into Kudu	210
Insert Kudu and SQL Server DataFrames into Oracle	214
Spark Streaming and Kudu	218
Kudu as a Feature Store for Spark MLlib	222
Summary	228
References	228

Chapter 7: Batch and Real-Time Data Ingestion and Processing 231

StreamSets Data Collector	231
Pipelines	232
Origins	232
Processors	232
Destinations	232
Executors	233
Data Collector Console	233
Deployment Options	237
Using StreamSets Data Collector	237
Ingesting XML to Kudu	238
Configure Pipeline	242
Configure the Directory Origin	243
Configure the XML Parser Processor	246
Validate and Preview Pipeline	247
Start the Pipeline	251
Stream Selector	255
Expression Evaluator	265
Using the JavaScript Evaluator	274
Ingesting into Multiple Kudu Clusters	281
REST API	286
Event Framework	289
Dataflow Performance Manager	289

xi

TABLE OF CONTENTS

Other Next-Generation Big Data Integration Tools 290
Data Ingestion with Kudu 290
Pentaho Data Integration 306
Ingest CSV into HDFS and Kudu 306
Data Ingestion to Kudu with Transformation 328
SQL Server to Kudu 331
Talend Open Studio 341
Ingesting CSV Files to Kudu 342
SQL Server to Kudu 349
Data Transformation 355

Other Big Data Integration Players 359
Informatica 360
Microsoft SQL Server Integration Services 360
Oracle Data Integrator for Big Data 360
IBM InfoSphere DataStage 361
Syncsort 361
Apache NIFI 361

Data Ingestion with Native Tools 362
Kudu and Spark 362
Sqoop 369

Kudu Client API 370
MapReduce and Kudu 370
Summary 371
References 371

Chapter 8: Big Data Warehousing 375
Enterprise Data Warehousing in the Era of Big Data 376
Structured Data Still Reigns Supreme 376
EDW Modernization 376
ETL Offloading 378
Analytics Offloading and Active Archiving 379
Data Consolidation 379

Replatforming the Enterprise Data Warehouse	380
Big Data Warehousing 101	381
Dimensional Modeling	381
Big Data Warehousing with Impala and Kudu	384
Summary	405
References	405

Chapter 9: Big Data Visualization and Data Wrangling 407

Big Data Visualization	407
SAS Visual Analytics	408
Zoomdata	408
Self-Service BI and Analytics for Big Data	408
Real-Time Data Visualization	409
Architecture	409
Deep Integration with Apache Spark	410
Zoomdata Fusion	411
Data Sharpening	411
Support for Multiple Data Sources	412
Real-Time IoT with StreamSets, Kudu, and Zoomdata	426
Create the Kudu Table	426
Data Wrangling	445
Trifacta	447
Alteryx	455
Datameer	466
Summary	474
References	475

Chapter 10: Distributed In-Memory Big Data Computing 477

Architecture	478
Why Use Alluxio?	479
Significantly Improve Big Data Processing Performance and Scalability	480
Multiple Frameworks and Applications Can Share Data at Memory Speed	480

TABLE OF CONTENTS

    Provides High Availability and Persistence in Case of Application Termination or Failure 482

    Optimize Overall Memory Usage and Minimize Garbage Collection 486

    Reduce Hardware Requirements 486

    Alluxio Components 487

    Installation 487

Apache Spark and Alluxio 489

Administering Alluxio 489

    Master 489

    Worker 490

Apache Ignite 490

Apache Geode 491

Summary 491

References 491

Chapter 11: Big Data Governance and Management 495

Data Governance for Big Data 496

Cloudera Navigator 496

    Metadata Management 498

    Data Classification 499

    Data Lineage and Impact Analysis 500

    Auditing and Access Control 500

    Policy Enforcement and Data Lifecycle Automation 501

    Cloudera Navigator REST API 502

Cloudera Navigator Encrypt 502

Other Data Governance Tools 503

    Apache Atlas 503

    Informatica Metadata Manager and Enterprise Data Catalog 503

    Collibra 503

    Waterline Data 504

    Smartlogic 504

Summary 504

References 505

Chapter 12: Big Data in the Cloud 507
Amazon Web Services (AWS) 507
Microsoft Azure Services 507
Google Cloud Platform (GCP) 508
Cloudera Enterprise in the Cloud 509
Hybrid and Multi-Cloud 509
Transient Clusters 510
Persistent Clusters 510
Cloudera Director 511
Summary 532
References 532

Chapter 13: Big Data Case Studies 537
Navistar 537
Use Cases 537
Solution 538
Technology and Applications 538
Outcome 539
Cerner 539
Use Cases 539
Solution 539
Technology and Applications 540
Outcome 541
British Telecom 541
Use Cases 541
Solution 542
Technology and Applications 542
Outcome 542
Shopzilla (Connexity) 543
Use Cases 543
Solution 543
Technology and Applications 544
Outcome 544

TABLE OF CONTENTS

Thomson Reuters 544
    Use Cases 545
    Solution 545
    Technology and Applications 545
    Outcome 546
Mastercard 546
    Use Cases 546
    Solution 547
    Technology and Applications 547
    Outcome 547
Summary 547
References 547

Index 549

About the Author

Butch Quinto is Chief Data Officer at Lykuid, Inc. an advanced analytics company that provides an AI-powered infrastructure monitoring platform. As Chief Data Officer, Butch serves as the head of AI and data engineering, leading product innovation, strategy, research and development. Butch was previously Director of Analytics at Deloitte where he led strategy, solutions development and delivery, technology innovation, business development, vendor alliance and venture capital due diligence. While at Deloitte, Butch founded and developed several key big data, IoT and artificial intelligence applications including Deloitte's IoT Framework, Smart City Platform and Geo-Distributed Telematics Platform. Butch was also the co-founder and lead lecturer of Deloitte's national data science and big data training programs.

Butch has more than 20 years of experience in various technical and leadership roles at start-ups and Global 2000 corporations in several industries including banking and finance, telecommunications, government, utilities, transportation, e-commerce, retail, technology, manufacturing, and bioinformatics. Butch is a recognized thought leader and a frequent speaker at conferences and events. Butch is a contributor to the Apache Spark and Apache Kudu open source projects, founder of the Cloudera Melbourne User Group and was Deloitte's Director of Alliance for Cloudera.

About the Technical Reviewer

Irfan Elahi has years of multidisciplinary experience in Data Science and Machine Learning. He has worked in a number of verticals such as consultancy firms, his own start-ups, and academia research lab. Over the years he has worked on a number of data science and machine learning projects in different niches such as telecommunication, retail, Web, public sector, and energy with the goal to enable businesses to derive immense value from their data-assets.

Acknowledgments

I would like to thank everyone at Apress, particularly Rita Fernando Kim, Laura C. Berendson, and Susan McDermott for all the help and support in getting this book published. It was a pleasure working with the Apress team. Several people have contributed to this book directly and indirectly. Thanks to Matei Zaharia, Sean Owen, Todd Lipcon, Grant Henke, William Berkeley, David Alves, Harit Iplani, Hui Ting Ong, Deborah Wiltshire, Steve Tuohy, Haoyuan Li, John Goode, Rick Sibley, Russ Cosentino, Jobi George, Rupal Shah, Pat Patterson, Irfan Elahi, Duncan Lee, Lee Anderson, Steve Janz, Stu Scotis, Chris Lewin, Julian Savaridas, and Tim Nugent. Thanks to the entire Hadoop, Kudu, Spark, and Impala community. Last but not the least, thanks to my wife, Aileen; and children, Matthew, Timothy, and Olivia.

Introduction

This book serves as a practical guide on how to utilize big data to store, process, and analyze structured data, focusing on three of the most popular Apache projects in the Hadoop ecosystem: Apache Spark, Apache Impala, and Apache Kudu (incubating). Together, these three Apache projects can rival most commercial data warehouse platforms in terms of performance and scalability at a fraction of the cost. Most next-generation big data and data science use cases are driven by structured data, and this book will serve as your guide.

I approach this book from an enterprise point of view. I cover not just the main technologies, I also examine advanced enterprise topics. This includes data governance and management, big data in the cloud, in-memory computing, backup and recovery, high availability, Internet of Things (IoT), data wrangling, and real-time data ingestion and visualization. I also discuss integration with popular third-party commercial applications and platforms from popular software vendors such as Oracle, Microsoft, SAS, Informatica, StreamSets, Zoomdata, Talend, Pentaho, Trifacta, Alteryx, Datameer, and Cask. For most of us, integrating big data with existing business intelligence and data warehouse infrastructure is a fact of life. Last but definitely not the least, I discuss several interesting big data case studies from some of the most innovative companies in the world including Navistar, Cerner, British Telecom, Shopzilla (Connexity), Thomson Reuters, and Mastercard.

It is not the goal of this book to provide a comprehensive coverage of every feature of Apache Kudu, Impala, and Spark. Instead, my goal is to provide real-world advice and practical examples on how to best leverage these components together to enable innovative enterprise use cases.

CHAPTER 1

Next-Generation Big Data

Despite all the excitement around big data, the large majority of mission-critical data is still stored in relational database management systems. This fact is supported by recent studies online and confirmed by my own professional experience working on numerous big data and business intelligence projects. Despite widespread interest in unstructured and semi-structured data, structured data still represents a significant percentage of data under management for most organizations, from the largest corporations and government agencies to small businesses and technology start-ups. Use cases that deals with unstructured and semi-structured data, while valuable and interesting, are few and far between. Unless you work for a company that does a lot of unstructured data processing such as Google, Facebook, or Apple, you are most likely working with structured data.

Big data has matured since the introduction of Hadoop more than 10 years ago. Take away all the hype, and it is evident that structured data processing and analysis has become the *next-generation* killer use case for big data. Most big data, business intelligence, and advanced analytic use cases deal with structured data. In fact, some of the most popular advances in big data such as Apache Impala, Apache Phoenix, and Apache Kudu as well as Apache Spark's recent emphasis on Spark SQL and DataFrames API are all about providing capabilities for structured data processing and analysis. This is largely due to big data finally being accepted as part of the enterprise. As big data platforms improved and gained new capabilities, they have become suitable alternatives to expensive data warehouse platforms and relational database management systems for storing, processing, and analyzing mission-critical structured data.

CHAPTER 1   NEXT-GENERATION BIG DATA

About This Book

This book is for business intelligence and data warehouse professionals who are interested in gaining practical and real-world insight into next-generation big data processing and analytics using Apache Kudu, Apache Impala, and Apache Spark. Experienced big data professionals who would like to learn more about Kudu and other advanced enterprise topics such as real-time data ingestion and complex event processing, Internet of Things (IoT), distributed in-memory computing, big data in the cloud, big data governance and management, real-time data visualization, data wrangling, data warehouse optimization, and big data warehousing will also benefit from this book.

I assume readers will have basic knowledge of the various components of Hadoop. Some knowledge of relational database management systems, business intelligence, and data warehousing is also helpful. Some programming experience is required if you want to run the sample code provided. I focus on three main Hadoop components: Apache Spark, Apache Impala, and Apache Kudu.

Apache Spark

Apache Spark is the next-generation data processing framework with advanced in-memory capabilities and a directed acyclic graph (DAG) engine. It can handle interactive, real-time, and batch workloads with built-in machine learning, graph processing, streaming, and SQL support. Spark was developed to address the limitation of MapReduce. Spark can be 10–100x faster than MapReduce in most data processing tasks. It has APIs for Scala, Java, Python, and R. Spark is one of the most popular Apache projects and is currently used by some of the largest and innovative companies in the world. I discuss Apache Spark in Chapter 5 and Spark and Kudu integration in Chapter 6.

Apache Impala

Apache Impala is a massively parallel processing (MPP) SQL engine designed to run on Hadoop platforms. The project was started by Cloudera and eventually donated to the Apache Software Foundation. Impala rivals traditional data warehouse platforms in terms of performance and scalability and was designed for business intelligence and OLAP workloads. Impala is compatible with some of the most popular BI and data visualization tools such as Tableau, Qlik, Zoomdata, Power BI, and MicroStrategy to mention a few. I cover Apache Impala in Chapter 3 and Impala and Kudu integration in Chapter 4.

Apache Kudu

Apache Kudu is a new mutable columnar storage engine designed to handle fast data inserts and updates and efficient table scans, enabling real-time data processing and analytic workloads. When used together with Impala, Kudu is ideal for Big Data Warehousing, EDW modernization, Internet of Things (IoT), real-time visualization, complex event processing, and feature store for machine learning. As a storage engine, Kudu's performance and scalability rivals other columnar storage format such as Parquet and ORC. It also performs significantly faster than Apache Phoenix with HBase. I discuss Kudu in Chapter 2.

Navigating This Book

This book is structured in easy-to-digest chapters that focus on one or two key concepts at a time. Chapters 1 to 9 are designed to be read in order, with each chapter building on the previous. Chapters 10 to 13 can be read in any order depending on your interest. The chapters are filled with practical examples and step-by-step instructions. Along the way, you'll find plenty of practical information on best practices and advice that will steer you to the right direction in your big data journey.

Chapter 1 - Next-Generation Big Data provides a brief introduction about the contents of this book.

Chapter 2 - Introduction to Kudu provides an introduction to Apache Kudu, starting with a discussion of Kudu's architecture. I talk about various topics such as how to access Kudu from Impala, Spark, and Python, C++ and Java using the client API. I provide details on how to administer, configure, and monitor Kudu, including backup and recovery and high availability options for Kudu. I also discuss Kudu's strength and limitations, including practical workarounds and advice.

Chapter 3 - Introduction to Impala provides an introduction to Apache Impala. I discuss Impala's technical architecture and capabilities with easy-to-follow examples. I cover details on how to perform system administration, monitoring, and performance tuning.

Chapter 4 - High Performance Data Analysis with Impala and Kudu covers Impala and Kudu integration with practical examples and real-world advice on how to leverage both components to deliver a high performance environment for data analysis. I discuss Impala and Kudu's strength and limitations, including practical workarounds and advice.

Chapter 5 - Introduction to Spark provides an introduction to Apache Spark. I cover Spark's architecture and capabilities, with practical explanations and easy-to-follow examples to help you get started with Spark development right away.

Chapter 6 - High Performance Data Processing with Spark and Kudu covers Spark and Kudu integration with practical examples and real-world advice on how to use both components for large-scale data processing and analysis.

Chapter 7 - Batch and Real-Time Data Ingestion and Processing covers batch and real-time data ingestion and processing using native and third-party commercial tools such as Flume, Kafka, Spark Streaming, StreamSets, Talend, Pentaho, and Cask. I provide step-by-step examples on how to implement complex event processing and the Internet of Things (IoT).

Chapter 8 - Big Data Warehousing covers designing and implementing star and snowflake dimensional models with Impala and Kudu. I talk about how to utilize Impala and Kudu for data warehousing including its strengths and limitations. I also discuss EDW modernization use cases such as data consolidation, data archiving, and analytics and ETL offloading.

Chapter 9 – Big Data Visualization and Data Wrangling discusses real-time data visualization and wrangling tools designed for extremely large data sets with easy-to-follow examples and advice.

Chapter 10 – Distributed In-Memory Big Data Computing covers Alluxio, previously known as Tachyon. I discuss its architecture and capabilities. I also discuss Apache Ignite and Geode.

Chapter 11 – Big Data Governance and Management covers big data governance and management. I discuss data lineage, metadata management, auditing, and policy enforcement using Cloudera Navigator. I also examine other popular data governance and metadata management applications.

Chapter 12 – Big Data in the Cloud covers deploying and using Apache Kudu, Spark, and Impala in the cloud with step-by-step instructions and examples.

Chapter 13 – Big Data Case Studies provides six innovative big data case studies, including details about challenges, implementation details, solutions, and outcomes. The case studies are provided with permission from Cloudera.

Summary

I suggest you set up your own Cloudera cluster as a development environment if you want to follow along the examples in this book. You can also use the latest version of the Cloudera Quickstart VM, freely downloadable from Cloudera's website. I do not recommend using a different data platform such as Hortonworks, MapR, EMR, or Databricks since they are not compatible with the other components discussed in this book such as Impala and Kudu.

CHAPTER 2

Introduction to Kudu

Kudu is an Apache-licensed open source columnar storage engine built for the Apache Hadoop platform. It supports fast sequential and random reads and writes, enabling real-time stream processing and analytic workloads.[i] It integrates with Impala, allowing you to insert, delete, update, upsert, and retrieve data using SQL. Kudu also integrates with Spark (and MapReduce) for fast and scalable data processing and analytics. Like other projects in the Apache Hadoop ecosystem, Kudu runs on commodity hardware and was designed to be highly scalable and highly available.

The Apache Kudu project was founded in 2012 by Todd Lipcon, a software engineer at Cloudera and PMC member and committer on the Hadoop, HBase, and Thrift projects.[ii] Kudu was developed to address the limitations of HDFS and HBase while combining both of its strengths. While HDFS supports fast analytics and large table scans, files stored in HDFS are immutable and can only be appended to after they are created.[iii] HBase makes it possible to update and randomly access data, but it's slow for analytic workloads. Kudu can handle both high velocity data and real-time analytics, allowing you to update Kudu tables and run analytic workloads at the same time. Batch processing and analytics on HDFS are still slightly faster than Kudu in some cases and HBase beats Kudu in random reads and writes performance. Kudu is somewhere in the middle. As shown in Figure 2-1, Kudu's performance is close enough to HDFS with Parquet (Kudu is faster in some cases) and HBase in terms of random reads and writes so that most of the time the performance difference is negligible.

CHAPTER 2 INTRODUCTION TO KUDU

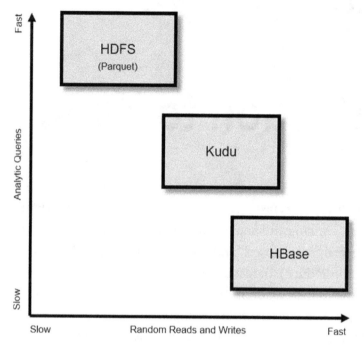

Figure 2-1. *High-level performance comparison of HDFS, Kudu, and HBase*

Prior to Kudu, some data engineers used a data processing architecture called the Lambda architecture to work around the limitations of HDFS and HBase. The Lambda architecture works by having a speed and batch layer (and technically, there's also a serving layer). Transaction data goes to the speed layer (usually HBase) where users get immediate access to the latest data. Data from the speed layer is copied at regular intervals (hourly or daily) to the batch layer (usually HDFS) in Parquet format, to be utilized for reporting and analytics. As you can see in Figure 2-2, data is copied twice and the data pipline is more complicated than necessary with the Lambda architecture. This is somewhat similar to a typical enterprise data warehouse environment with OLTP databases representing the "speed layer" and the data warehouse acting as the "batch layer."

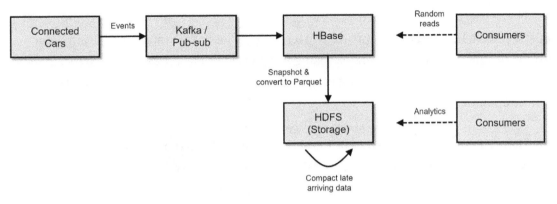

Figure 2-2. *Lambda Architecture*

Kudu makes the Lambda architecture obsolete due to its ability to simultaneously handle random reads and writes and analytic workloads. With Kudu, there is no data duplication and the data pipeline is considerably simpler, as shown in Figure 2-3.

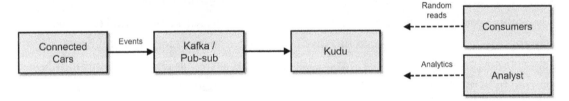

Figure 2-3. *Modern data ingest pipeline using Kudu*

Kudu Is for Structured Data

Kudu was designed to store structured data similar to relational databases. In fact, Kudu (when used with Impala) is often used for relational data management and analytics. Kudu rivals commercial data warehouse platforms in terms of capabilities, performance, and scalability. We'll discuss Impala and Kudu integration later in the chapter and more thoroughly in Chapter 4.

Use Cases

Before we begin, let's talk about what Kudu is not. Kudu is not meant to replace HBase or HDFS. HBase is a schema-less NoSQL-style data store that makes it suitable for sparse data or applications that requires variable schema. HBase was designed for OLTP-type workloads that requires random reads and writes. For more information on HBase, see the HBase online documentation.

HDFS was designed to store all types of data: structured, semi-structured, and unstructured. If you need to store data in a highly scalable file system, HDFS is a great option. As mentioned earlier, HDFS (using Parquet) is still faster in some cases than Kudu when it comes to running analytic workloads. For more on HDFS, see the HDFS online documentation.

As discussed earlier, Kudu excels at storing structured data. It doesn't have an SQL interface, therefore you need to pair Kudu with Impala. Data that you would normally think of storing in a relational or time series database can most likely be stored in Kudu as well. Below are some use cases where Kudu can be utilized.[iv]

Relational Data Management and Analytics

Kudu (when used with Impala) exhibits most of the characteristics of a relational database. It stores data in rows and columns and organizes them in databases and tables. Impala provides a highly scalable MPP SQL engine and allows you to interact with Kudu tables using ANSI SQL commands just as you would with a relational database. Relational database use cases can be classified into two main categories, online transactional processing (OLTP) and decision support systems (DSS) or as commonly referred to in modern nomenclature, data warehousing. Kudu was not designed for OLTP, but it can be used for data warehousing and other enterprise data warehouse (EDW) modernization use cases.

Data Warehousing

Kudu can be used for dimensional modeling – the basis of modern data warehousing and online analytic processing (OLAP). Kudu lacks foreign key constraints, auto-increment columns, and other features that you would normally find in a traditional data warehouse platform; however these limitations do not preclude you from organizing your data in facts and dimensions tables. Impala can be accessed using your favorite BI and OLAP tools via ODBC/JDBC. I discuss data warehousing using Impala and Kudu in Chapter 8.

ETL Offloading

ETL offloading is one of the many EDW optimization use cases that you can use Kudu for. Critical reports are unavailable to the entire organization due to ETL processes running far beyond its processing window and pass into the business hours. By offloading time-consuming ETL processing to an inexpensive Kudu cluster, ETL jobs can finish before business hours, making critical reports and analytics available to business users when they need it. I discuss ETL offloading using Impala and Kudu in Chapter 8.

Analytics Offloading and Active Archiving

Impala is an extremely fast and scalable MPP SQL engine. You can reduce the load on your enterprise data warehouse by redirecting some of your ad hoc queries and reports to Impala and Kudu. Instead of spending millions of dollars upgrading your data warehouse, analytics offloading and active archiving is the smarter and more cost-effective way to optimize your EDW environment. I discuss analytics offloading and active archiving using Impala and Kudu in Chapter 8.

Data Consolidation

It's not unusual for large organizations to have hundreds or thousands of legacy databases scattered across its enterprise, paying millions of dollars in licensing, administration and infrastructure cost. By consolidating these databases into a single Kudu cluster and using Impala to provide SQL access, you can significantly reduce cost while improving performance and scalability. I discuss data consolidation using Impala and Kudu in Chapter 8.

Internet of Things (IoT) and Time Series

Kudu is perfect for IoT and time series applications where real-time data ingestion, visualization, and complex event processing of sensor data is critical. Several large companies and government agencies such as Xiaomi, JD.com,[v] and Australia Department of Defense[vi] are successfully using Kudu for IoT use cases. I discuss IoT, real-time data ingestion, and complex event processing using Impala, Kudu, and StreamSets in Chapter 7. I discuss real-time data visualization with Zoomdata in Chapter 9.

CHAPTER 2 INTRODUCTION TO KUDU

Feature Store for Machine Learning Platforms

Data science teams usually create a centralized feature store where they can publish and share highly selected sets of authoritative features with other teams for creating machine learning models. Creating and maintaining feature stores using immutable data formats such as ORC and Parquet is time consuming, cumbersome, and requires too much unnecessary hard work, especially for large data sets. Using Kudu as a fast and highly scalable mutable feature store, data scientists and engineers can easily update and add features using familiar SQL statements. The ability to update feature stores in seconds or minutes is critical in an Agile environment where data scientists are constantly iterating in building, testing, and improving the accuracy of their predictive models. In Chapter 6, we use Kudu as a feature store for building a predictive machine learning model using Spark MLlib.

Note Kudu allows up to a maximum of 300 columns per table. HBase is a more appropriate storage engine if you need to store more than 300 features. HBase tables can contain thousands or millions of columns. The downside in using HBase is that it is not as efficient in handling full table scans compared to Kudu. There is discussion within the Apache Kudu community to address the 300-column limitation in future versions of Kudu.

Strictly speaking, you can bypass Kudu's 300-column limit by setting an unsafe flag. For example, if you need the ability to create a Kudu table with 1000 columns, you can start the Kudu master with the following flags: --unlock-unsafe-flags --max-num-columns=1000. This has not been thoroughly tested by the Kudu development team and is therefore not recommended for production use.

Key Concepts

Kudu introduces a few concepts that describe different parts of its architecture.

Table A table is where data is stored in Kudu. Every Kudu table has a primary key and is divided into segments called tablets.

Tablet A tablet, or partition, is a segment of a table.

Tablet Server A tablet server stores and serves tablets to clients.

Master A master keeps track of all cluster metadata and coordinates metadata operations.

Catalog Table Central storage for all of cluster metadata. The catalog table stores information about the location of tables and tablets, their current state, and number of replicas. The catalog table is stored in the master.

Architecture

Similar to the design of other Hadoop components such as HDFS and HBase (and their Google counterparts, BigTable and GFS), Kudu has a master-slave architecture. As shown in Figure 2-4, Kudu comprises one or more Master servers responsible for cluster coordination and metadata management. Kudu also has one or more tablet servers, storing data and serving them to client applications.[vii] For a tablet, there can only be one acting master, the leader, at any given time. If the leader becomes unavailable, another master is elected to become the new leader. Similar to the master, one tablet server acts as a leader, and the rest are followers. All write request go to the leader, while read requests go to the leader or replicas. Data stored in Kudu is replicated using the Raft Consensus Algorithm, guaranteeing the availability of data will survive the loss of some of the replica as long as the majority of the total number of replicas is still available. Whenever possible, Kudu replicates logical operations instead of actual physical data, limiting the amount of data movement across the cluster.

Note The Raft Consensus Algorithm is described in detail in "The Raft Paper": In Search of an Understandable Consensus Algorithm (Extended Version) by Diego Ongaro and John Ousterhout.[viii] Diego Ongaro's PhD dissertation, "Consensus: Bridging Theory and Practice," published by Stanford University in 2014, expands on the content of the paper in more detail.[ix]

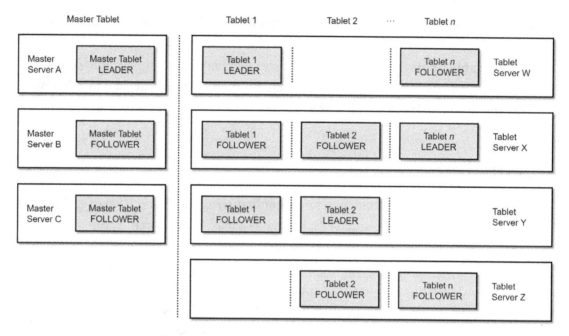

Figure 2-4. *Kudu Architecture*

Multi-Version Concurrency Control (MVCC)

Most modern databases use some form of concurrency control to ensure read consistency instead of traditional locking mechanisms. Oracle has a multi-version consistency model since version 6.0.[x] Oracle uses data maintained in the rollback segments to provide read consistency. The rollback segments contain the previous data that have been modified by uncommitted or recently committed transactions.[xi] MemSQL and SAP HANA manages concurrency using MVCC as well. Originally, SQL Server only supported a pessimistic concurrency model, using locking to enforce concurrency. As a result, readers block writers and writers block readers. The likelihood of blocking problems and lock contention increase as the number of concurrent users and operations rise, leading to performance and scalability issues. Things became so bad in SQL Server-land that developers and DBAs were forced to use the NOLOCK hint in their queries or set the READ UNCOMITTED isolation level, tolerating dirty reads in exchange for a minor performance boost. Starting in SQL Server 2005, Microsoft introduced its own version of multi-version concurrency control known as row-level versioning.[xii] SQL Server doesn't have the equivalent of rollback segments so it uses tempdb to store previously committed data. Teradata does not have multi-version consistency model and relies on transactions and locks to enforce concurrency control.[xiii]

Similar to Oracle, MemSQL, and SAP HANA, Kudu uses multi-version concurrency control to ensure read consistency.[xiv] Readers don't block writers and writers don't block readers. Kudu's optimistic concurrency model means that operations are not required to acquire locks during large full table scans, considerably improving query performance and scalability.

Impala and Kudu

Impala is the default MPP SQL engine for Kudu. Impala allows you to interact with Kudu using SQL. If you have experience with traditional relational databases where the SQL and storage engines are tightly integrated, you might find it unusual that Kudu and Impala are decoupled from each other. Impala was designed to work with other storage engines such as HDFS, HBase, and S3, not just Kudu. There's also work underway to integrate other SQL engines such as Apache Drill (DRILL-4241) and Hive (HIVE-12971) with Kudu. Decoupling storage, SQL, and processing engines are common practices in the open source community.

The Impala-Kudu integration works great but there is still work to be done. While it matches or exceeds traditional data warehouse platforms in terms of performance and scalability, Impala-Kudu lacks some of the enterprise features found in most traditional data warehouse platforms. We discuss some of these limitations later in the chapter.

Primary Key

Every Kudu table needs to have a primary key. Kudu's primary key is implemented as a clustered index. With a clustered index, the rows are stored physically in the tablet in the same order as the index. Also note that Kudu doesn't have an auto-increment feature so you will have to include a unique primary key value when inserting rows to a Kudu table. If you don't have a primary key value, you can use Impala's built-in uuid() function or another method to generate a unique value.

CHAPTER 2 INTRODUCTION TO KUDU

Data Types

Like other relational databases, Kudu supports various data types (Table 2-1).

Table 2-1. List of Data Types, with Available and Default Encoding

Data Type	Encoding	Default
boolean	plain, run length	run length
8-bit signed integer	plain, bitshuffle, run length	bitshuffle
16-bit signed integer	plain, bitshuffle, run length	bitshuffle
32-bit signed integer	plain, bitshuffle, run length	bitshuffle
64-bit signed integer	plain, bitshuffle, run length	bitshuffle
unixtime_micros (64-bit microseconds since the Unix epoch)	plain, bitshuffle, run length	bitshuffle
single-precision (32-bit) IEEE-754 floating-point number	plain, bitshuffle	bitshuffle
double-precision (64-bit) IEEE-754 floating-point number	plain, bitshuffle	bitshuffle
UTF-8 encoded string (up to 64KB uncompressed)	plain, prefix, dictionary	dictionary
binary (up to 64KB uncompressed)	plain, prefix, dictionary	dictionary

You may notice that Kudu currently does not support the decimal data type. This is a key limitation in Kudu. The float and double data types only store a very close approximation of the value instead of the exact value as defined in the IEEE 754 specification.[xv] Because of this behaviour, float and double are not appropriate for storing financial data. At the time of writing, support for decimal data type is still under development (Apache Kudu 1.5 / CDH 5.13). Decimal support is coming in Kudu 1.7. Check KUDU-721 for more details. There are various workarounds available. You can store financial data as string then use Impala to cast the value to decimal every time you need to read the data. Since Parquet supports decimals, another workaround would be to use Parquet for your fact tables and Kudu for dimension tables.

As shown in Table 2-1, Kudu columns can use different encoding types depending on the type of column. Supported encoding types includes Plain, Bitshuffle, Run Length, Dictionary, and Prefix. By default, Kudu columns are uncompressed. Kudu supports column compression using Snappy, zlib, or LZ4 compression codecs. Consult Kudu's documentation for more details on Kudu encoding and compression support.

> **Note** In earlier versions of Kudu, date and time are represented as a BIGINT. You can use the TIMESTAMP data type in Kudu tables starting in Impala 2.9/CDH 5.12. However, there are several things to keep in mind. Kudu represents date and time columns using 64-bit values, while Impala represents date and time as 96-bit values. Nanosecond values generated by Impala are rounded when stored in Kudu. When reading and writing TIMESTAMP columns, there is an overhead converting between Kudu's 64-bit representation and Impala's 96-bit representation. There are two workarounds: use the Kudu client API or Spark to insert data, or continue using BIGINT to represent date and time.[xvi]

Partitioning

Table partitioning is a common way to enhance performance, availability, and manageability of Kudu tables. Partitioning allows tables to be subdivided into smaller segments, or tablets. Partitioning enables Kudu to take advantage of partition pruning by allowing access to tables at a finer level of granularity. Table partitioning is required for all Kudu tables and is completely transparent to applications. Kudu supports Hash, Range, and Composite Hash-Range and Hash-Hash partitioning. Below are a few examples of partitioning in Kudu.

Hash Partitioning

There are times when it is desirable to evenly distribute data randomly across partitions to avoid IO bottlenecks. With hash partitioning, data is placed in a partition based on a hashing function applied to the partitioning key. Not that you are not allowed to add partitions on hash partitioned tables. You will have to rebuild the entire hash partitioned table if you wish to add more partitions.

```
CREATE TABLE myTable (
 id BIGINT NOT NULL,
 name STRING,
 PRIMARY KEY(id)
)
PARTITION BY HASH PARTITIONS 4
STORED AS KUDU;
```

Range Partitioning

Range partitioning stores data in partitions based on predefined ranges of values of the partitioning key for each partition. Range partitioning enhances the manageability of the partitions by allowing new partitions to be added to the table. It also improves performance of read operations via partition pruning. One downside: range partitioning can cause hot spots if you insert data in partition key order.

```
CREATE TABLE myTable (
  year INT,
  deviceid INT,
  totalamt INT,
  PRIMARY KEY (deviceid, year)
)
PARTITION BY RANGE (year) (
  PARTITION VALUE = 2016,
  PARTITION VALUE = 2017,
  PARTITION VALUE = 2018
)
STORED AS KUDU;
```

Hash-Range Partitioning

Hash-Range partitioning combines the benefits while minimizing the limitations of hash and range partitioning. Using hash partitioning ensures write IO is spread evenly across tablet servers, while using range partitions ensure new tablets can be added to accommodate future growth.

```
CREATE TABLE myTable (
 id BIGINT NOT NULL,
 sensortimestamp BIGINT NOT NULL,
 sensorid INTEGER,
 temperature INTEGER,
 pressure INTEGER,
 PRIMARY KEY(rowid,sensortimestamp)
)
```

```
PARTITION BY HASH (id) PARTITIONS 16,
RANGE (sensortimestamp)
(
PARTITION unix_timestamp('2017-01-01') <= VALUES < unix_timestamp('2018-01-01'),
PARTITION unix_timestamp('2018-01-01') <= VALUES < unix_timestamp('2019-01-01'),
PARTITION unix_timestamp('2019-01-01') <= VALUES < unix_timestamp('2020-01-01')
)
STORED AS KUDU;
```

I discuss table partitioning in more detail in Chapter 4.

Spark and Kudu

Spark is the ideal data processing and ingestion tool for Kudu. Spark SQL and the DataFrame API makes it easy to interact with Kudu. I discuss Spark and Kudu integration in more detail in Chapter 6.

You use Spark with Kudu using the DataFrame API. You can use the --packages option in spark-shell or spark-submit to include kudu-spark dependency. You can also manually download the jar file from central.maven.org and include it in your --jars option. Use the kudu-spark2_2.11 artifact if you are using Spark 2 with Scala 2.11. For example:

```
spark-shell --packages org.apache.kudu:kudu-spark2_2.11:1.1.0
spark-shell --jars kudu-spark2_2.11-1.1.0.jar
```

Kudu Context

You use a Kudu context in order to execute DML statements against a Kudu table.[xvii] For example, if we need to insert data into a Kudu table:

```
import org.apache.kudu.spark.kudu._
val kuduContext = new KuduContext("kudumaster01:7051")
case class CustomerData(id: Long, name: String, age: Short)

val data = Array(CustomerData(101,"Lisa Kim",60), CustomerData(102,"Casey Fernandez",45))
```

```
val insertRDD = sc.parallelize(data)
val insertDF = sqlContext.createDataFrame(insertRDD)

insertDF.show
```

```
+----------+---------------+---+
|customerid|           name|age|
+----------+---------------+---+
|       101|       Lisa Kim| 60|
|       102|Casey Fernandez| 45|
+----------+---------------+---+
```

Insert the DataFrame into Kudu table. I assume the table already exists.

```
kuduContext.insertRows(insertDF, "impala::default.customers")
```

Confirm that the data was successfully inserted.

```
val df = sqlContext.read.options(Map("kudu.master" -> "kuducluster:7051","kudu.table" -> "impala::default.customers")).kudu
df.select("id","name","age").show()
```

```
+---+---------------+---+
| id|           name|age|
+---+---------------+---+
|102|Casey Fernandez| 45|
|101|       Lisa Kim| 60|
+---+---------------+---+
```

I discuss Spark and Kudu integration in more detail in Chapter 6.

Note Starting in Kudu 1.6, Spark performs better by taking advantage of scan locality. Spark will scan the closest tablet replica instead of scanning the leader, which could be in a different tablet server.

Spark Streaming and Kudu

In our example shown in Listing 2-1, we will use Flafka (Flume and Kafka) and Spark Streaming to read data from a Flume spooldir source, store it in Kafka, and processing and writing the data to Kudu with Spark Streaming.

A new stream processing engine built on Spark SQL was included in Spark 2.0 called Structured Streaming. Starting with Spark 2.2.0, the experimental tag from Structured Streaming has been removed. However, Cloudera still does not support Structured Streaming as of this writing (CDH 5.13). Chapter 7 describes Flafka and Spark Streaming in more detail.

Listing 2-1. Spark Streaming and Kudu

```
import org.apache.kudu.client.CreateTableOptions;
import org.apache.kudu.spark.kudu._
import org.apache.spark._
import org.apache.spark.rdd.NewHadoopRDD
import org.apache.spark.SparkConf
import org.apache.spark.storage.StorageLevel
import org.apache.spark.streaming.flume._
import org.apache.spark.streaming.Seconds
import org.apache.spark.streaming.StreamingContext
import org.apache.spark.util.IntParam
import org.apache.spark.sql.SQLContext

object FlumeStreaming {

  case class MySensorData(tableid: String, deviceid: String, thedate:
  String, thetime: String, temp: Short, status: String)

  def readSensorData(str: String): MySensorData = {
    val col = str.split(",")

    val thetableid = col(0)
    val thedeviceid = col(1)
    val thedate = col(2)
    val thetime = col(3)
    val thetemp = col(4)
    val thestatus = col(5)
```

```scala
    MySensorData(col(0), col(1), col(2), col(3), col(4).toShort, col(5))
  }

  def main(args: Array[String]) {

    val sparkConf = new SparkConf().setMaster("local[2]").
    setAppName("FlumeStreaming")
    val sc = new SparkContext(sparkConf)
    val ssc = new StreamingContext(sc, Seconds(1))

      // the arguments are for host name and port number
    val flumeStream = FlumeUtils.createPollingStream(ssc,args(0),args(1).toInt)

    val sensorDStream = flumeStream.map (x => new String(x.event.getBody.
    array)).map(readSensorData)

    sensorDStream.foreachRDD {rdd =>

            val sqlContext = new SQLContext(sc)
            import sqlContext.implicits._
            val kuduContext = new KuduContext("kudumaster01:7051")
// convert the RDD into a DataFrame and insert it into the Kudu table
            val DataDF = rdd.toDF
            kuduContext.insertRows(DataDF, "impala::default.sensortable")

            DataDF.registerTempTable("currentDF")

            // Update the table based on the thresholds

            val WarningFilteredDF = sqlContext.sql("select * from
            currentDF where temp > 50 and temp <= 60")
            WarningFilteredDF.registerTempTable("warningtable")
            val UpdatedWarningDF = sqlContext.sql("select tableid,deviceid,
            thedate,thetime,temp,'WARNING' as status from warningtable")
            kuduContext.updateRows(UpdatedWarningDF, "impala::default.
            sensortable")

            val CriticalFilteredDF = sqlContext.sql("select * from
            currentDF where temp > 61")
```

```
            CriticalFilteredDF.registerTempTable("criticaltable")
            val UpdatedCriticalDF = sqlContext.sql("select tableid,deviceid,
            thedate,thetime,temp,'CRITICAL' as status from criticaltable")
            kuduContext.updateRows(UpdatedCriticalDF, "impala::default.
            sensortable")
      }
   ssc.start()
   ssc.awaitTermination()
 }

}
```

Listing 2-2 shows the flume configuration file, with Kafka used as a flume channel.

Listing 2-2. Flume configuration file

```
agent1.sources  = source1
agent1.channels = channel1
agent1.sinks = spark

agent1.sources.source1.type = spooldir
agent1.sources.source1.spoolDir = /tmp/streaming
agent1.sources.source1.channels = channel1

agent1.channels.channel1.type = org.apache.flume.channel.kafka.KafkaChannel
agent1.channels.channel1.brokerList = kafkabroker01:9092,
kafkabroker02:9092, kafkabroker03:9092
agent1.channels.channel1.zookeeperConnect = server03:2181
agent1.channels.channel1.topic = mytopic

agent1.sinks.spark.type = org.apache.spark.streaming.flume.sink.SparkSink
agent1.sinks.spark.hostname = 127.0.0.1
agent1.sinks.spark.port =  9999
agent1.sinks.spark.channel = channel1
agent1.sinks.spark.batchSize=5
```

After compiling the package, submit the application to the cluster to execute it.

```
spark-submit \
--class FlumeStreaming \
--jars kudu-spark_2.10-0.10.0.jar \
--master yarn-client \
--driver-memory=512m \
--executor-memory=512m \
--executor-cores 4  \
/mydir/spark/flume_streaming_kudu/target/scala-2.10/test-app_2.10-1.0.jar \
localhost 9999
```

The Flafka pipeline with Spark Streaming and Kudu should look like Figure 2-5.

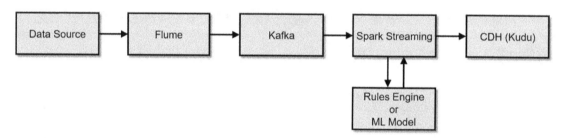

Figure 2-5. *A Flafka pipeline with Spark Streaming and Kudu*

Kudu C++, Java, and Python Client APIs

Kudu provides NoSQL-style Java, C++, and Python client APIs. Applications that require the best performance from Kudu should use the client APIs. In fact, some of the data ingestion tools discussed in Chapter 7, such as StreamSets, CDAP, and Talend utilize the client APIs to ingest data into Kudu. DML changes via the API are available for querying in Impala immediately without the need to execute INVALIDATE METADATA.

Kudu Java Client API

Listing 2-3 provides an example using the Java client API.

Listing 2-3. Sample Java code using the Kudu client API

```java
import org.apache.kudu.ColumnSchema;
import org.apache.kudu.Schema;
import org.apache.kudu.Type;
import org.apache.kudu.client.*;

import java.util.ArrayList;
import java.util.List;

public class JavaKuduClient {

  public static void main(String[] args) {

// Create Kudu client object

KuduClient myKuduClient = new KuduClient.KuduClientBuilder("kudumaster01").build();

// Create the schema

        List<ColumnSchema> myColumns = new ArrayList(3);
      myColumns.add(new ColumnSchema.ColumnSchemaBuilder("rowid", Type.INT32)
         .key(true)
         .build());
      myColumns.add(new ColumnSchema.ColumnSchemaBuilder("customername",
      Type.STRING)
         .build());
         myColumns.add(new ColumnSchema.ColumnSchemaBuilder("customerage",
         Type.INT8)
          .build());
      List<String> partKeys = new ArrayList<>();
      partKeys.add("key");

// Create the table based on the schema

        Schema mySchema = new Schema(myColumns);
        client.createTable("CustomersTbl", mySchema, new
        CreateTableOptions().setRangePartitionColumns(partKeys));
```

CHAPTER 2 INTRODUCTION TO KUDU

```java
// Open the Kudu table

    KuduTable myTable = myKuduClient.openTable("CustomersTbl");
    KuduSession mySession = myKuduClient.newSession();

// Insert new rows

    Insert myInsert = myTable.newInsert();
    myInsert.getRow().addInt("rowid", 1);
    myInsert.getRow().addString("customername", "Jerry Walsh");
        myInsert.getRow().addInt("customerage", 64)
        mySession.apply(myInsert);

// Update existing rows

    Update myUpdate = myTable.newUpdate();
    myUpdate.getRow().addInt("rowid", 1);
    myUpdate.getRow().addString("customername", "Jerome Walsh");
        myUpdate.getRow().addInt("customerage", 65)
        mySession.apply(myUpdate);

// Upsert rows

    Upsert myUpsert = myTable.newUpsert();
    myUpsert.getRow().addInt("rowid", 2);
    myUpsert.getRow().addString("customername", "Tim Stein");
        myUpsert.getRow().addInt("customerage", 49)
        myUpsert.apply(myUpdate);

// Delete row

        Delete myDelete = myTable.newDelete()
        myDelete.getrow().addString("rowid", 1);
        mySession.apply(myDelete)

// Display rows

    List<String> myColumns = new ArrayList<String>();
    myColumns.add("rowid");
      myColumns.add("customername");
      myColumns.add("customerage");
```

```
        KuduScanner myScanner = myClient.newScannerBuilder(myTable)
            .setProjectedColumnNames(myColumns)
            .build();

        while (myScanner.hasMoreRows()) {
          RowResultIterator myResultIterator = myScanner.nextRows();
          while (myResultIterator.hasNext()) {
            RowResult myRow = myResultIterator.next();
            System.out.println(myRow.getInt("rowid"));
                System.out.println(myRow.getString("customername"));
                System.out.println(myRow.getString("customerage"));
          }
        }
// Delete table

        myKuduClient.deleteTable(myTable);

// Close the connection

        myKuduClient.shutdown();

  }
}
```

Maven Artifacts

You will need the following in your pom.xml file.

```
<dependency>
  <groupId>org.apache.kudu</groupId>
  <artifactId>kudu-client</artifactId>
  <version>1.1.0</version>
</dependency>
```

Kudu Python Client API

The Python client API provides an easy way to interact with Kudu. The Python API is still in experimental stage and might change at any time. See Listing 2-4 for an example.

CHAPTER 2 INTRODUCTION TO KUDU

Listing 2-4. Sample Python code using the Kudu client API

```python
import kudu
from kudu.client import Partitioning
from datetime import datetime

# Connect to Kudu
myclient = kudu.connect(host='kudumaster01', port=7051)

# Define the columns
mybuilder = kudu.schema_builder()
mybuilder.add_column('rowid').type(kudu.int64).nullable(False).primary_key()
mybuilder.add_column('customername', type_=kudu.string, nullable=False)
mybuilder.add_column('customerage', type_=kudu.int8, nullable=False)
myschema = mybuilder.build()

# Define partitioning method
mypartitioning = Partitioning().add_hash_partitions(column_names=['rowid'],
num_buckets=24)

# Create new table
myclient.create_table('customers', myschema, mypartitioning)

# Open a table
mytable = myclient.table('customers')

# Create a new session
mysession = client.new_session()

# Insert a row
myinsert = mytable.new_insert({'rowid': 1, 'customername': "Jason
Weinstein", 'customerage': 62})
mysession.apply(myinsert)

# Upsert a row
myupsert = mytable.new_upsert({'rowid': 2, 'customername': "Frank Nunez",
'customerage': 47})
session.apply(myupsert)
```

```
# Updating a row
myupdate = table.new_update({'rowid': 1, 'customername': "Jason Dean Weinstein"})
session.apply(myupdate)

# Delete a row
mydelete = table.new_delete({'rowid': 1})
session.apply(mydelete`)

# Flush the session
mysession.flush()

# Create a scanner with a predicate
myscanner = mytable.scanner()
myscanner.add_predicate(table['customerage'] < 50)

# Read the data. Note that this method doesn't scale well for large table scans
myresult = myscanner.open().read_all_tuples()
```

Kudu C++ Client API

Kudu also provides a C++ client API. See Listing 2-5 for an example.

Listing 2-5. Sample C++ code using the Kudu client API

```
#include <ctime>
#include <iostream>
#include <sstream>

#include "kudu/client/callbacks.h"
#include "kudu/client/client.h"
#include "kudu/client/row_result.h"
#include "kudu/client/stubs.h"
#include "kudu/client/value.h"
#include "kudu/client/write_op.h"
#include "kudu/common/partial_row.h"
#include "kudu/util/monotime.h"
```

CHAPTER 2 INTRODUCTION TO KUDU

```cpp
using kudu::client::KuduClient;
using kudu::client::KuduClientBuilder;
using kudu::client::KuduColumnSchema;
using kudu::client::KuduError;
using kudu::client::KuduInsert;
using kudu::client::KuduPredicate;
using kudu::client::KuduRowResult;
using kudu::client::KuduScanner;
using kudu::client::KuduSchema;
using kudu::client::KuduSchemaBuilder;
using kudu::client::KuduSession;
using kudu::client::KuduStatusFunctionCallback;
using kudu::client::KuduTable;
using kudu::client::KuduTableAlterer;
using kudu::client::KuduTableCreator;
using kudu::client::KuduValue;
using kudu::client::sp::shared_ptr;
using kudu::KuduPartialRow;
using kudu::MonoDelta;
using kudu::Status;

using std::string;
using std::vector;

int main(int argc, char* argv[]) {

  // Enable verbose debugging for the client library.
  // Set parameter to 0 to disable
  kudu::client::SetVerboseLogLevel(2);

  // Create and connect a client.
  shared_ptr<KuduClient> client;
  KUDU_CHECK_OK(KuduClientBuilder().add_master_server_
  addr("kudumaster01:7051").Build(&client));
  KUDU_LOG(INFO) << "Client connection created.";
```

```cpp
// Create a schema.
// Available data types:   INT8 = 0, INT16 = 1, INT32 = 2,
// INT64 = 3, STRING = 4, BOOL = 5, FLOAT = 6, DOUBLE = 7,
// BINARY = 8, UNIXTIME_MICROS = 9, TIMESTAMP = UNIXTIME_MICROS
KuduSchema mytable_schema;
KuduSchemaBuilder mytable_builder;
categories_builder.AddColumn("rowid")->Type(KuduColumnSchema::INT32)->NotNull();
categories_builder.AddColumn("name")->Type(KuduColumnSchema::STRING)->NotNull();
categories_builder.AddColumn("age")->Type(KuduColumnSchema::INT8)->NotNull();
categories_builder.AddColumn("salary")->Type(KuduColumnSchema::DOUBLE)
->NotNull();
categories_builder.SetPrimaryKey({"rowid"});
KUDU_CHECK_OK(categories_builder.Build(&mytable_schema));
KUDU_LOG(INFO) << "Created a schema for mytable";

// Delete table if it exists
bool exists;
KUDU_CHECK_OK(client->TableExists("mytable", &exists));
if (exists) {
  KUDU_CHECK_OK(client->DeleteTable("mytable"));
    KUDU_LOG(INFO) << "Deleting table if it exists.";
}

 // Generate the split keys for the table.
vector<const KuduPartialRow*> splits;
int32_t num_tablets = 20;
int32_t increment = 1000 / num_tablets;
for (int32_t i = 1; i < num_tablets; i++) {
  KuduPartialRow* row = mytable_schema.NewRow();
  KUDU_CHECK_OK(row->SetInt32(0, i * increment));
  splits.push_back(row);
}

vector<string> column_names;
column_names.push_back("rowid");
```

```cpp
// Create the table.
KuduTableCreator* table_creator = client->NewTableCreator();

KUDU_CHECK_OK(table_creator->table_name("mytable")
    .schema(&mytable_schema)
  .set_range_partition_columns(column_names)
  .split_rows(splits)
  .Create());

// Confirm if the table was successfully created
bool created;
KUDU_CHECK_OK(client->TableExists("mytable", &created));
created ? KUDU_LOG(INFO) << "Created table mytable." :
          KUDU_LOG(INFO) << "Failed to create table mytable.";

// Insert two rows into the table.
shared_ptr<KuduTable> table;
client->OpenTable("mytable", &table);

KuduInsert* my_insert = table->NewInsert();
KuduPartialRow* row = categories_insert->mutable_row();
KUDU_CHECK_OK(row->SetInt32("rowid", 100));
KUDU_CHECK_OK(row->SetStringCopy("name", "Fred Smith"));
KUDU_CHECK_OK(row->SetInt8("age", 56));
KUDU_CHECK_OK(row->SetDouble("salary", 110000));
KUDU_CHECK_OK(session->Apply(my_insert));

KuduInsert* my_insert = table->NewInsert();
KuduPartialRow* row = categories_insert->mutable_row();
KUDU_CHECK_OK(row->SetInt32("rowid", 101));
KUDU_CHECK_OK(row->SetStringCopy("name", "Linda Stern"));
KUDU_CHECK_OK(row->SetInt8("age", 29));
KUDU_CHECK_OK(row->SetDouble("salary", 75000));
KUDU_CHECK_OK(session->Apply(my_insert));

KUDU_CHECK_OK(session->Flush());
KUDU_LOG(INFO) << "Inserted two rows into mytable";
```

```
  // Scan one row based on a predicate
  KuduScanner scanner(table.get());

  // Add a predicate: WHERE name = "Linda Stern"
  KuduPredicate* pred = table->NewComparisonPredicate(
      "name", KuduPredicate::EQUAL, KuduValue::FromString("Linda Stern"));
  KUDU_RETURN_NOT_OK(scanner.AddConjunctPredicate(pred));

  KUDU_RETURN_NOT_OK(scanner.Open());
  vector<KuduRowResult> results;

  while (scanner.HasMoreRows()) {
    KUDU_RETURN_NOT_OK(scanner.NextBatch(&results));
    for (vector<KuduRowResult>::iterator iter = results.begin();
         iter != results.end();
         iter++) {
      const KuduRowResult& result = *iter;
      string_t myname;
      KUDU_RETURN_NOT_OK(result.GetString("name", &myname));
      KUDU_LOG(INFO) << "Scanned some rows out of a table" << myname;
    }
    results.clear();
  }
  // Delete the table.
  KUDU_CHECK_OK(client->DeleteTable("mytable"));
  KUDU_LOG(INFO) << "Deleted mytable.";
}
```

More examples[xviii] can be found on Kudu's official website[xix] and github repository.[xx] The sample code available online was contributed by the Kudu development team and served as reference for this chapter.

CHAPTER 2 INTRODUCTION TO KUDU

Backup and Recovery

Kudu doesn't have a backup and recovery utility. However, there are a few ways to back up (and recover) Kudu tables using Impala, Spark, and third-party tools such as StreamSets and Talend.

Note HDFS snapshots cannot be used to back up Kudu tables since Kudu data does not reside in HDFS.[xxi]

Backup via CTAS

The simplest way to back up a Kudu table is to use CREATE TABLE AS (CTAS). You're basically just creating another copy of the Kudu table in HDFS, preferably in Parquet format (or other compressed format), so you can copy the file to a remote location such as another cluster or S3.

```
CREATE TABLE AS DimCustomer_copy AS SELECT * FROM DimCustomer;

+-----------------------+
| summary               |
+-----------------------+
| Inserted 18484 row(s) |
+-----------------------+
```

You can create the table first so you can customize table options if needed, and then use INSERT INTO to insert data from the Kudu table.

```
CREATE TABLE DimCustomer_Parquet (
ID STRING,
CustomerKey BIGINT,
FirstName STRING,
LastName STRING,
BirthDate STRING,
YearlyIncome FLOAT,
TotalChildren INT,
EnglishEducation STRING,
EnglishOccupation STRING,
```

```
HouseOwnerFlag INT,
NumberCarsOwned INT
)
STORED AS PARQUET;

set COMPRESSION_CODEC=gzip;

insert into DimCustomer_Parquet Select * from DimCustomer;
Modified 18484 row(s) in 4.52s
```

> **Note** The CREATE TABLE LIKE syntax is not supported on Kudu tables. If you try to use the syntax to create a table, you will receive an error message similar to the following: "ERROR: AnalysisException: Cloning a Kudu table using CREATE TABLE LIKE is not supported."

Check the files with the HDFS command.

```
hadoop fs -du -h /user/hive/warehouse/dimcustomer_parquet

636.8 K  1.9 M  /user/hive/warehouse/dimcustomer_parquet/f948582ab9f8dfbb-
5e57d0ca00000000_1052047868_data.0.parq
```

Copy the Parquet Files to Another Cluster or S3

You can now copy the Parquet files to another cluster using distcp.

```
hadoop distctp -pb hftp://kuducluster:50070/user/hive/warehouse/
dimcustomer_parquet hdfs://kuducluster2/backup_files
```

You can also copy the files to S3.

```
hadoop distcp –pb -Dfs.s3a.access.key=s3-access-key -Dfs.s3a.secret.key=
s3-secret-key hdfs://user/hive/warehouse/dimcustomer_parquet
s3a://myWarehouseBucket/backup_files
```

To maintain consistency, note that I used the -pb option to guarantee that the special block size of the Parquet data files is preserved.[xxii]

> **Note** Cloudera has a cluster replication feature called Cloudera Enterprise Backup and Disaster Recovery (BDR). BDR provides an easy-to-use graphical user interface that lets you schedule replication from one cluster to another. BDR does not work with Kudu, but you can replicate the destination Parquet files residing in HDFS.[xxiii]

Export Results via impala-shell to Local Directory, NFS, or SAN Volume

Impala-shell can generate delimited files that you can then compress and copy to a remote server or NFS/SAN volume. Note that this method is not appropriate for large tables.

```
impala-shell -q "SELECT * FROM DimCustomer" --delimited --output_
delimiter=, --output_file /backup_nfs/dimcustomer_bak.csv
```

Export Results Using the Kudu Client API

The Kudu client API can be used to export data as well. See Listing 2-6 for an example.

Listing 2-6. Sample Java code using the Kudu client API to export data

```java
import org.apache.kudu.ColumnSchema;
import org.apache.kudu.Schema;
import org.apache.kudu.Type;
import org.apache.kudu.client.*;

import java.util.ArrayList;
import java.util.List;

public class TableBackup {

  public static void main(String[] args) {

// Create Kudu client object
```

```java
KuduClient myKuduClient = new KuduClient.KuduClientBuilder("kudumaster").
build();

KuduTable myTable = myKuduClient.openTable("CustomersTbl");
KuduSession mySession = myKuduClient.newSession();

// Display rows

        List<String> myColumns = new ArrayList<String>();
        myColumns.add("rowid");
          myColumns.add("customername");
          myColumns.add("customerage");

        KuduScanner myScanner = myKuduClient.newScannerBuilder(myTable)
            .setProjectedColumnNames(myColumns)
            .build();

        while (myScanner.hasMoreRows()) {
          RowResultIterator myResultIterator = myScanner.nextRows();
          while (myResultIterator.hasNext()) {
            RowResult myRow = myResultIterator.next();
            System.out.println(myRow.getInt("rowid"));
                System.out.println(myRow.getString("customername"));
                System.out.println(myRow.getString("customerage"));
          }
        }
 }
}
```

Compile the java code and run it from the command line. Redirect the results to a file. This method is appropriate for small data sets.

```
java TableBackup >> /backup_nfs/mybackup.txt
```

Export Results with Spark

You can also back up data using Spark. This is more appropriate for large tables since you can control parallelism, the number of executors, executor cores, and executor memory.

Start by creating a Data Frame.

```
val df = sqlContext.read.options(Map("kudu.master" -> "localhost:7051","kudu.table" -> "impala::default.DimCustomer")).kudu
```

Save the data in CSV format.

```
df.coalesce(1).write.format("com.databricks.spark.csv").option("header", "true").save("/backup/dimcustomer_bak")
```

Or you can save it as Parquet.

```
df.coalesce(1).write.mode("append").parquet("/backup/dimcustomer_p_bak")
```

Using coalesce to limit the amount of files generated when writing to HDFS may cause performance issues. I discuss coalesce in more details in Chapter 5.

Replication with Spark and Kudu Data Source API

We can use Spark to copy data from one Kudu cluster to another.

Start the spark-shell.

```
spark-shell --packages org.apache.kudu:kudu-spark_2.10:1.1.0 --driver-class-path mysql-connector-java-5.1.40-bin.jar --jars mysql-connector-java-5.1.40-bin.jar
```

Connect to the Kudu master and check the data in the users table. We're going to sync this Kudu table with another Kudu table in another cluster.

```
import org.apache.kudu.spark.kudu._

val kuduDF = sqlContext.read.options(Map("kudu.master" -> "kuducluster:7051","kudu.table" -> "impala::default.users")).kudu

kuduDF.select("userid","name","city","state","zip","age").sort($"userid".asc).show()
```

```
+------+----------------+---------------+-----+-----+---+
|userid|            name|           city|state|  zip|age|
+------+----------------+---------------+-----+-----+---+
|   100|    Wendell Ryan|      San Diego|   CA|92102| 24|
|   101| Alicia Thompson|       Berkeley|   CA|94705| 52|
|   102| Felipe Drummond|      Palo Alto|   CA|94301| 33|
|   103|   Teresa Levine|   Walnut Creek|   CA|94507| 47|
|   200|   Jonathan West|         Frisco|   TX|75034| 35|
|   201|  Andrea Foreman|         Dallas|   TX|75001| 28|
|   202|    Kirsten Jung|          Plano|   TX|75025| 69|
|   203| Jessica Nguyen|          Allen|   TX|75002| 52|
|   300|    Fred Stevens|       Torrance|   CA|90503| 23|
|   301|     Nancy Gibbs|       Valencia|   CA|91354| 49|
|   302|      Randy Park|Manhattan Beach|   CA|90267| 21|
|   303|    Victoria Loma|  Rolling Hills|   CA|90274| 75|
+------+----------------+---------------+-----+-----+---+
```

Let's go ahead and insert the data to a table in another Kudu cluster. The destination table needs to be present in the other Kudu cluster.

val kuduContext = new KuduContext("kuducluster2:7051")

kuduContext.insertRows(kuduDF, "impala::default.users2")

Verify the data in the destination table.

impala-shell

```
select * from users2 order by userid;
+------+----------------+---------------+-----+-----+---+
|userid|            name|           city|state|  zip|age|
+------+----------------+---------------+-----+-----+---+
|   100|    Wendell Ryan|      San Diego|   CA|92102| 24|
|   101| Alicia Thompson|       Berkeley|   CA|94705| 52|
|   102| Felipe Drummond|      Palo Alto|   CA|94301| 33|
|   103|   Teresa Levine|   Walnut Creek|   CA|94507| 47|
|   200|   Jonathan West|         Frisco|   TX|75034| 35|
```

```
|   201| Andrea Foreman|               Dallas| TX|75001| 28|
|   202|   Kirsten Jung|                Plano| TX|75025| 69|
|   203| Jessica Nguyen|                Allen| TX|75002| 52|
|   300|   Fred Stevens|             Torrance| CA|90503| 23|
|   301|    Nancy Gibbs|             Valencia| CA|91354| 49|
|   302|     Randy Park|     Manhattan Beach| CA|90267| 21|
|   303|  Victoria Loma|        Rolling Hills| CA|90274| 75|
+------+---------------+---------------------+-----+-----+---+
```

The rows were successfully replicated.

Real-Time Replication with StreamSets

StreamSets is a powerful real-time and batch ingestion tool used mostly in real-time streaming and Internet of Things (IoT) use cases. You can use StreamSets to replicate data from JDBC sources such as Oracle, MySQL, SQL Server, or Kudu to another destination in real time or near real time. StreamSets offers two origins to facilitate replication through a JDBC connection: JDBC Query Consumer and JDBC Multitable Consumer. The JDBC Query Consumer origin uses a user-defined SQL query to read data from tables. See Figure 2-6.

Figure 2-6. *StreamSets and Kudu*

CHAPTER 2 INTRODUCTION TO KUDU

The JDBC Multitable Consumer origin reads multiple tables in the same database. The JDBC Multitable Consumer origin is appropriate for database replication. StreamSets includes a Kudu destination; alternatively the JDBC Producer is a (slower) option and can be used to replicate data to other relational databases. Chapter 7 covers StreamSets in more detail.

Replicating Data Using ETL Tools Such as Talend, Pentaho, and CDAP

Talend (Figure 2-9) and Cask Data Platform (Figure 2-7) offer native support for Kudu. Both provide Kudu source and sinks and can be used to replicate data from one Kudu cluster to one or more Kudu clusters; another destination such as S3; or an RDBMS such as SQL Server, MySQL, or Oracle. Other tools such as Pentaho PDI (Figure 2-8) does not have native Kudu support. However, it can transfer data to Kudu via Impala, albeit slower. Chapter 7 covers batch and real-time ingestion tools in including StreamSets, Talend, Pentaho, and CDAP in detail.

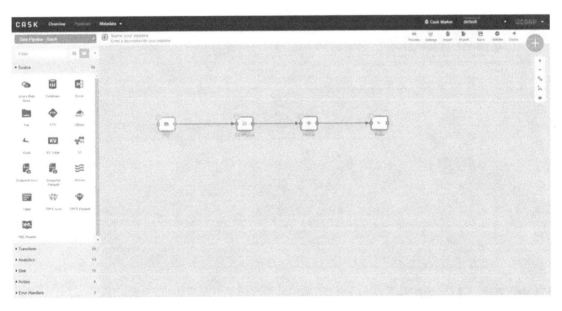

Figure 2-7. ETL with CDAP and Kudu

41

CHAPTER 2 INTRODUCTION TO KUDU

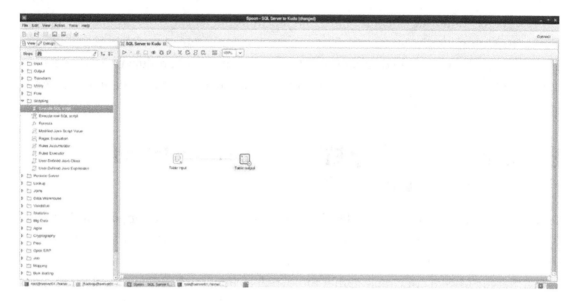

Figure 2-8. *ETL with Pentaho and Kudu*

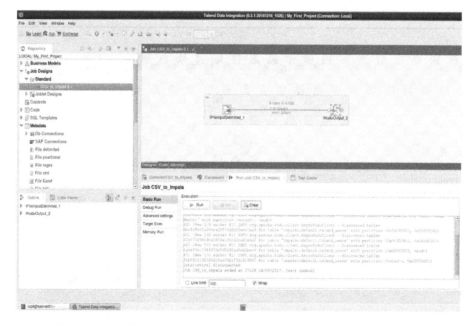

Figure 2-9. *ETL with Talend and Kudu*

> **Note** Talend Kudu components are provided by a third-party company, One point Ltd. These components are free and downloadable from the Talend Exchange at – `https://exchange.talend.com/`. The Kudu Output and Input components need to be installed before you can use Talend with Kudu.

Python and Impala

Using Python is not the fastest or most scalable way to back up large Kudu tables, but it should be adequate for small- to medium-sized data sets. Below is a list of the most common ways to access Kudu tables from Python.

Impyla

Cloudera built a Python package known as Impyla.[xxiv] Impyla simply communicates with Impala using standard ODBC/JDBC. One of the nice features of Impyla is its ability to easily convert query results into pandas DataFrame (not to be confused with Spark DataFrames). Here's an example.

```
>>> from impala.util import as_pandas
>>> cur.execute('SELECT id, name, salary FROM employees')
>>> df = as_pandas(cur)
>>> type(df)
<class 'pandas.core.frame.DataFrame'>
>>> df
          id    name                salary
0         001   James Chan          100000
1         002   Roger Lim            75000
2         003   Dan Tanner       v   65000
3         004   Lilian Russo         90000
4         005   Edith Sarkisian     110000
```

pyodbc

pyodbc is a popular open source Python package that you can use to access databases via ODBC/JDBC.[xxv] Here's an example on how to use pyodbc. To learn more about pyodbc, visit its github page at github.com/mkleehammer/pyodbc.

```
import pyodbc

myconnection_str = '''Driver=/mypath/libclouderaimpalaodbc.dylib;HOST=local
host;PORT=21050'''

myconnection = pyodbc.connect("myconnection_str")
cursor = myconnection.cursor()
cursor.execute("select id, name, salary  from employees")
```

SQLAlchemy

SQLAlchemy is an SQL toolkit for Python that features an object-relational mapper (ORM). To learn more about SQLAlchemy, visit its website at sqlalchemy.org. Here's an example on how to use SQLAlchemy to connect to Impala.

```
import sqlalchemy
from sqlalchemy.orm import sessionmaker
from sqlalchemy import create_engine

myconnection_str = '''Driver=/mypath/libclouderaimpalaodbc.dylib;HOST=local
host;PORT=21050'''

myconnection = create_engine("myconnection_str")
session = sessionmaker(bind=db)
user = session.query(Employees).filter_by(salary > '65000').first()
```

High Availability Options

Aside from Kudu, having a default tablet replica factor of 3 (which can be increased to 5 or 7), there are no built-in high availability tools or features available for Kudu. Fortunately, you can use built-in components in Cloudera Enterprise and third-party tools such as StreamSets to provide high availability capabilities to Kudu. High availability can protect you from complete site failure by having two or more Kudu clusters. The clusters can be in geographically distributed data centers or cloud providers.[xxvi] Just be aware that the amount of data that is being replicated could impact performance and cost. An added benefit of having an active-active environment is being able to use both clusters for different use cases. For example, a second cluster can be used for ad hoc queries, building machine learning models, and other data science workloads, while the first cluster is used for real-time analytics or use cases with well-defined SLAs. Let's explore a few high availability options for Kudu.

Active-Active Dual Ingest with Kafka and Spark Streaming

In this option, all data is published to a Kafka cluster. Spark streaming is used to read data from the Kafka topics. Using Spark Streaming, you have the option to perform data transformation and cleansing before writing the data to Kudu. Figure 2-10 shows two Kudu destinations, but you could have more depending on your HA requirements.

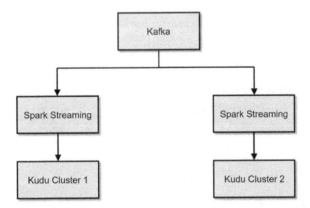

Figure 2-10. *Active-Active Dual Ingest with Kafka and Spark Streaming*

Active-Active Kafka Replication with MirrorMaker

Another high availability option is to replicate Kafka using MirrorMaker. As shown in Figure 2-11, data ingested into the source Kafka cluster is replicated using MirrorMaker to the destination Kafka cluster. From there, the data is read by Spark Streaming and written to the Kudu destinations similar to Figure 2-10. If your goal is dual ingest, using MirrorMaker to replicate your Kudu cluster might be overkill. However, it provides better data protection since data is replicated on two or more Kafka and Kudu clusters.

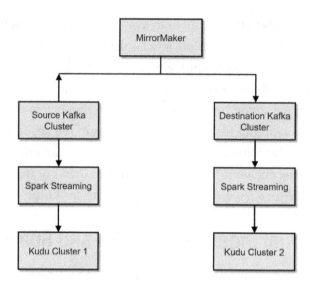

Figure 2-11. *Active-Active Kafka Replication with MirrorMaker*

Active-Active Dual Ingest with Kafka and StreamSets

This option is very similar to Figure 2-10 but uses StreamSets instead of Spark Streaming. StreamSets is easier to use and administer than Spark Streaming and provides built-in monitoring, alerting, and exception handling. It also offers an event framework that makes it easy to kick off a task based on events. I usually recommend StreamSets over Spark Streaming in most projects (Figure 2-12).

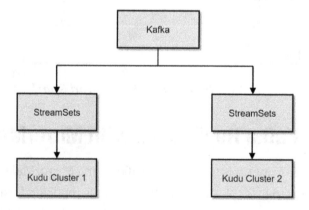

Figure 2-12. *Active-Active Dual Ingest with Kafka and StreamSets*

Active-Active Dual Ingest with StreamSets

Technically, you don't need Kafka in order to have active-active dual ingest. StreamSets allows you to configure multiple destinations. Using the Stream Selector processor, it can even route data to different destinations based on certain conditions (Figure 2-13). Using Kafka provides added high availability and scalability and is still recommended in most cases. I cover StreamSets in Chapter 7 in more detail.

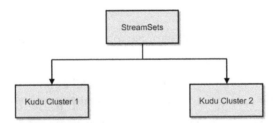

Figure 2-13. *Active-Active Dual Ingest with Kafka and StreamSets*

Administration and Monitoring

Just like other data management platforms, Kudu provides tools to aid in system administration and monitoring.

Cloudera Manager Kudu Service

Cloudera Manager is Cloudera's cluster management tool, providing a single pane of glass for administering and managing Cloudera Enterprise clusters. With Cloudera Manager, you can perform common administration tasks such as starting and stopping the Kudu service, updating configuration, monitoring performance, and checking logs.

Kudu Master Web UI

Kudu Masters provide a web interface (available on port 8051) that provides information about the cluster. It displays information about tablet servers, heartbeat, hostnames, tables, and schemas. You can also view details on available logs, memory usage, and resource consumption.

Kudu Tablet Server Web UI

Each table server also provides a web interface (available on port 8050) that provides information about the tablet servers cluster. It displays more detailed information about each tablet hosted on the tablet servers, debugging and state information, resource consumption, and available logs.

Kudu Metrics

Kudu provides several metrics that you can use to monitor and troubleshoot your cluster. You can get a list of available Kudu metrics by executing $ kudu-tserver --dump_metrics_json or kudu-master --dump_metrics_json. Once you know the metric you want to check, you can collect the actual value via HTTP by visiting /metrics end-point. These metrics are also collected and aggregated by Cloudera Manager. For example:

```
curl -s 'http://tabletserver01:8050/metrics?include_schema=1&metrics=read_bytes_rate'
```

Kudu Command-Line Tools

In addition to Cloudera Manager and the web user interfaces accessible provided by the master and tablet servers, Kudu includes command-line tools for common system administration tasks.

Validate Cluster Health

ksck: Checks that the cluster metadata is consistent and that the masters and tablet servers are running. Ksck checks all tables and tablets by default, but you can specify a list of tables to check using the tables flag, or a list of tablet servers using the tablets flag. Use the checksum_scan and checksum_snapshot to check for inconsistencies in your data.

Usage:

```
kudu cluster ksck --checksum_scan [--tables <tables>] <master_address>
```

File System

check: Check a Kudu filesystem for inconsistencies
  Usage:

```
kudu fs check [-fs_wal_dir=<dir>] [-fs_data_dirs=<dirs>] [-repair]
```

  list: Show list of tablet replicas in the local filesystem
  Usage:

```
kudu local_replica list [-fs_wal_dir=<dir>] [-fs_data_dirs=<dirs>]
[-list_detail]
```

  data_size: Summarize the data size/space usage of the given local replica(s).
  Usage:

```
kudu local_replica data_size <tablet_id_pattern> [-fs_wal_dir=<dir>]
[-fs_data_dirs=<dirs>] [-format=<format>]
```

Master

status: Get the status of a Kudu Master
  Usage:

```
kudu master status <master_address>
```

  timestamp: Get the current timestamp of a Kudu Master
  Usage:

```
kudu master timestamp <master_address>
```

  list: List masters in a Kudu cluster
  Usage:

```
kudu master list <master_addresses> [-columns=<columns>] [-format=<format>]
[-timeout_ms=<ms>]
```

Measure the Performance of a Kudu Cluster

loadgen: Run load generation with optional scan afterward
  loadgen inserts auto-generated random data into an existing or auto-created table as fast as the cluster can execute it. Loadgen can also check whether the actual count of inserted rows matches the original row count.

CHAPTER 2　INTRODUCTION TO KUDU

Usage:

```
kudu perf loadgen <master_addresses> [-buffer_flush_watermark_pct=<pct>]
[-buffer_size_bytes=<bytes>] [-buffers_num=<num>] [-error_buffer_size_
bytes=<bytes>] [-flush_per_n_rows=<rows>] [-keep_auto_table] [-num_
rows_per_thread=<thread>] [-num_threads=<threads>] [-run_scan] [-seq_
start=<start>] [-show_first_n_errors=<errors>] [-string_fixed=<fixed>]
[-string_len=<len>] [-table_name=<name>] [-table_num_buckets=<buckets>]
[-table_num_replicas=<replicas>] [-use_random]
```

Table

delete: Delete a table
Usage:

```
kudu table delete <master_addresses> <table_name>
```

list: List all tables
Usage:

```
kudu table list <master_addresses> [-list_tablets]
```

Tablets

leader_step_down: Force the tablet's leader replica to step down
Usage:

```
kudu tablet leader_step_down <master_addresses> <tablet_id>
```

add_replica: Add a new replica to a tablet's Raft configuration
Usage:

```
kudu tablet change_config add_replica <master_addresses> <tablet_id> <ts_
uuid> <replica_type>
```

move_replica: Move a tablet replica from one tablet server to another
The replica move tool effectively moves a replica from one tablet server to another by adding a replica to the new server and then removing it from the old one.
Usage:

```
kudu tablet change_config move_replica <master_addresses> <tablet_id>
<from_ts_uuid> <to_ts_uuid>
```

Tablet Server

status: Get the status of a Kudu Tablet Server
Usage:

```
kudu tserver status <tserver_address>
```

timestamp: Get the current timestamp of a Kudu Tablet Server
Usage:

```
kudu tserver timestamp <tserver_address>
```

list: List tablet servers in a Kudu cluster
Usage:

```
kudu tserver list <master_addresses> [-columns=<columns>]
[-format=<format>] [-timeout_ms=<ms>]
```

Consult Kudu's online command-line reference guide[xxvii] for a complete list and description of Kudu's command-line tools.

Known Issues and Limitations

There are several issues and limitations in Kudu. Depending on your use case, they can be considered minor or major issues. Most of them have workarounds but some do not. You have to be aware of these limitations when building new applications or migrating workloads to Kudu. I list some of the major ones below. Kudu committers and contributors are hard at work fixing these limitations.

- Kudu does not support DECIMAL, CHAR, VARCHAR, DATE, and complex types like ARRAY, MAP, and STRUCT.
- Kudu tables can have a maximum of 300 columns.
- Kudu does not have secondary indexes.
- Kudu does not have foreign keys.
- Multi-row and multi-table transactions are not supported.
- Kudu does not have a built-in backup and recovery and high availability feature.

- Kudu does not support row, column, and table-level role-based access control.

- Kudu recommends 100 as a maximum number of tablet servers.

- Kudu recommends 3 as a maximum number of masters.

- Kudu recommends 8TB as the maximum amount of stored data, post-replication and post-compression, per tablet server.

- Kudu recommended 2000 as a maximum number of tablets per tablet server, post-replication.

- Kudu recommends 60 as a maximum number of tablets per table for each tablet server, post-replication, at table-creation time.

- Kudu does not support rack-awareness, multiple data centers, and rolling restarts.

For a more complete and up-to-date list of Kudu's limitations, consult Cloudera's online documentation.[xxviii]

Security

Kudu supports Kerberos for strong authentication. Communication between Kudu clients and servers is encrypted with TLS. Kudu does not support table, row, or column-level access control. Instead it uses a white-list style access control list to implement coarse-grained authorization. Two levels of access include Superuser and User. Unauthenticated users will be unable to access the Kudu cluster.[xxix]

There's still a lot of work to be done from a security standpoint. For the meantime, additional suggestions to tighten up security include restricting direct access to the Kudu tables and implementing role-based access control via the business intelligence tool's semantic layer. Implementing databases views to mimic row and column-level role-based access control is another option. Configuring IP access lists to restrict access from certain IP addresses to the port used by the master for RPC (the default port is 7051) can also be explored.

Consult Cloudera's online documentation for a more up-to-date development on Kudu security.

Summary

Although Hadoop is known for its ability to handle structured, unstructured, and semi-structured data, structured relational data remains the focus of most companies' data management and analytic strategies and will continue to be in the foreseeable future.[xxx] In fact, a majority of the big data use cases involves replicating workloads from relational databases. Kudu is the perfect storage engine for structured data. Throughout the book, we will focus on Kudu and how it integrates with other projects in the Hadoop ecosystem and third-party applications to enable useful business use cases.

References

i. Globenewswire; "Cloudera Announces General Availability of Apache Kudu with Release of Cloudera Enterprise 5.10," Cloudera, 2017, `https://globenewswire.com/news-release/2017/01/31/912363/0/en/Cloudera-Announces-General-Availability-of-Apache-Kudu-with-Release-of-Cloudera-Enterprise-5-10.html`

ii. Todd Lipcon; "A brave new world in mutable big data: Relational storage," O'Reilly, 2017, `https://conferences.oreilly.com/strata/strata-ny/public/schedule/speaker/75982`

iii. Jimmy Xiang; "Apache HBase Write Path," Cloudera, 2012, `https://blog.cloudera.com/blog/2012/06/hbase-write-path/`

iv. Apache Software Foundation; "Introducing Apache Kudu," ASF, 2017, `https://kudu.apache.org/docs/#kudu_use_cases`

v. Apache Software Foundation; "The Apache Software Foundation Announces Apache® Kudu™ v1.0," ASF, 2017, `https://blogs.apache.org/foundation/entry/the_apache_software_foundation_announces100`

vi. Pat Patterson; "Innovation with @ApacheKafka, #StreamSets, @ApacheKudu & @Cloudera at the Australian @DeptDefence - spotted on the Kudu Slack channel," Twitter, 2017, `https://twitter.com/metadaddy/status/843842328242634754`

vii. Todd Lipcon; "Kudu: Storage for Fast Analytics on Fast Data," Cloudera, 2015, https://kudu.apache.org/kudu.pdf

viii. Diego Ongaro and John Ousterhout; "In Search of an Understandable Consensus Algorithm (Extended Version)," Stanford University, 2014, https://raft.github.io/raft.pdf

ix. Diego Ongaro; "Consensus: Bridging Theory and Practice," Stanford University, 2014, https://github.com/ongardie/dissertation#readme

x. Neil Chandler; "Oracle's Locking Model – Multi Version Concurrency Control," Neil Chandler, 2013, https://chandlerdba.com/2013/12/01/oracles-locking-model-multi-version-concurrency-control/

xi. Oracle; "Multiversion Concurrency Control," Oracle, 2018, https://docs.oracle.com/cd/B19306_01/server.102/b14220/consist.htm#i17881

xii. Microsoft; "Database Concurrency and Row Level Versioning in SQL Server 2005," Microsoft, 2018, https://technet.microsoft.com/en-us/library/cc917674.aspx

xiii. Teradata; "About Concurrency Control," Teradata, 2018, https://info.teradata.com/HTMLPubs/DB_TTU_16_00/index.html#page/General_Reference%2FB035-1091-160K%2Fvju1472241438286.html%23

xiv. David Alves and James Kinley; "Apache Kudu Read & Write Paths," Cloudera, 2017, https://blog.cloudera.com/blog/2017/04/apache-kudu-read-write-paths/

xv. Microsoft; "Using decimal, float, and real Data," Microsoft, 2018, https://technet.microsoft.com/en-us/library/ms187912(v=sql.105).aspx

xvi. Apache Impala; "TIMESTAMP Data Type," Apache Impala, 2017, https://impala.apache.org/docs/build/html/topics/impala_timestamp.html

xvii. Cloudera; "Example Impala Commands With Kudu," Cloudera, 2017, https://kudu.apache.org/docs/developing.html#_kudu_integration_with_spark

xviii. William Berkeley, "scantoken_noncoveringrange.cc," Cloudera, 2017, https://gist.github.com/wdberkeley/50e2e47548a0daa3d3bff68e388da37a

xix. Apache Kudu, "Developing Applications With Apache Kudu," Apache Kudu, 2017, http://kudu.apache.org/docs/developing.html

xx. Apache Kudu; "Kudu C++ client sample," Apache Kudu, 2018, https://github.com/cloudera/kudu/tree/master/src/kudu/client/samples

xxi. Apache Kudu; "Apache Kudu FAQ," Apache Kudu, 2018, https://kudu.apache.org/faq.html

xxii. Cloudera; "Using the Parquet File Format with Impala Tables", Cloudera, 2018, https://www.cloudera.com/documentation/enterprise/latest/topics/impala_parquet.html

xxiii. Cloudera; "How To Back Up and Restore HDFS Data Using Cloudera Enterprise BDR," Cloudera, 2018, https://www.cloudera.com/documentation/enterprise/latest/topics/cm_bdr_howto_hdfs.html

xxiv. Cloudera; "A New Python Client for Impala", Cloudera, 2018, http://blog.cloudera.com/blog/2014/04/a-new-python-client-for-impala/

xxv. Cloudera; "Importing Data into Cloudera Data Science Workbench," Cloudera, 2018, https://www.cloudera.com/documentation/data-science-workbench/latest/topics/cdsw_import_data.html#impala_impyla

xxvi. Cloudera; "Implementing Active/Active Multi-Cluster Deployments with Cloudera Enterprise," Cloudera, 2018, https://www.cloudera.com/content/dam/www/marketing/resources/whitepapers/implementing-active-deployments-with-cloudera-enterprise-whitepaper.pdf.landing.html

xxvii. Cloudera; "Apache Kudu Command Line Tools Reference," Cloudera, 2018, https://kudu.apache.org/docs/command_line_tools_reference.html

xxviii. Cloudera; "Impala Integration Limitations," Cloudera, 2018, https://www.cloudera.com/documentation/kudu/latest/topics/kudu_known_issues.html#impala_kudu_limitations

xxix. Cloudera; "Apache Kudu Security," Cloudera, 2018, https://www.cloudera.com/documentation/kudu/latest/topics/kudu_security.html

xxx. Cloudera; "Dell Survey: Structured Data Remains Focal Point Despite Rapidly Changing Information Management Landscape," Cloudera, 2018, http://www.dell.com/learn/us/en/uscorp1/press-releases/2015-04-15-dell-survey

CHAPTER 3

Introduction to Impala

Impala is a massively parallel processing (MPP) SQL engine designed and built from the ground up to run on Hadoop platforms.[i] Impala provides fast, low-latency response times appropriate for business intelligence applications and ad hoc data analysis. Impala's performance matches, and in most cases, surpasses commercial MPP engines.

Impala started out as a project within Cloudera. It was donated to the Apache Software Foundation and accepted into the Apache incubator on December 2, 2015. Prior to joining Cloudera, Impala's architect and tech lead, Marcel Kornacker, was a software engineer at Google who led the development of the distributed query engine of Google's F1 Project,[ii] a distributed relational database system used to support Google's highly critical and massively popular AdWords business,[iii] Cloudera released a beta version of Impala in October of 2012 and announced its general availability in May of 2013.

This chapter will not provide you an exhaustive list of every feature of Impala. My goal with this chapter is to give you enough information about Impala to get you started immediately.

Architecture

Impala has a distributed architecture just like other Hadoop components (Figure 3-1). It utilizes daemon processes for direct local execution of queries on worker nodes. Impala is just a query engine designed to integrate with multiples storage engines such as HDFS, HBase, S3, and Kudu. This is a different architecture compared to traditional relational database systems such as Oracle and SQL Server where the query and storage engine are tightly coupled together.

CHAPTER 3   INTRODUCTION TO IMPALA

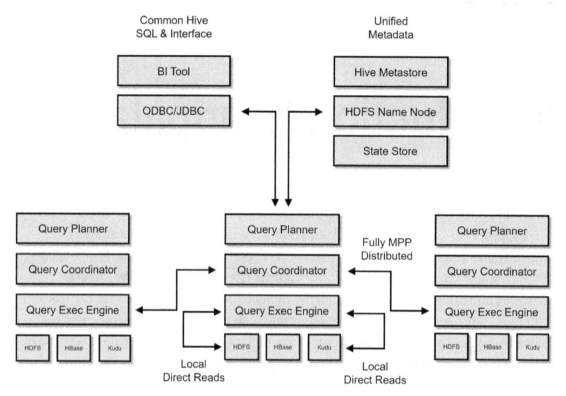

Figure 3-1. *Impala Architecture*

Impala can be accessed with popular business intelligence and data visualization tools via ODBC or JDBC such as Microsoft Power BI, Qlik, Tableau, Zoomdata, Business Objects, and Microstrategy to name a few.

Impala Server Components

Impala's architecture consists of three services: the Impala Daemon, Catalog Service, and Statestore.

The Impala Daemon

The Impala daemon acts as a query coordinator that is responsible for accepting and managing execution of queries across the cluster. The Impala daemons are usually installed on the data nodes, taking advantage of data locality, allowing it direct access to the data blocks stored in HDFS.

The Impala Catalog Service

Impala's catalog service stores and aggregates metadata information collected from other metadata repositories such as the Hive Metastore and HDFS Namenode.

The Impala Statestore

The Statestore distributes and synchronizes metadata information to all Impala processes across the cluster. The Statestore is a pub-sub Impala service that pushes updates to the Impala daemons.

Impala and the Hadoop Ecosystem

Impala is part of the Hadoop Ecosystem and was designed to work with other Hadoop components.

Impala and Hive

Hive was the first SQL interface for Hadoop. Hive was originally built on top of MapReduce but also runs on other data processing engines such as Tez and Spark. Even though Hive is notoriously slow, a lot of people were already using it, so building an incompatible SQL interface made no sense. When Cloudera engineers were designing Impala, one of their main goals was to make it as interoperable as possible with Hive. Utilizing existing Hive infrastructure such as the Hive Metastore was key to enabling Impala access to Hive tables. Impala (and Hive) can be accessed via HUE (the Hadoop User Experience), a web-based SQL workbench for Hadoop. Additionally, Impala has its own shell called impala-shell, which can be accessed via the command line. I discuss the impala-shell and HUE later in the chapter.

Impala and HDFS

Tables created in Hive and Impala are usually stored in HDFS using various file formats. To create a CSV table in Impala:

```
CREATE TABLE my_csv_table (
id BIGINT,
name STRING,
gender STRING
)
ROW FORMAT DELIMITED FIELDS TERMINATED BY ',';
```

To bulk insert data you can use the LOAD DATA or INSERT...SELECT statement below. Note that you have to create the table first.

```
LOAD DATA INPATH '/mydata/mytextfile.txt' INTO TABLE my_csv_table;INSERT INTO my_csv_table SELECT customerkey, concat(firstname,' ',lastname), gender FROM dimcustomer;
```

Inserting one value at a time is possible but not recommended.

```
INSERT INTO my_csv_table VALUES (200,'Norman Bates','M');
```

Each INSERT...VALUES statement executed will produce a small file in HDFS and may cause fragmentation and performance issues. Using LOAD DATA or INSERT...SELECT is recommended when performing bulk inserts. If you have to perform a lot of single row inserts, it may be best to use a different storage engine such as HBase or Kudu.

You can also manually copy CSV files in the data directory of the table (/user/hive/warehouse/my_csv_table). Data added this way will not be visible to Impala until you run a refresh or invalidate metadata statement. We'll cover refresh and invalidate metadata later in the chapter.

Impala and HBase

Impala can query tables stored in HBase via a Hive external table. Impala can also insert data to the Hive external table with the data stored in the underlying HBase table. Here's an example.

Start the hbase shell. Create an HBase table and insert some test data.

```
create 'table1', 'colfam1'

put 'table1','firstrow','colfam1:name', 'Jeff Frazier'
put 'table1','firstrow','colfam1:city', 'Los Angeles'
put 'table1','firstrow','colfam1:age', '75'

put 'table1','secondrow','colfam1:name', 'Susan Fernandez'
put 'table1','secondrow','colfam1:city', 'New York'
put 'table1','secondrow','colfam1:age', '72'

put 'table1','thirdrow','colfam1:name', 'Tony Cheng'
put 'table1','thirdrow','colfam1:city','San Francisco'
put 'table1','thirdrow','colfam1:age','77'
```

Create the external table in Hive.

```
CREATE EXTERNAL TABLE hive_on_hbase (key string,name string, age int, city
string)
STORED BY
'org.apache.hadoop.hive.hbase.HBaseStorageHandler'
WITH SERDEPROPERTIES
('hbase.columns.mapping'=':key, colfam1:name,colfam1:age,colfam1:city')
TBLPROPERTIES('hbase.table.name'='table1');
```

Start the impala-shell. Since the table was created outside of Impala, it does not know anything about the new table. Execute invalidate metadata before selecting data from the new table.

```
invalidate metadata;

select * from hive_on_hbase;
```

key	age	city	name
firstrow	75	Los Angeles	Jeff Frazier
secondrow	72	New York	Susan Fernandez
thirdrow	77	San Francisco	Tony Cheng

```
INSERT INTO hive_on_hbase VALUES ('fourthrow', 65,'Palo Alto','Kevin Lee');

select * from hive_on_hbase;
```

key	age	city	name
firstrow	75	Los Angeles	Jeff Frazier
secondrow	72	New York	Susan Fernandez
thirdrow	77	San Francisco	Tony Cheng
fourthrow	65	Palo Alto	Kevin Lee

Impala and S3

Amazon S3 is a popular object store frequently used as data store for transient clusters. It's also a cost-effective storage for backups and cold data. Reading data from S3 is just like reading data from HDFS or any other file system. Impala supports queries as well as inserts for data stored on S3 starting in CDH 5.8 (Impala 2.6) and higher.

```
CREATE TABLE my_s3_table (
id BIGINT,
name STRING,
gender STRING
)
ROW FORMAT DELIMITED FIELDS TERMINATED BY ','
LOCATION 's3://impala01/csvfiles/';
```

Impala and Kudu

Kudu is the new relational data store from Cloudera. Kudu is the recommended storage engine for structured data and works seamlessly with Impala. I'll provide in-depth coverage of Impala and Kudu in Chapter 4.

File Formats

Impala supports several file formats such as Text, Avro, RCFile, SequenceFile, and Parquet. Impala does not support the ORC file format supported by Hortonworks. Impala can create tables in all file formats and query from them. Impala doesn't allow inserts to Avro, RCFile, and SequenceFile tables. You have to run your insert statement from Hive or use LOAD DATA to import your data.

Impala's default file format when creating tables is Text with values separated by the ASCII 0x01 character (Ctrl-A). Several compression codecs are supported such as Snappy (the right balance of speed and compression rate), GZIP (highest level of compression but slower), Deflate, BZip2, and LZO. Generally, Snappy is used in most cases.

Parquet is the recommended file format for Impala. Parquet started out as a joint project between Twitter and Cloudera. Parquet is a columnar file format specifically designed for large full-table scans and analytics queries that Impala was designed for. To create a Parquet table in Impala, you simply include STORED AS PARQUET in your CREATE TABLE statement such as:

```
CREATE TABLE customer (
id BIGINT,
name STRING)
STORED AS PARQUET;
```

CHAPTER 3 INTRODUCTION TO IMPALA

Impala SQL

Impala behaves just like a traditional relational database management system. Impala uses SQL to retrieve, manage, and modify data (when used with Kudu). Tables, views, and other objects are logically organized inside a database. Data stored in tables is organized in rows and columns. Each one of these columns has an associated data type.

Data Types

Just like a typical RDBMS, Impala supports standard data types to store different kinds of data as shown in Table 3-1.

Table 3-1. *List of Data Types Supported by Impala*

Data Type	Description
Array	A complex data type that can contain a group of elements.
Bigint	An 8-byte integer data type. Range: -9223372036854775808 .. 9223372036854775807.
Boolean	A data type representing a true or false choice.
Char	A fixed-length character type. The maximum length you can specify is 255.
Decimal	A numeric data type with fixed scale and precision. Suitable for financial and other arithmetic calculations. DECIMAL with no precision or scale values is equivalent to DECIMAL(9,0).
Double	A double precision floating-point data type. Range: 4.94065645841246544e-324d .. 1.79769313486231570e+308, positive or negative
Float	A single precision floating-point data type. Range: 1.40129846432481707e-45 .. 3.40282346638528860e+38, positive or negative.
Int	A 4-byte integer data type. Range: -2147483648 .. 2147483647. There is no UNSIGNED subtype.
Map	A complex data type that can contain a set of key-value pairs.
Real	An alias for the DOUBLE data type.
Smallint	A 2-byte integer data type. Range: -32768 .. 32767.

(*continued*)

63

Table 3-1. (*continued*)

Data Type	Description
String	Length: Maximum of 32,767 bytes.
Struct	A complex data type that can contain multiple fields of a single item.
Timestamp	A data type. Range: Allowed date values range from 1400-01-01 to 9999-12-31.
Tinyint	A 1-byte integer data type. Range: -128 .. 127.
Varchar	A variable-length character type. The maximum length you can specify is 65,535.

SQL Statements

As previously mentioned, Impala uses SQL as the language to query, manage and manipulate data. Impala adheres to the SQL-92 standard, which includes many standard features and built-in functions you would find in a modern SQL implementation. Below is a partial list of common SQL statements supported by Impala. Consult the Impala SQL Language Reference[iv] for a complete coverage of Impala SQL.

CREATE DATABASE

CREATE DATABASE creates a database in Impala. Just like other RDBMS, Impala uses the notion of a database to store its tables and other objects.

```
CREATE DATABASE salesdb;
```

CREATE TABLE

Impala creates "internal" tables by default. Internal tables are managed by Impala. If you drop an internal table, the physical files that comprise that table is also deleted.

```
CREATE TABLE salesdb.customers (
id BIGINT,
name STRING,
gender STRING
)
ROW FORMAT DELIMITED FIELDS TERMINATED BY ','
```

CREATE EXTERNAL TABLE

An external table lets you specify the location of the data that's already in stored in a storage engine such as HDFS, HBase, S3, or Kudu. For example, I can create an external table called products against CSV data stored in the HDFS directory /user/bob/products. External tables are like views. If you drop an external table, the physical files that comprise that table is not deleted, only the table definition is dropped. You need to manually delete the data from the underlying storage.

```
CREATE EXTERNAL TABLE salesdb.products (
(
id BIGINT,
productname STRING,
productdesc STRING,
cost DOUBLE
)
ROW FORMAT DELIMITED FIELDS TERMINATED BY ','
LOCATION '/user/bob/products/';
```

SELECT

SELECT allows you to retrieve data from tables or views. You can specify the column names manually to only return specific columns, or specify an asterisk if you want to return all columns.

Example

```
SELECT customerkey, firstname, lastname, birthdate
FROM dimcustomer;
```

Customerkey	firstname	lastname	birthdate
11001	Eugene	Huang	1976-05-10
11044	Adam	Flores	1954-11-21
11063	Angela	Murphy	1980-10-04
11075	Felicia	Jimenez	1963-05-16
11099	Adam	Ross	1966-09-05
11107	Bianca	Lin	1976-03-04
11118	Alvin	Zeng	1962-12-31

11156	Maria	Roberts	1981-08-06
11169	Bryce	Richardson	1973-12-20
11171	Jonathan	Hill	1978-10-05

WHERE

WHERE Filter data based on conditions.
   Example

```
SELECT customerkey, firstname, lastname, birthdate
FROM dimcustomer
WHERE firstname='Eugene';
```

Customerkey	firstname	lastname	birthdate
11001	Eugene	Huang	1976-05-10

AND and OR

AND and OR lets users combine more than one condition in the WHERE clause.
   Example

```
SELECT customerkey, firstname, lastname, birthdate
FROM dimcustomer
WHERE firstname = 'Adam'
AND lastname='Flores';
```

customerkey	firstname	lastname	birthdate
11044	Adam	Flores	1954-11-21

```
SELECT customerkey, firstname, lastname, birthdate
FROM dimcustomer
WHERE firstname = 'Adam'
OR firstname='Alvin';
```

customerkey	firstname	lastname	birthdate
11044	Adam	Flores	1954-11-21
11099	Adam	Ross	1966-09-05
11118	Alvin	Zeng	1962-12-31

LIKE

LIKE allows users to specify wildcards to search for patterns in columns.
    Example

```
SELECT customerkey, firstname, lastname, birthdate
FROM dimcustomer
WHERE lastname
LIKE 'R%'
```

Customerkey	firstname	lastname	birthdate
11099	Adam	Ross	1966-09-05
11156	Maria	Roberts	1981-08-06
11169	Bryce	Richardson	1973-12-20

LIMIT

LIMIT limits the number of rows returned by an SQL query.
    Example

```
SELECT customerkey, firstname, lastname, birthdate
FROM dimcustomer
LIMIT 5;
```

customerkey	firstname	lastname	birthdate
11001	Eugene	Huang	1976-05-10
11044	Adam	Flores	1954-11-21
11063	Angela	Murphy	1980-10-04
11075	Felicia	Jimenez	1963-05-16
11099	Adam	Ross	1966-09-05

ORDER BY

ORDER BY sorts the results in descending or ascending order.
    Example

```
SELECT customerkey, firstname, lastname, birthdate
FROM dimcustomer
ORDER BY customerkey;
```

Customerkey	firstname	lastname	birthdate
11001	Eugene	Huang	1976-05-10
11002	Ruben	Torres	1971-02-09
11003	Christy	Zhu	1973-08-14
11004	Elizabeth	Johnson	1979-08-05
11005	Julio	Ruiz	1976-08-01
11006	Janet	Alvarez	1976-12-02
11007	Marco	Mehta	1969-11-06
11008	Rob	Verhoff	1975-07-04
11009	Shannon	Carlson	1969-09-29

GROUP BY and HAVING

The GROUP BY is used with aggregate functions such as COUNT, AVG, and SUM to group results by one or more columns. HAVING filters data after the results have been aggregated. It is used in conjunction with GROUP BY.

Example

```
SELECT firstname, count(*)
FROM dimcustomer
GROUP BY firstname
HAVING firstname='Adam';
```

firstname	count(*)
Adam	51

DISTINCT

DISTINCT returns distinct values in a column.

Example

```
SELECT DISTINCT firstname
FROM dimcustomer;
```

UNION and UNION DISTINCT

UNION and UNION DISTINCT combines the results of two or more SELECT statements. Both only return distinct values.

Example

```
SELECT firstname
FROM dimcustomer
UNION
SELECT firstname
FROM dimcustomer2;
```

UNION ALL

UNION ALL is similar to UNION except that it allows duplicate values in the results.

Example

```
SELECT firstname
FROM dimcustomer
UNION ALL SELECT
firstname
FROM dimcustomer2;
```

JOIN

JOIN combines rows from two or more tables using common values in both tables. Impala supports different types of joins. Consult the Impala SQL Language Reference for more details.

Example

```
SELECT t1.firstname, t1.lastname, t2.salesamount
FROM
dimcustomer AS t1
INNER JOIN
salesfact AS t2
WHERE
t1.customerkey = t2.customerkey;
```

SUBQUERIES

A subquery is a SQL query within another SQL query.
   Example

```
SELECT firstname
FROM
dimcustomer
WHERE firstname
IN
(SELECT firstname FROM dimcustomer2);
```

DESCRIBE

DESCRIBE returns metadata information about a table.
   Example

```
DESCRIBE dimcustomer;

DESCRIBE FORMATTED dimcustomer;
```

INVALIDATE METADATA

INVALIDATE METADATA invalidates the metadata for one or more tables. It forces Impala to reload the table's metadata before a query is executed against it.
   Example

```
INVALIDATE METADATA dimcustomer;
```

LOAD DATA

LOAD DATA loads data stored in HDFS file into a table.
   Example

```
LOAD DATA INPATH '/home/bob/customers.txt' INTO TABLE dimcustomer;
```

REFRESH

Refresh reloads the metadata for a table immediately but only loads partial metadata for newly added data, making it a less-expensive operation compared to INVALIDATE METADATA.

Example

```
REFRESH dimcustomer;
```

SET Statements

SET Statements allow users to change the behavior of SQL statements The settings only affect queries executed within that an impala-shell session.

NUM_NODES

NUM_NODES limits the number of nodes that will execute a SQL query.

Example

```
SET NUM_NODES=1
```

MEM_LIMIT

MEN_LIMIT defines the maximum amount of memory an SQL query can utilize on each node.

Example

```
set mem_limit=3G;

set mem_limit=3000M;
```

LIVE_PROGRESS

LIVE_PROGRESS displays a live progress bar showing what percentage of execution has been completed.

Example

```
SET LIVE_PROGRESS=TRUE;
```

BATCH_SIZE

BATCH_SIZE shows the number of rows processed at a time by an SQL query. A large number improves SQL query's performance, but consumes more memory. The default is set to 1024.

Example

```
SET BATCH_SIZE=2048
```

SHOW Statements

The SHOW statement enables users to obtain information about Impala objects. Below is a list of the most common SHOW statements supported by Impala.

SHOW DATABASES

SHOW DATABASES displays all available databases.

Example

```
SHOW DATABASES;

name                comment
_impala_builtins    System database for Impala builtin functions
default             Default Hive database
```

SHOW TABLES

SHOW TABLES displays all available tables within a database.

Example

```
SHOW TABLES;

name
customers
dimcustomer
dimgeography
dimproductcategory
hbase_users
```

sample_07
sample_08
sensordata
testtable
testuser
users
web_logs

SHOW FILES

SHOW FILES displays all the HDFS files that comprise a table.
   Example

```
SHOW FILES IN web_logs;
```

Path	Size	Partition
hdfs://kuducluster:8020/user/cloudera/2015_11_18/web_logs_1.csv	113.00KB	date=2015-11-18
hdfs://kuducluster:8020/user/cloudera/2015_11_19/web_logs_2.csv	103.18KB	date=2015-11-19
hdfs://kuducluster:8020/user/cloudera/2015_11_20/web_logs_3.csv	105.59KB	date=2015-11-20
hdfs://kuducluster:8020/user/cloudera/2015_11_21/web_logs_4.csv	82.68KB	date=2015-11-21

SHOW TABLE STATS

SHOW TABLE TATS displays low-level information about a table.

Example

```
SHOW TABLE STATS web_logs;
date                      #Rows      #Files        Size        Bytes Cached
Cache Replication         Format     Incremental stats Location
2015-11-18                -1         1             113.00KB    NOT CACHED
NOT CACHED                TEXT       false         hdfs://kuducluster:8020/user/
                                                   cloudera/2015_11_18
2015-11-19                -1         1             103.18KB    NOT CACHED
NOT CACHED                TEXT       false         hdfs://kuducluster:8020/user/
                                                   cloudera/2015_11_19
2015-11-20                -1         1             105.59KB    NOT CACHED
NOT CACHED                TEXT       false         hdfs://kuducluster:8020/user/
                                                   cloudera/2015_11_20
2015-11-21                -1         1             82.68KB     NOT CACHED
NOT CACHED                TEXT       false         hdfs://kuducluster:8020/user/
                                                   cloudera/2015_11_21
Total                     -1         4             404.45KB    0B
```

Built-In Functions

Impala includes built-in functions similar to those you would find in a modern RDBMS. These functions can be used in SQL statements to transform data, manipulate strings, and perform date and mathematical calculations. Impala includes aggregate functions, string functions, date and time functions, and mathematical functions to mention a few. Listed are some of the frequently used functions included in Impala. Consult the Impala SQL Language Reference[v] for complete coverage of Impala's built-in functions.

uuid

The uuid function returns a UUID or a universal unique identifier as a string. This function is useful for generating unique values to identify objects. It can also be used as primary key values.

```
select uuid();
uuid()
c14f1469-233b-4d95-9efe-8cf1d2205d7f
```

now

The now function returns the current date and time.

```
select now();
```

```
now()
2017-11-06 22:44:19.926002000
```

regexp_like

The regexp function returns true or false to signify if the string contains the regular expression specified by the pattern.

```
select regexp_like('water','wat');
```

```
regexp_like('water', 'wat')
true
```

abs

The abs function returns the absolute value of the argument.

```
select abs(-100);
```

```
abs(-100)
100
```

fnv_hash(type v)

Returns a 64-bit BIGINT value based on the input argument. This function implements a version of the Fowler–Noll–Vo hash function, which is not cryptographically secure. It can be used for applications requiring simple hashing functionality such as load balancing and hash partitioning.

```
select abs(fnv_hash((now())));
abs(fnv_hash((now())))
5911639952301673123
```

User-Defined Functions

User-defined functions or UDFs lets you write your own Impala function. This is useful in cases where you need a feature that is not included in Impala. Impala supports UDFs written in Java, but for maximum performance it is recommended that you write your UDFs in C++. You can also create user-defined aggregate functions (UDAFs) that return a value based on a group of values. Consult Cloudera's documentation on how to create user-defined functions.

Complex Types in Impala

Impala supports complex types such as STRUCT, ARRAY, and MAP.[vi] Complex data types have several benefits such as improving query performance by eliminating the need for table joins and simpler ETL and data modeling. Currently, support for complex types only works with Parquet tables created through Hive or Spark. Impala currently cannot create Parquet tables containing complex types. Additionally, Kudu does not yet support complex types; therefore traditional dimensional modeling or denormalization is needed when working with Kudu tables. CDH includes sample tables with nested types that you can install via HUE. Let's show some examples on how to work with complex types.

```
DESCRIBE customers;
+-------------------+-------------------------+---------+
| name              | type                    | comment |
+-------------------+-------------------------+---------+
| id                | int                     |         |
| name              | string                  |         |
| email_preferences | struct<                 |         |
|                   |   email_format:string,  |         |
|                   |   frequency:string,     |         |
|                   |   categories:struct<    |         |
|                   |     promos:boolean,     |         |
|                   |     surveys:boolean     |         |
|                   |   >                     |         |
|                   | >                       |         |
| addresses         | map<string,struct<      |         |
|                   |   street_1:string,      |         |
```

```
|                   |       street_2:string,   |         |
|                   |       city:string,       |         |
|                   |       state:string,      |         |
|                   |       zip_code:string    |         |
|                   |    >>                    |         |
| orders            | array<struct<            |         |
|                   |    order_id:string,      |         |
|                   |    order_date:string,    |         |
|                   |    items:array<struct<   |         |
|                   |       product_id:int,    |         |
|                   |       sku:string,        |         |
|                   |       name:string,       |         |
|                   |       price:double,      |         |
|                   |       qty:int            |         |
|                   |    >>                    |         |
|                   | >>                       |         |
+-------------------+--------------------------+---------+
```

Querying Struct Fields

The nested fields are accessed via dot notation.

```
SELECT
id,
name,
email_preferences.email_format ef,
email_preferences.categories.promos ecp
FROM customers
LIMIT 10;

+-------+---------------------+------+-------+
| id    | name                | ef   | ecp   |
+-------+---------------------+------+-------+
| 75012 | Dorothy Wilk        | html | true  |
| 17254 | Martin Johnson      | text | true  |
| 12532 | Melvin Garcia       | html | true  |
```

```
| 42632 | Raymond S. Vestal    | html | true  |
| 77913 | Betty J. Giambrone   | text | false |
| 38807 | Rebecca T. Johnson   | html | true  |
| 71843 | David B. Allison     | text | true  |
| 67099 | Jay N. Weaver        | text | false |
| 83510 | Carol B. Houser      | html | false |
| 48072 | Octaviana Guiterrez  | text | false |
+-------+----------------------+------+-------+
```

Querying Deeply Nested Collections

You specify the nested array in your FROM clause using dot notation.

```
SELECT
order_id,
order_date,
items.name,
items.qty
FROM
customers.orders,
customers.orders.items
LIMIT 10;
+----------------------+-----+
| name                 | qty |
+----------------------+-----+
| Evening Clutch       | 1   |
| Large Tassel Pouch   | 1   |
| Flameless Candle     | 2   |
| Tea for One          | 1   |
| Towel Set            | 4   |
| Maple Dining Table   | 1   |
| Paloma Accent Table  | 1   |
| Trunk Coffee Table   | 1   |
| Simple Scallop Table | 2   |
| Murano Glass Vase    | 4   |
+----------------------+-----+
```

Querying Using ANSI-92 SQL Joins with Nested Collections

Impala supports ANSI-92 SQL joins with nested collections. Note that you don't need to use the ON clause due to the implicit join on the parent-child relationship.

```
SELECT
id,
name,
co.order_id,
co.order_date
FROM
customers c
INNER JOIN
customers.orders co
LIMIT 10;
```

```
+-------+--------------+----------+---------------------------+
| id    | name         | order_id | order_date                |
+-------+--------------+----------+---------------------------+
| 75012 | Dorothy Wilk | 4056711  | 2015-05-01T14:22:25-04:00 |
| 75012 | Dorothy Wilk | J882C2   | 2015-06-10T11:00:00-05:00 |
| 75012 | Dorothy Wilk | I72T39   | 2015-03-14T11:00:00-05:00 |
| 75012 | Dorothy Wilk | PB6268   | 2015-02-11T14:22:25-04:00 |
| 75012 | Dorothy Wilk | B8623C   | 2015-04-21T11:00:00-05:00 |
| 75012 | Dorothy Wilk | R9S838   | 2015-07-09T11:00:00-05:00 |
| 75012 | Dorothy Wilk | HS3124   | 2015-10-14T00:00:00       |
| 75012 | Dorothy Wilk | BS5902   | 2014-10-23T00:00:00       |
| 75012 | Dorothy Wilk | DN8815   | 2015-09-07T00:00:00       |
| 75012 | Dorothy Wilk | XR2771   | 2015-12-25T00:00:00       |
+-------+--------------+----------+---------------------------+
```

Impala Shell

The Impala shell (impala-shell) is a command-line utility that you can use to interact with Impala. You can run SQL statements, save your query's output to a file, or display low-level performance information about a query. You can use impala-shell in interactive and non-interactive mode.

Several command-line options are available to you when executing the impala-shell command. Listed in Table 3-2 are some of the most important options. For a more exhaustive list, consult the Impala documentation.[vii]

Table 3-2. Impala Shell Command-Line Options

Command-Line Option	Description
-i *hostname* or --impalad=*hostname*	impala-shell will connect to the host running the impala daemon.
-o *filename* or --output_file *filename*	Stores the query result in the specified text file.
-d *default_db* or --database=*default_db*	Specifies the default database to use upon starting up the impala-shell.
-q *query* or --query=*query*	Executes a query from the command line.
-f *query_file* or --query_file=*query_file*	Executes a query stored in a file.
-h or --help	Show help information.

Let's start using impala-shell in interactive mode. Let's make the output a little easier to display by using impala-shell options.

```
impala-shell
Starting Impala Shell without Kerberos authentication
Connected to quickstart.cloudera:21000
Server version: impalad version 2.9.0-cdh5.12.0 RELEASE (build 03c6ddbdcec39238be4f5b14a300d5c4f576097e)
***********************************************************************
Welcome to the Impala shell.
(Impala Shell v2.9.0-cdh5.12.0 (03c6ddb) built on Thu Jun 29 04:17:31 PDT 2017)
Press TAB twice to see a list of available commands.
***********************************************************************
Pressing tab twice will list available impala-shell commands.

alter      create     describe   explain    insert     quit       shell      src
unset      use        with       compute    delete     drop       help       load
select     show       summary    update     values     connect    desc       exit
history    profile    set        source     tip        upsert     version
```

CHAPTER 3　INTRODUCTION TO IMPALA

You can use help to list the commands anytime you're inside impala-shell.

help;

Documented commands (type help <topic>):
==
compute   describe explain profile select shell tip   use   version
connect   exit   history quit   set   show   unset values with

Undocumented commands:
======================
alter   delete drop insert source summary upsert
create   desc   help load   src   update

If you want a description of what a specific command does, you can type help "command."

help tip;
Print a random tip

Execute the tip command.

tip;

When you set a query option it lasts for the duration of the Impala shell session.

You can view available databases and tables and query them.

show databases;

```
+------------------+---------------------------------------------+
| name             | comment                                     |
+------------------+---------------------------------------------+
| _impala_builtins | System database for Impala builtin functions|
| default          | Default Hive database                       |
+------------------+---------------------------------------------+
```

use default;

show tables;

81

CHAPTER 3 INTRODUCTION TO IMPALA

```
+-----------------+
| name            |
+-----------------+
| customer_backup |
| customers       |
| sample_07       |
| sample_08       |
| testtable       |
| web_logs        |
+-----------------+
```

select count(*) from customers;

```
+----------+
| count(*) |
+----------+
| 53       |
+----------+
```

select client_ip,city,country_name from web_logs limit 10;

```
+-----------------+---------------+---------------+
| client_ip       | city          | country_name  |
+-----------------+---------------+---------------+
| 128.199.234.236 | Singapore     | Singapore     |
| 128.199.234.236 | Singapore     | Singapore     |
| 128.199.234.236 | Singapore     | Singapore     |
| 128.199.234.236 | Singapore     | Singapore     |
| 128.199.234.236 | Singapore     | Singapore     |
| 66.249.76.236   | Mountain View | United States |
| 222.85.131.87   | Guiyang       | China         |
| 222.85.131.87   | Guiyang       | China         |
| 101.226.168.225 | Shanghai      | China         |
| 66.249.76.225   | Mountain View | United States |
+-----------------+---------------+---------------+
```

describe web_logs;

name	type	comment
version	bigint	
app	string	
bytes	smallint	
city	string	
client_ip	string	
code	tinyint	
country_code	string	
country_code3	string	
country_name	string	
device_family	string	
extension	string	
latitude	float	
longitude	float	
method	string	
os_family	string	
os_major	string	
protocol	string	
record	string	
referer	string	
region_code	bigint	
request	string	
subapp	string	
time	string	
url	string	
user_agent	string	
user_agent_family	string	
user_agent_major	string	
id	string	
date	string	

A handy feature of the impala-shell is its ability to run a shell command.

shell iostat;

```
Linux 2.6.32-642.15.1.el6.x86_64 (kuducluster)          11/03/2017
_x86_64_        (4 CPU)

avg-cpu:   %user   %nice %system %iowait  %steal   %idle
           40.91    0.00    2.75    0.12    0.00   56.22

Device:              tps   Blk_read/s   Blk_wrtn/s   Blk_read   Blk_wrtn
sda                16.11       643.23       619.74    6647326    6404562
dm-0               83.61       631.40       597.55    6525034    6175256
dm-1                0.03         0.25         0.00       2600          0
dm-2                3.74        10.91        22.18     112770     229256
```

History lets you list a history of recent commands you've executed.

history;

```
[1]
[2]: show tables;
[3]: select * from users;
[4]: select count(*) from dimcustomer;
[5]: select customerkey, firstname, lastname, birthdate from dimcustomer
     limit 10;
[6]: describe dimcustomer;
[7]: show databases;
[8]: show tables;
[9]: help;
[10]: history;
```

Performance Tuning and Monitoring

Impala is a high-performance SQL MPP engine able to query 1 trillion rows or more.[viii] However, some performance tuning and monitoring is sometimes still required to ensure that Impala is performing at its maximum potential. Fortunately, Cloudera Enterprise includes tools and commands that can assist you tune and monitor Impala.

Explain

Explain displays the execution plan for an SQL query. The execution plan shows the sequence of low-level operations that Impala will perform to execute your SQL query. You can use the execution plan to evaluate a query's performance. You read the execution plan from bottom to top.

explain select count(*) from web_logs;

Query: explain select count(*) from web_logs
Explain String
Estimated Per-Host Requirements: Memory=90.00MB VCores=1
WARNING: The following tables are missing relevant table and/or column statistics.
default.web_logs
" "
PLAN-ROOT SINK
|
03:AGGREGATE [FINALIZE]
| output: count:merge(*)
|
02:EXCHANGE [UNPARTITIONED]
|
01:AGGREGATE
| output: count(*)
|
00:SCAN HDFS [default.web_logs]
   partitions=4/4 files=4 size=404.45KB
Fetched 16 row(s) in 3.50s

Summary

Provide timings for the different stages of query execution. It helps you determine where the potential bottlenecks are. The results below indicate that most of the activity was spent performing an HDFS scan.

summary;

Operator	#Hosts	Avg Time	Max Time	#Rows
03:AGGREGATE	1	0ns	0ns	1
02:EXCHANGE	1	0ns	0ns	1
01:AGGREGATE	1	0ns	0ns	1
00:SCAN HDFS	1	44.00ms	44.00ms	1.00K

Est. #Rows	Peak Mem	Est. Peak Mem	Detail
1	20.00 KB	-1 B	FINALIZE
1	0 B	-1 B	UNPARTITIONED
1	16.00 KB	10.00 MB	
-1	280.00 KB	80.00 MB	default.web_log

Profile

Profile shows a more detailed performance report about your SQL query such as memory usage, network, and storage wait times to name a few. Shown below is an excerpt of the result.

profile;

```
Query Runtime Profile:
Query (id=26414d8a59a8fff4:fd59c4f500000000):
  Summary:
    Session ID: 8e4d099336a7cabe:83d7327345384cb4
    Session Type: BEESWAX
    Start Time: 2017-11-05 22:21:59.218421000
    End Time: 2017-11-05 22:21:59.345006000
    Query Type: QUERY
    Query State: FINISHED
    Query Status: OK
    Impala Version: impalad version 2.7.0-cdh5.10.0 RELEASE (build 785a073cd07e2540d521ecebb8b38161ccbd2aa2)
    User: hadoop
    Connected User: hadoop
    Delegated User:
```

CHAPTER 3 INTRODUCTION TO IMPALA

```
Network Address: ::ffff:10.0.1.101:53796
Default Db: default
Sql Statement: select count(*) from web_logs
Coordinator: kuducluster:22000
Query Options (non default):
Plan:
```

Cloudera Manager

Cloudera Manager provides a graphical user interface that you can use to monitor Impala (Figure 3-2).

Figure 3-2. *Cloudera Manager*

CHAPTER 3   INTRODUCTION TO IMPALA

You can enlarge the charts for a more detailed view of different performance metrics such as Query Duration (Figure 3-3).

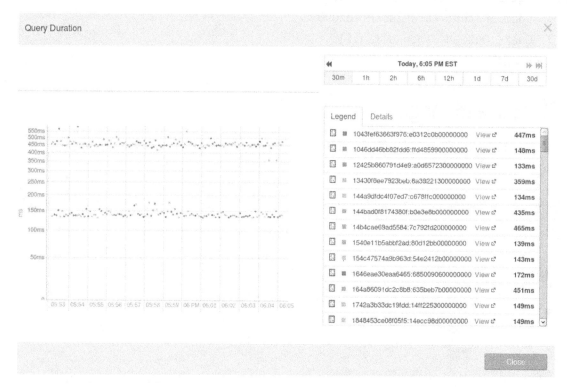

Figure 3-3. *Query Duration*

CHAPTER 3 INTRODUCTION TO IMPALA

You can also monitor queries. In Cloudera Manager, navigate to Clusters ➤ Impala Queries as shown in Figure 3-4.

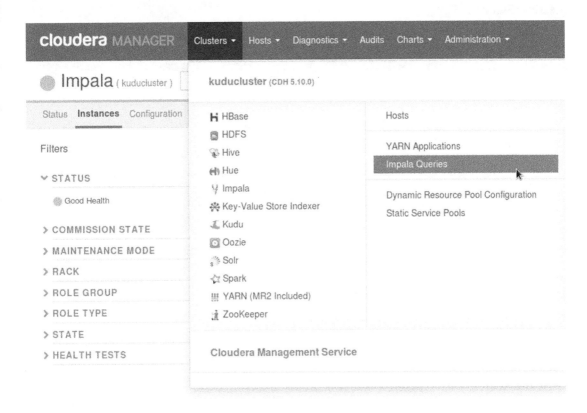

Figure 3-4. *Cloudera Manager – Impala Queries*

CHAPTER 3 INTRODUCTION TO IMPALA

You'll be shown a history of Impala queries that has executed or are executing in the cluster. Cloudera Manager also shows information such as duration, user, and other metrics (Figure 3-5).

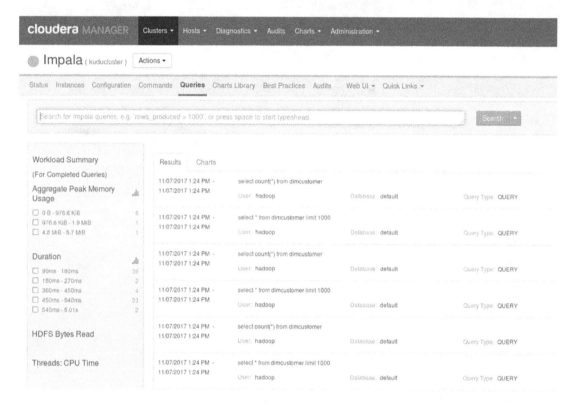

Figure 3-5. Impala Queries

CHAPTER 3 INTRODUCTION TO IMPALA

Impala has its own web interface that provides low-level details about its StateStore and Catalog server (Figure 3-6).

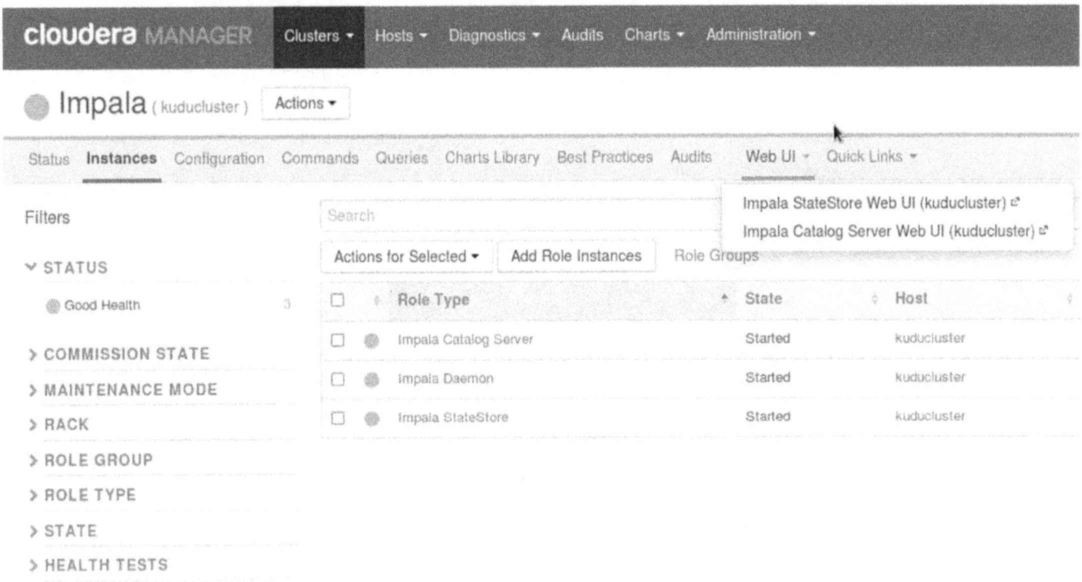

Figure 3-6. *Impala Web UI*

CHAPTER 3 INTRODUCTION TO IMPALA

The Impala StateStore WebUI provides detailed performance and diagnostic information about the Impala StateStore (Figure 3-7).

Figure 3-7. Impala StateStore WebUI

CHAPTER 3　INTRODUCTION TO IMPALA

The Impala Catalog Server WebUI provides detailed performance and diagnostic information about the Impala Catalog Server (Figure 3-8).

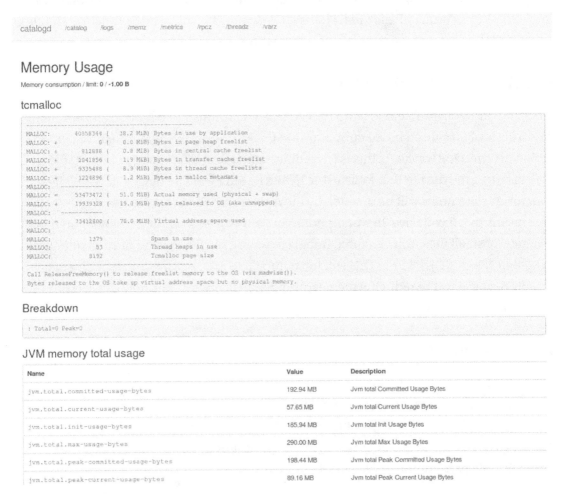

Figure 3-8. Impala Catalog Server WebUI

Impala Performance Recommendations

Some of the recommendations are specific to Impala, while some are common to most data warehouses.

Use Parquet with Impala

Parquet is a highly optimized columnar file format that was designed to work well with Impala. Use the Parquet file format whenever you can when creating your tables (If Kudu is not available in your environment). I'll discuss Impala and Kudu in the next chapter.

Address "Small Files" Problem

Small files are a big problem for Impala. HDFS is not optimized to handle lots of small files. This is often caused by executing several INSERT…VALUES against a table over a period of time. One way to fix this is to combine small files by creating another table and bulk inserting the data to that table using INSERT…SELECT. The degree of parallelism of your insert statements will be dictated by the number of worker nodes running Impala. This means that if you have 20 worker nodes in your cluster, your insert statement might generate 20 small files. You can force Impala to write a big file instead of 20 smaller files by setting NUM_NODES to 1. This will force Impala to execute the insert statement on 1 node. Because you're reducing the degree of parallelism to 1, this may cause other performance issues, so please test accordingly.

Table Partitioning

Partitioning is a requirement for large tables. If you have large tables partitioning, it should cause queries to process individual smaller partitions rather than reading the entire table. This is known as partition pruning. In most cases, partition pruning can dramatically shorten execution time and reduce the amount of data retrieved by Impala. I discuss table partitioning in Chapter 4.

Denormalization

Denormalization is another optimization option. Table joins are usually some of the most inefficient database operations, particularly if you're joining several large tables. By denormalizing your tables, you increase read performance at the expense of duplicate data and decreasing write performance. For most DSS and reporting systems this is an acceptable trade-off.

Create Aggregate or Summary Tables

Technically, creating aggregate or summary tables is a form of denormalization. In most RDBMS this is implemented using materialized views. Impala doesn't support materialized views but you can mimic them by pre-building summary tables containing aggregate data.

Gather Statistics for Tables Used in Large Table Joins

Table statistics needs to be up to date so that Impala can figure out the most efficient join order for your tables included in your query. You can use COMPUTE STATS to regularly gather table statistics for frequently joined tables.

For more details on how to tune Impala, consult the Impala Performance Guidelines and Best Practices.[ix]

Workload and Resource Management

Impala has its own resource and workload management features and doesn't integrate with YARN like other Hadoop components such as Spark and MapReduce.

Note There was a component called Llama in earlier versions of CDH that provided YARN integration with Impala, but it was discontinued starting in CDH 5.5 / Impala 2.3.

In Impala, resource limits on CPU, network, and disk IO are enforced via the use of cgroups if Static Partitioning is enabled. In terms of memory usage, Impala's memory limit (setting MEM_LIMIT) is used to limit memory usage. Dynamic resource pools allow dynamic resource management for Impala queries based on resources that are available on pools.

Admission Control

Admission Control is a resource management feature in Impala that limits concurrency on SQL queries as a safeguard for overutilization due to a large number of simultaneous users. Admission control queues user queries until resources are freed up and become available.

Consult Impala's online documentation for more information on Impala's workload and resource management features.[x]

Hadoop User Experience

The Hadoop User Experience or HUE is a web-based user interface that provides an easy-to-use SQL Editor and Metastore Manager for Impala and Hive. HUE also includes other features that lets you interact with HBase, a File Browser, and a Job Browser to mention a few.

The SQL Editor lets users run Impala queries and is a more user-friendly tool than the impala-shell (Figure 3-9).

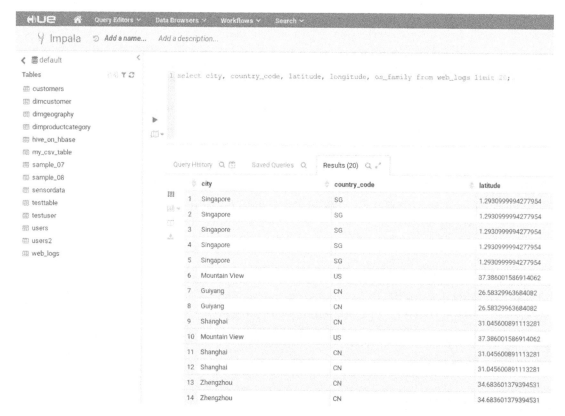

Figure 3-9. *HUE – Impala*

CHAPTER 3 INTRODUCTION TO IMPALA

The Metastore Manager lets users manage Impala and Hive tables and other objects (Figure 3-10).

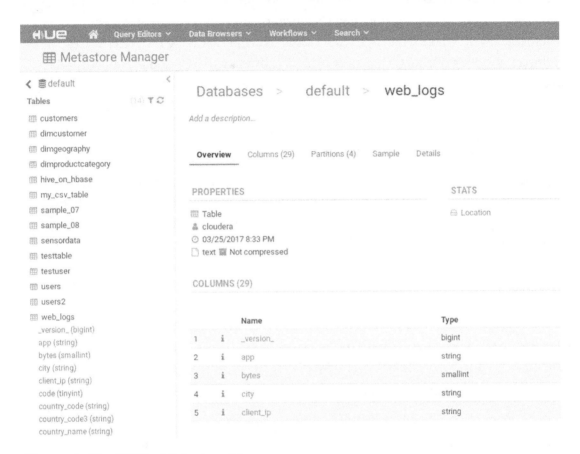

Figure 3-10. HUE – Metastore Manager

CHAPTER 3 INTRODUCTION TO IMPALA

Impala in the Enterprise

Thanks to its JDBC and ODBC drivers, Impala is compatible with popular business intelligence, data wrangling, and data visualization tools in the market such as Microstrategy, Business Objects, Cognos, Oracle Business Intelligence Enterprise Edition (OBIEE), Tableau, Qlik, Power BI, and Zoomdata to mention a few. I cover data visualization and business intelligence with Impala in in Chapter 9.

Summary

Impala is one of the most important components of Cloudera Enterprise. While Hive made Hadoop accessible to the masses, it was SQL MPP engines such as Impala who made it possible for Hadoop to invade the enterprise. We'll cover Impala and Kudu in the next chapter.

References

 i. Marcel Kornacker; "Impala: A Modern, Open-Source SQL Engine for Hadoop," Cloudera, 2015, `http://cidrdb.org/cidr2015/Papers/CIDR15_Paper28.pdf`.

 ii. Marcel Kornacker; "Meet the Engineer: Marcel Kornacker," Cloudera, 2013, `https://blog.cloudera.com/blog/2013/01/meet-the-engineer-marcel-kornacker/`.

 iii. Jeff Shute; F1: "A Distributed SQL Database That Scales," Google, 2013, `https://research.google.com/pubs/pub41344.html`.

 iv. Cloudera; "Impala SQL Language Reference," Cloudera, 2018, `https://www.cloudera.com/documentation/enterprise/latest/topics/impala_langref.html`.

 v. Cloudera; "Impala SQL Language Reference," Cloudera, 2018, `https://www.cloudera.com/documentation/enterprise/latest/topics/impala_langref.html`.

vi. http://blog.cloudera.com/blog/2015/11/new-in-cloudera-enterprise-5-5-support-for-complex-types-in-impala/.

vii. Cloudera; "impala-shell Command-Line Options," Cloudera, 2018, http://www.cloudera.com/documentation/cdh/5-1-x/Impala/Installing-and-Using-Impala/ciiu_shell_options.html.

viii. Cloudera; "Impala Frequently Asked Questions," Cloudera, 2018, https://www.cloudera.com/documentation/enterprise/latest/topics/impala_faq.html.

ix. Cloudera; "Impala Performance Guidelines and Best Practices," Cloudera, 2018, https://www.cloudera.com/documentation/enterprise/latest/topics/impala_perf_cookbook.html#perf_cookbook__perf_cookbook_stats.

x. Cloudera; "Resource Management for Impala," Cloudera, 2018, https://www.cloudera.com/documentation/enterprise/latest/topics/impala_resource_management.html.

CHAPTER 4

High Performance Data Analysis with Impala and Kudu

Impala is the default MPP SQL engine for Kudu. Impala allows you to interact with Kudu using SQL. If you have experience with traditional relational databases where the SQL and storage engines are tightly integrated, you might find it unusual that Kudu and Impala are decoupled from each other. Impala was designed to work with other storage engines such as HDFS, HBase, and S3, not just Kudu. There's also work underway to integrate other SQL engines such as Apache Drill (DRILL-4241) and Hive (HIVE-12971) with Kudu. Decoupling storage, SQL, and processing engines is common in the open source community.

The Impala-Kudu integration works great but there is still work to be done. While it matches or exceeds traditional data warehouse platforms in terms of performance and scalability, Impala-Kudu still lacks some of the enterprise features found in most traditional data warehouse platforms. Kudu is a young project. We discuss some of these limitations later in the chapter.

Primary Key

Every Kudu table needs to have a primary key. When creating Kudu tables, the column or columns used as the primary key must be listed first. Kudu's primary key is implemented as a clustered index. With a clustered index the rows are stored physically in the tablet in the same order as the index. Also note that Kudu doesn't have an auto-increment feature so you will have to include a unique primary key value when inserting rows to a Kudu table. If you don't have a primary key value, you can use Impala's built-in uuid() function or a more efficient method of generating a unique value.

Data Types

Like other relational databases, Kudu supports various data types (Table 4-1).

You may notice that Kudu does not support the decimal data type. This is a key limitation in Kudu. The float and double data types only store a very close approximation of the value instead of the exact value as defined in the IEEE 754 specification.[i]

Table 4-1. List of Data Types, with Available and Default Encoding

Data Type	Encoding	Default
boolean	plain, run length	run length
8-bit signed integer	plain, bitshuffle, run length	bitshuffle
16-bit signed integer	plain, bitshuffle, run length	bitshuffle
32-bit signed integer	plain, bitshuffle, run length	bitshuffle
64-bit signed integer	plain, bitshuffle, run length	bitshuffle
unixtime_micros (64-bit microseconds since the Unix epoch)	plain, bitshuffle, run length	bitshuffle
single-precision (32-bit) IEEE-754 floating-point number	plain, bitshuffle	bitshuffle
double-precision (64-bit) IEEE-754 floating-point number	plain, bitshuffle	bitshuffle
UTF-8 encoded string (up to 64KB uncompressed)	plain, prefix, dictionary	dictionary
binary (up to 64KB uncompressed)	plain, prefix, dictionary	dictionary

Because of this, behavior float and double are not appropriate for storing financial data. At the time of writing, support for decimal data type is still under development (Apache Kudu 1.5 / CDH 5.13). Check KUDU-721 for more details. There are various work arounds. You can store financial data as string and then use Impala to cast the value to decimal every time you need to read the data. Since Parquet supports decimals, another workaround would be to use Parquet for your fact tables and Kudu for dimension tables. Kudu committers are working on adding decimal support and may be included in newer versions of Kudu.

As shown in Table 4-1, Kudu columns can use different encoding types depending on the type of column. Supported encoding types includes Plain, Bitshuffle, Run Length, Dictionary, and Prefix. By default Kudu columns are uncompressed. Kudu supports column compression using Snappy, zlib, or LZ4 compression codecs. Compression and encoding can significantly reduce space overhead and improve performance. Refer to Kudu's online documentation for more information on encoding and compression.

> **Note** In earlier versions of Kudu, date and time are represented as a BIGINT. You can use the TIMESTAMP data type in Kudu tables starting in Impala 2.9/CDH 5.12. However, there are several things to keep in mind. Kudu represents date and time columns using 64-bit values, while Impala represents date and time as 96-bit values. Nanosecond values generated by Impala is rounded when stored in Kudu. When reading and writing TIMESTAMP columns, there is an overhead converting between Kudu's 64-bit representation and Impala's 96-bit representation. There are two workarounds: use the Kudu client API or Spark to insert data, or continue using BIGINT to represent date and time.[ii]

Internal and External Impala Tables

You can create internal and external tables in Impala.

Internal Tables

Internal tables are created and managed by Impala. Internal tables are immediately visible to Impala as soon as they are created. Administration tasks such as dropping and renaming tables are performed using Impala. Here's an example on how to create an internal table in Impala.

```
CREATE TABLE users
(
  id BIGINT,
  name STRING,
  age TINYINT,
  salary FLOAT,
  PRIMARY KEY(id)
)
PARTITION BY HASH PARTITIONS 8
STORED AS KUDU;
```

External Tables

Kudu tables created via the Kudu API and Spark are not immediately visible to Impala. The table exists in Kudu but since it was not created via Impala, it does not know anything about the table. An external table must be created in Impala that references the Kudu table. Dropping an external table only removes the mapping between Impala and the Kudu table, it does not drop the physical table in Kudu. Below is an example on how to create an external table in Impala to an existing Kudu table.

```
CREATE EXTERNAL TABLE users
STORED AS KUDU
TBLPROPERTIES (
  'kudu.table_name' = 'kudu_users'
);
```

Changing Data

You can change the data stored in Kudu tables using SQL statements via Impala, just like traditional relational databases. This is one of the main reasons why you would use Kudu over immutable storage formats such as ORC or Parquet.

Inserting Rows

You can use a standard SQL insert statement.

```
INSERT INTO users VALUES (100, "John Smith", 25, 50000);
```

Inserting rows with multiple VALUES subclauses.

```
INSERT INTO users VALUES (100, "John Smith", 25, 50000), (200, "Cindy Nguyen", 38, 120000), (300, "Steve Mankiw", 60, 75000);
```

Bulk inserts are also supported.

```
INSERT INTO users SELECT * from users_backup;
```

Updating Rows

Standard SQL update statement.

```
UPDATE users SET age=39 where name = 'Cindy Nguyen';
```

Bulk updates are also supported.

```
UPDATE users SET salary=150000 where id > 100;
```

Upserting Rows

Upserts are supported. If the primary key does not exist in the table, the entire row is inserted.

```
UPSERT INTO users VALUES (400, "Mike Jones", 21, 80000);
```

However, if the primary key already exists, the columns are updated with the new values.

```
UPSERT INTO users VALUES (100, "John Smith", 27, 70000);
```

Deleting Rows

Standard SQL Delete statement.

```
DELETE FROM users WHERE id < 3;
```

More complicated Delete statements are also supported.

```
DELETE FROM users WHERE id in (SELECT id FROM users WHERE name = 'Steve Mankiw');
```

Note As discussed in Chapter 2, Kudu doesn't support ACID-compliant transactions. Updates will not be rolled back if it fails halfway through. Additional data validation must be performed after an update to ensure data integrity.

CHAPTER 4 HIGH PERFORMANCE DATA ANALYSIS WITH IMPALA AND KUDU

Changing Schema

Typical database administration tasks such as renaming tables, adding and dropping range partitions, and dropping tables are supported with Kudu. For more information, please consult Kudu's online documentation.

Partitioning

Table partitioning is a common way to enhance performance, availability, and manageability of Kudu tables. Partitioning allows tables to be subdivided into smaller segments, or tablets. Partitioning enables Kudu to take advantage of partition pruning by allowing access to tablets at a finer level of granularity. Table partitioning is required for all Kudu tables and is completely transparent to applications. Kudu supports Hash, Range, and Composite Hash-Range and Hash-Hash partitioning. Below are a few examples of partitioning in Kudu. Partitioning is discussed in more detail in Chapter 2.

Hash Partitioning

Hash partitioning uses a hash key to distribute rows evenly across the different tablets.

```
CREATE TABLE myTable (
 id BIGINT NOT NULL,
 name STRING,
 PRIMARY KEY(id)
)
PARTITION BY HASH PARTITIONS 4
STORED AS KUDU;
```

Range Partitioning

Range partitioning stores ranges of data separately in different tablets.

```
CREATE TABLE myTable (
  year INT,
  deviceid INT,
  totalamt INT,
  PRIMARY KEY (deviceid, year)
)
```

```
PARTITION BY RANGE (year) (
  PARTITION VALUE = 2016,
  PARTITION VALUE = 2017,
  PARTITION VALUE = 2018
)
STORED AS KUDU;
```

Hash-Range Partitioning

Hash partitioning spread writes evenly across tablets. Range partitioning enables partition pruning and allows adding and removing partitions. Composite partitioning combines the strengths of both types of partitioning schemes while limiting its weaknesses. This is a good partitioning scheme for IoT use cases.

```
CREATE TABLE myTable (
 id BIGINT NOT NULL,
 sensortimestamp BIGINT NOT NULL,
 sensorid INTEGER,
 temperature INTEGER,
 pressure INTEGER,
 PRIMARY KEY(id, sensortimestamp)
)
PARTITION BY HASH (id) PARTITIONS 16,
RANGE (sensortimestamp)
(
PARTITION unix_timestamp('2017-01-01') <= VALUES < unix_timestamp('2018-01-01'),
PARTITION unix_timestamp('2018-01-01') <= VALUES < unix_timestamp('2019-01-01'),
PARTITION unix_timestamp('2019-01-01') <= VALUES < unix_timestamp('2020-01-01')
)
STORED AS KUDU;
```

Hash-Hash Partitioning

You can have multiple levels of hash partitions in your table. Each partition level should use a different hashed column.

```
CREATE TABLE myTable (
  id BIGINT,
  city STRING,
  name STRING
  age TINYINT,
  PRIMARY KEY (id, city)
)
PARTITION BY HASH (id) PARTITIONS 8,
            HASH (city) PARTITIONS 8
STORED AS KUDU;
```

List Partitioning

You may have used list partitions if you are familiar with other relational database management systems such as Oracle. While Kudu doesn't technically have list partitioning, you can imitate its behavior using range partitioning.

```
CREATE TABLE myTable (
  city STRING
  name STRING,
  age TINYINT,
  PRIMARY KEY (city, name)
)
PARTITION BY RANGE (city)
(
  PARTITION VALUE = 'San Francisco',
  PARTITION VALUE = 'Los Angeles',
  PARTITION VALUE = 'San Diego',
  PARTITION VALUE = 'San Jose'
)
STORED AS KUDU;
```

Note Partitioning is not a panacea. It is just one of the many ways to optimize Kudu. You may still run into performance issues when ingesting data into Kudu if you are using default Kudu settings. Make sure you adjust the parameters maintenance_manager_num_threads,[iii] which is the number of maintenance threads and can help to speed up compactions and flushes. You can monitor the bloom_lookups_per_op metric and memory pressure rejection to see if compactions and flushes are affecting performance. Another setting you may need to adjust is memory_limit_hard_bytes, which controls the total amount of memory allocated to the Kudu daemon.[iv] Refer to the online Kudu documentation for more details.

Using JDBC with Apache Impala and Kudu

Popular BI and data visualization tools such as Power BI, Tableau, Qlik, OBIEE, and MicroStrategy (to mention a few) can access Apache Impala and Kudu using JDBC/ODBC. The Impala JDBC drivers can be downloaded from Cloudera's website. The Progress DataDirect JDBC driver is another alternative.[v] In some cases, some JDBC drivers from different companies have additional performance features. Your experience may vary. Figure 4-1 shows a sample screenshot of a Zoomdata dashboard visualizing data stored in Kudu via Impala JDBC/ODBC.

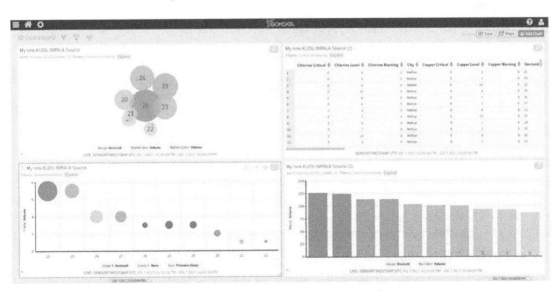

Figure 4-1. Kudu data accessible from ZoomData in real time

Federation with SQL Server Linked Server and Oracle Gateway

You can create database links from SQL Server and Oracle to Impala. This is sometimes useful if copying data back and forth is too burdensome and data is relatively small. Additionally, if you need access to the latest data stored in Kudu in near real time from SQL Server or Oracle and ETL is too slow, then accessing data via a database link is an option. Here's an example on how to create a database link in SQL Server. Users can then access data stored in a remote Impala environment (as shown in Listing 4-1). Note that it is generally recommended to use OpenQuery when executing queries to ensure the query is executed on the remote server.

Listing 4-1. Creating and using linked server from SQL Server to Impala

```
EXEC master.dbo.sp_addlinkedserver
 @server = 'ClouderaImpala', @srvproduct='ClouderaImpala',
 @provider='MSDASQL', @datasrc='ClouderaImpala',
 @provstr='Provider=MSDASQL.1;Persist Security Info=True;User ID=;Password=';

SELECT * FROM OpenQuery(ClouderaImpala, 'SELECT * FROM order_items');

SELECT * FROM OpenQuery(ClouderaImpala, 'SELECT * FROM orders');

SELECT count(*)
FROM [ClouderaImpala].[IMPALA].[default].order_items

SELECT count(*)
FROM [ClouderaImpala].[IMPALA].[default].orders

create view OrderItems_v as
SELECT * FROM OpenQuery(ClouderaImpala, 'SELECT * from order_items');

create view Orders_v as
SELECT * FROM OpenQuery(ClouderaImpala, 'SELECT * from orders');

create view OrderDetail_v as
SELECT * FROM OpenQuery(ClouderaImpala, 'SELECT o.order_id,oi.order_item_
id, o.order_date,o.order_status
FROM [IMPALA].[default].orders o, [IMPALA].[default].order_items oi
where o.order_id=oi.order_item_order_id')
```

You can create a database link from Oracle to Impala using Oracle's ODBC Heterogeneous Gateway. Refer to Oracle's documentation for more details.

Database links are appropriate if accessing small- to medium-sized data sets. If you need a full-blown data virtualization tool, you might want to look at using Polybase, Denodo, or TIBCO data virtualization (previously owned by Cisco).

Summary

Impala provides Kudu with a powerful MPP SQL Engine. Together, they rival traditional data warehouse platforms in terms of performance and scalability. Kudu committers and contributors are hard at work at adding more features and capabilities. Chapter 2 provides an in-depth discussion of Kudu, including its limitations. For more information on Impala, I refer you to Chapter 3. Chapter 8 covers using Impala and Kudu for big data warehousing. Chapter 9 shows users how to use Impala and Kudu for real-time data visualization.

References

i. Microsoft; "Complete tutorial to understand IEEE floating point errors," Microsoft, 2018, `https://support.microsoft.com/en-us/help/42980/-complete-tutorial-to-understand-ieee-floating-point-errors`

ii. Apache Impala; "TIMESTAMP Data Type," Apache Impala, 2018, `https://impala.apache.org/docs/build/html/topics/impala_timestamp.html`

iii. Cloudera; "Apache Kudu Background Maintenance Tasks," Cloudera, 2018, `https://www.cloudera.com/documentation/kudu/latest/topics/kudu_background_tasks.html`

iv. Todd Lipcon; "Re: How to calculate the optimal value of `maintenance_manager_num_threads`," Todd Lipcon, 2017, `https://www.mail-archive.com/user@kudu.apache.org/msg00358.html`

v. Saikrishna Teja Bobba; "Tutorial: Using Impala JDBC and SQL with Apache Kudu," Progress, 2017, `https://www.progress.com/blogs/tutorial-using-impala-jdbc-and-sql-with-apache-kudu`

CHAPTER 5

Introduction to Spark

Spark is the next-generation big data processing framework for processing and analyzing large data sets. Spark features a unified processing framework that provides high-level APIs in Scala, Python, Java, and R and powerful libraries including Spark SQL for SQL support, MLlib for machine learning, Spark Streaming for real-time streaming, and GraphX for graph processing.[i] Spark was founded by Matei Zaharia at the University of California, Berkeley's AMPLab and was later donated to the Apache Software Foundation, becoming a top-level project in February 24, 2014.[ii] The first version was released on May 30, 2014.[iii]

Entire books have been written about Spark. This chapter will give you a quick introduction to Spark, enough to give you the skills needed to perform common data processing tasks. My goal is to make you productive as quickly as possible. For a more thorough treatment, *Learning Spark* by Holden Karau, Andy Konwinski, Patrick Wendell and Matei Zaharia, (O'Reilly, 2015) remains the best introduction to Spark. *Advanced Analytics with Spark 2nd edition* by Sandy Ryza, Uri Laserson, Sean Owen, and Josh Wills (O'Reilly, 2015) covers advanced Spark topics and is also highly recommended. I assume no previous knowledge of Spark. However, some knowledge of Scala is helpful. *Learning Scala* by Jason Swartz (O'Reilly, 2014) and *Programming in Scala 2nd edition* by Martin Odersky, Lex Spoon, and Bill Venners (Artima, 2011) are good books to help you learn Scala. For a primer on Hadoop and Hadoop components such as HDFS and YARN, visit the Apache Hadoop website. Integration with Kudu is discussed in Chapter 6.

Overview

Spark was developed to address the limitations of MapReduce, Hadoop's original data processing framework. Matei Zaharia saw a lot of MapReduce's limitations at UC Berkeley and Facebook (where he did his internship) and sought to create a faster and more generalized, multipurpose data processing framework that can handle iterative

CHAPTER 5 INTRODUCTION TO SPARK

and interactive applications.[iv] Matei succeeded in his goal in making Spark better than MapReduce in almost every way. Spark is more accessible and easier to use due to its simple but powerful APIs. Spark provides a unified platform (Figure 5-1) that supports more types of workloads such as streaming, interactive, graph processing, machine learning, and batch.[v] Spark jobs can run 10-100x faster than equivalent MapReduce jobs due to its fast in-memory capabilities and advanced DAG (Directed Acyclic Graph) execution engine. Data scientists and engineers are just generally more productive with Spark than MapReduce.

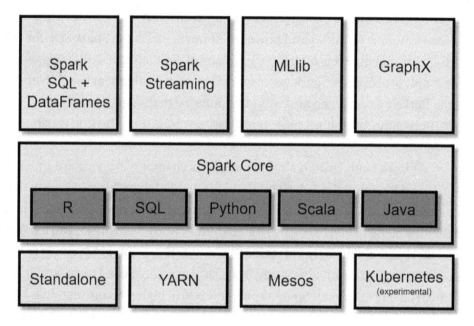

Figure 5-1. Apache Spark Ecosystem

Cluster Managers

Cluster managers manage and allocate cluster resources applications. Spark supports the stand-alone cluster manager that comes with Spark (Standalone Scheduler), YARN, and Mesos. There's an experimental project to bring native support for Spark to utilize Kubernetes as a cluster manager. Check SPARK-18278 for more details.

CHAPTER 5 INTRODUCTION TO SPARK

Architecture

At a high level, Spark distributes the execution of Spark applications tasks across cluster nodes (Figure 5-2). Every Spark application has a SparkContext object within its driver program. The SparkContext represents a connection to your cluster manager, which provides computing resources to your Spark applications. After connecting to the cluster, Spark acquires executors on your worker nodes. Spark then sends your application code to the executors. An application will usually run one or more jobs in response to a Spark action. Each job is then divided by Spark into smaller directed acyclic graphs (DAGs) of stages of tasks. Each task is then distributed and sent to executors across the worker nodes for execution.

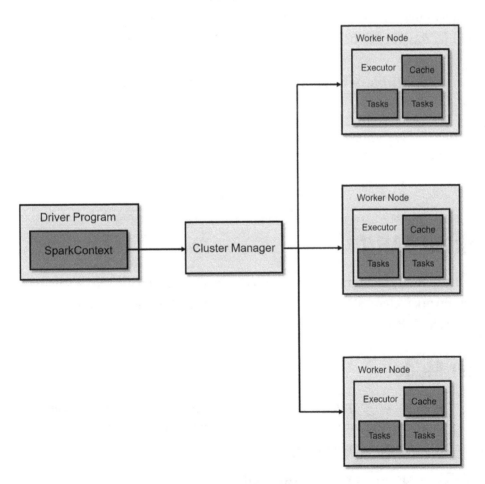

Figure 5-2. Apache Spark Architecture

Each Spark application gets its own set of executors. Because tasks from different applications run in different JVMs, a Spark application cannot interfere with another Spark application. This also means that it's difficult for Spark applications to share data without using a slow external data source such as HDFS or S3 [38]. Using an off-heap memory storage such as Tachyon (a.k.a Alluxio) can make data sharing faster and easier. I discuss Alluxio in Chapter 10.

Executing Spark Applications

I assume you're using Cloudera Enterprise as your big data platform. In CDH, Spark 1.x and 2.x can coexist on the same cluster without any issues. You use an interactive shell (spark-shell or pyspark) or submit an application (spark-submit) to execute Spark 1.x applications. You need to use different commands to access Spark 2.x (Table 5-1).

Table 5-1. Spark 1.x and Spark 2.x Commands

Spark 1.x	Spark 2.x
spark-submit	spark2-submit
spark-shell	spark2-shell
pyspark	pyspark2

Spark on YARN

YARN is the default cluster manager for most Hadoop-based platforms such as Cloudera and Hortonworks. There are two deploy modes that can be used to launch Spark applications in YARN.

Cluster Mode

In cluster mode, the driver program runs inside an application master managed by YARN. The client can exit without affecting the execution of the application. To launch applications or the spark-shell in cluster mode:

spark-shell --master yarn --deploy-mode cluster

spark-submit --class mypath.myClass --master yarn --deploy-mode cluster

Client Mode

In client mode, the driver program runs in the client machine. The application master is only used for requesting resources from YARN. To launch applications or the spark-shell in client mode:

```
spark-shell --master yarn --deploy-mode client

spark-submit --class mypath.myClass --master yarn --deploy-mode client
```

Introduction to the Spark-Shell

You typically use an interactive shell for ad hoc data analysis or exploration. It's also a good tool to learn the Spark API. Spark's interactive shell is available in Scala or Python. In our example below, we'll create a list of cities and convert them all to uppercase.

Listing 5-1. Introduction to spark-shell

```
spark2-shell

Setting default log level to "WARN".
To adjust logging level use sc.setLogLevel(newLevel). For SparkR, use
setLogLevel(newLevel).
Spark context Web UI available at http://10.0.1.101:4040
Spark context available as 'sc' (master = yarn,
app id = application_1513771857144_0002).
Spark session available as 'spark'.
Welcome to
      ____              __
     / __/__  ___ _____/ /__
    _\ \/ _ \/ _ `/ __/  '_/
   /___/ .__/\_,_/_/ /_/\_\   version 2.2.0.cloudera1
      /_/

Using Scala version 2.11.8 (Java HotSpot(TM) 64-Bit Server VM, Java 1.8.0_151)
Type in expressions to have them evaluated.
Type :help for more information.

scala> val myCities = sc.parallelize(List("tokyo","new york","sydney",
"san francisco"))
```

```
myCities: org.apache.spark.rdd.RDD[String] = ParallelCollectionRDD[0]
at parallelize at <console>:24

scala> val uCities = myCities.map {x => x.toUpperCase}
uCities: org.apache.spark.rdd.RDD[String] = MapPartitionsRDD[1] at map
at <console>:26

scala> uCities.collect.foreach(println)
TOKYO
NEW YORK
SYDNEY
SAN FRANCISCO
```

You will use the spark-shell throughout the chapter. A SparkSession named "spark" is automatically created when you start the spark2-shell as shown in Listing 5-1.

SparkSession

As you can see in Figure 5-2, SparkContext enables access to all Spark features and capabilities. The driver program uses the SparkContext to access other contexts such as StreamingContext, SQLContext, and HiveContext. Starting in Spark 2.0, SparkSession provides a single point of entry to interact with Spark. All features available through SparkContext such as SQLContext, HiveContext, and StreamingContext in Spark 1.x are now accessible via SparkSession.[vi]

In Spark 1.x you would write something like this.

```
val sparkConf = new SparkConf().setAppName("MyApp").setMaster("local")

val sc = new SparkContext(sparkConf).set("spark.executor.cores", "4")

val sqlContext = new org.apache.spark.sql.SQLContext(sc)
```

In Spark 2.x, you don't have to explicitly create SparkConf, SparkContext, or SQLContext since all of their functionalities are already included in SparkSession.

```
val spark = SparkSession.
builder().
appName("MyApp").
config("spark.executor.cores", "4").
getOrCreate()
```

Accumulator

Accumulators are variables that are only "added" to. They are usually used to implement counters. In the example, I add up the elements of an array using an accumulator:

```
val accum = sc.longAccumulator("Accumulator 01")

sc.parallelize(Array(10, 20, 30, 40)).foreach(x => accum.add(x))

accum.value
res2: Long = 100
```

Broadcast Variables

Broadcast variables are read-only variable stored on each executor node's memory. Spark uses high-speed broadcast algorithms to reduce network latency of copying broadcast variables. Instead of storing data in a slow storage engine such as HDFS or S3, using broadcast variables is a faster way to store a copy of a dataset on each node.

```
val broadcastVar = sc.broadcast(Array(10, 20, 30))

broadcastVar.value
res0: Array[Int] = Array(10, 20, 30)
```

RDD

An RDD is a resilient immutable distributed collection of objects partitioned across one or more nodes in your cluster. RDD's can be processed and operated in parallel by two types of operations: transformations and actions.

> **Note** The RDD was Spark's primary programming interface in Spark 1.x. Dataset has replaced RDD as the main API starting in Spark 2.0. Users are highly recommended to switch from RDD to Dataset due to richer programming interface and better performance. We discuss Dataset and DataFrame later in the chapter.

Creating an RDD

Creating an RDD is straightforward. You can create an RDD from an existing Scala collection or from reading from an external file stored in HDFS or S3.

parallelize

Parallelize creates an RDD from a Scala collection.

```
val myList = (1 to 5).toList
val myRDD = sc.parallelize(myList)
val myCitiesRDD = sc.parallelize(List("tokyo","new york","sydney",
"san francisco"))
```

textFile

Textfile creates an RDD from a text file stored in HDFS or S3.

```
val myRDD = sc.textFile("hdfs://master01:9000/files/mydirectory")
```

```
val myRDD = sc.textFile("s3a://mybucket/files/mydata.csv")
```

Note that an RDD is immutable. Operations that need to perform any type of data transformation will need to create another RDD. RDD operations can be classified into two categories: transformation and action.

Transformations

A transformation is an operation that creates a new RDD. I describe some of the most common transformations. Refer to the online Spark documentation for a complete list.

Map

Map executes a function against each element in the RDD. It creates and returns a new RDD of the result. Map's return type doesn't necessarily have to be the same type of the original RDD.

```
val myCities = sc.parallelize(List("tokyo","new york","paris","san francisco"))
val uCities = myCities.map {x => x.toUpperCase}
```

```
uCities.collect.foreach(println)
TOKYO
NEW YORK
PARIS
SAN FRANCISCO
```

Let's show another example of map.

```
val lines = sc.parallelize(List("Michael Jordan", "iPhone"))

val words = lines.map(line => line.split(" "))

words.collect

res2: Array[Array[String]] = Array(Array(Michael, Jordan), Array(iPhone))
```

Flatmap

Flatmap executes a function against each element in the RDD and then flattens the results.

```
val lines = sc.parallelize(List("Michael Jordan", "iPhone"))

val words = lines.flatMap(line => line.split(" "))

words.collect

res3: Array[String] = Array(Michael, Jordan, iPhone)
```

Filter

Returns RDD that only includes elements that match the condition specified.

```
val lines = sc.parallelize(List("Michael Jordan", "iPhone","Michael Corleone"))

val words = lines.map(line => line.split(" "))

val results = words.filter(w => w.contains("Michael"))

results.collect

res9: Array[Array[String]] = Array(Array(Michael, Jordan), Array(Michael, Corleone))
```

Distinct

Returns only distinct values.

```
val myCities1 = sc.parallelize(List("tokyo","tokyo","paris","sydney"))

val myCities2 = sc.parallelize(List("perth","tokyo","canberra","sydney"))
```

Combine results using union.

```
val citiesFromBoth = myCities1.union(myCities2)
```

Display only distinct values.

```
citiesFromBoth.distinct.collect.foreach(println)

sydney
perth
canberra
tokyo
paris
```

ReduceByKey

Combine values with the same key. using the specified reduce function.

```
val pairRDD = sc.parallelize(List(("a", 1), ("b",2), ("c",3), ("a", 30),
("b",25), ("a",20)))
val sumRDD = pairRDD.reduceByKey((x,y) => x+y)
sumRDD.collect
res15: Array[(String, Int)] = Array((b,27), (a,51), (c,3))
```

Keys

Return an RDD containing just the keys.

```
val myRDD = sc.parallelize(List(("a", "Larry"), ("b", "Curly"), ("c", "Moe")))

val keysRDD = myRDD.keys
```

```
keysRDD.collect.foreach(println)

a
b
c
```

Values

Return an RDD containing just the values.

```
val myRDD = sc.parallelize(List(("a", "Larry"), ("b", "Curly"), ("c", "Moe")))

val valRDD = myRDD.values

valRDD.collect.foreach(println)

Larry
Curly
Moe
```

Inner Join

Returns an RDD of all elements from both RDDs based on the join predicate.

```
val employee = Array((100,"Jim Hernandez"), (101,"Shane King"))
val employeeRDD = sc.parallelize(employee)

val employeeCity = Array((100,"Glendale"), (101,"Burbank"))
val employeeCityRDD = sc.parallelize(employeeCity)

val employeeState = Array((100,"CA"), (101,"CA"), (102,"NY"))
val employeeStateRDD = sc.parallelize(employeeState)

val employeeRecordRDD = employeeRDD.join(employeeCityRDD).
join(employeeStateRDD)

employeeRecordRDD.collect.foreach(println)

(100,((Jim Hernandez,Glendale),CA))
(101,((Shane King,Burbank),CA))
```

RightOuterJoin / LeftOuterJoin

Returns an RDD of all elements from the right RDD even if there are no matching rows on the left RDD. A Left Outer Join is equivalent to the Right Outer Join with the columns in a different order.

```
val employeeRecordRDD = employeeRDD.join(employeeCityRDD).rightOuterJoin(employeeStateRDD)

employeeRecordRDD.collect.foreach(println)

(100,(Some((Jim Hernandez,Glendale)),CA))
(102,(None,NY))
(101,(Some((Shane King,Burbank)),CA))
```

Union

Returns an RDD that contains the combination of two or more RDDs.

```
val  employee2 = Array((103,"Mark Choi","Torrance","CA"),
(104,"Janet Reyes","Rolling Hills","CA"))
val employee2RDD = sc.parallelize(employee2)
val  employee3 = Array((105,"Lester Cruz","Van Nuys","CA"),
(106,"John White","Inglewood","CA"))
val employee3RDD = sc.parallelize(employee3)
employeesRDD.collect.foreach(println)

(103,Mark Choi,Torrance,CA)
(104,Janet Reyes,Rolling Hills,CA)
(105,Lester Cruz,Van Nuys,CA)
(106,John White,Inglewood,CA)
```

Subtract

Returns an RDD that contains only the elements that are in the first RDD.

```
val  listEmployees = Array((103,"Mark Choi","Torrance","CA"),
(104,"Janet Reyes","Rolling Hills","CA"),(105,"Lester Cruz","Van Nuys","CA"))
val listEmployeesRDD = sc.parallelize(listEmployees)
```

```
val  exEmployees = Array((103,"Mark Choi","Torrance","CA"))
val exEmployeesRDD = sc.parallelize(exEmployees)

val currentEmployeesRDD = listEmployeesRDD.subtract(exEmployeesRDD)

currentEmployeesRDD.collect.foreach(println)

(105,Lester Cruz,Van Nuys,CA)
(104,Janet Reyes,Rolling Hills,CA)
```

Coalesce

Coalesce reduces the number of partitions in an RDD. You might want to use coalesce after performing a filter on a large RDD. While filtering reduces the amount of data consumed by the new RDD, it inherits the number of partitions of the original RDD. If the new RDD is significantly smaller than the original RDD, it may have hundreds or thousands of small partitions, which could cause performance issues.

Using coalesce is also useful when you want to reduce the number of files generated by Spark when writing to HDFS, preventing the dreaded "small file" problem. Each partition gets written as separate files to HDFS. Note that you might run into performance issues when using coalesce since you are effectively reducing the degree of parallelism while writing to HDFS. Try increasing the number of partitions if that happens. This is applicable to DataFrames as well, which I will discuss later. In the example below, we're writing only one Parquet file to HDFS.

```
DF.coalesce(1).write.mode("append").parquet("/user/hive/warehouse/Mytable")
```

Repartition

Repartition can both decrease and increase the number of partitions in an RDD. You would generally use coalesce when reducing partitions since it's more efficient than repartition. Increasing the number of partitions might be useful by increasing the degree of parallelism when writing to HDFS. This is applicable to DataFrames as well, which I will discuss later. In the example below, we're writing 6 Parquet files to HDFS.

```
DF.repartition (6).write.mode("append").parquet("/user/hive/warehouse/Mytable")
```

> **Note** Coalesce is generally faster than repartition. Repartition will perform a full shuffle, creating new partitions and equally distributing data across worker nodes. Coalesce minimizes data movement and avoids a full shuffle by using existing partitions.

Actions

An action is an RDD operation that returns a value to the driver program. I list some of the most common actions. Refer to the online Spark documentation for a complete list of RDD actions.

Collect

Returns the entire content of the RDD to the driver program. Not advisable for large data sets.

```
val myCities = sc.parallelize(List("tokyo","new york","paris","san francisco"))
myCities.collect
res2: Array[String] = Array(tokyo, new york, paris, san francisco)
```

Take

Returns a subset of the RDD to the driver program.

```
val myCities = sc.parallelize(List("tokyo","new york","paris","san francisco"))
myCities.take(2)
res4: Array[String] = Array(tokyo, new york)
```

Count

Returns the number of items in the RDD.

```
val myCities = sc.parallelize(List("tokyo","new york","paris","san francisco"))
myCities.count
res3: Long = 4
```

Foreach

Execute the provided function to each item of the RDD.

```
val myCities = sc.parallelize(List("tokyo","new york","paris","san francisco"))

myCities.collect.foreach(println)

tokyo
new york
paris
san Francisco
```

Lazy Evaluation

Spark supports lazy evaluation, which is critical for big data processing. All transformations in Spark are lazily evaluated. Spark does not execute transformations immediately. You can continue to define more transformations. When you finally want the final results, you execute an action, which causes the transformations to be executed.

Caching

Each transformation is re-executed each time you run an action by default. You can cache an RDD in memory using the cache or persist method to avoid re-executing the transformation multiple times. There are several persistence levels to choose from such as MEMORY_ONLY, MEMORY_ONLY_SER, MEMORY_AND_DISK, MEMORY_AND_DISK_SER. and DISK_ONLY. Consult Apache Spark's online documentation for more details on caching. Off-heap caching with Alluxio (previously known as Tachyon) is discussed in Chapter 10.

Spark SQL, Dataset, and DataFrames API

While most of the excitement these days is focused on use cases that involve unstructured and semi-structured data (video analytics, image processing, and text mining to mention a few), a majority of the actual data analysis and processing is still done on structured, relational data. Most of the important business decisions by companies are still based on analysis done on relational data.

CHAPTER 5   INTRODUCTION TO SPARK

SparkSQL was developed to make it easier for Spark data engineers and data science to process and analyze structured data. Dataset is similar to an RDD in that it supports strong typing, but under the hood it has a much more efficient engine. Starting in Spark 2.0, the Dataset API is now the primary programming interface. The DataFrame is just a Dataset with named columns, very similar to relational table. Together, Spark SQL and DataFrames provide a powerful programming interface for processing and analyzing structured data. Here's a quick example on how to use the DataFrames API. I'll discuss the DataFrames API in more detail later.

```
val jsonDF = spark.read.json("/jsondata")
jsonDF.show
+---+------+--------------+-----+------+-----+
|age|  city|          name|state|userid|  zip|
+---+------+--------------+-----+------+-----+
| 35|Frisco| Jonathan West|   TX|   200|75034|
| 28|Dallas|Andrea Foreman|   TX|   201|75001|
| 69| Plano|  Kirsten Jung|   TX|   202|75025|
| 52| Allen|Jessica Nguyen|   TX|   203|75002|
+---+------+--------------+-----+------+-----+

jsonDF.select ("age", "city").show
+---+------+
|age|  city|
+---+------+
| 35|Frisco|
| 28|Dallas|
| 69| Plano|
| 52| Allen|
+---+------+

jsonDF.filter($"userid" < 202).show()
+---+------+--------------+-----+------+-----+
|age|  city|          name|state|userid|  zip|
+---+------+--------------+-----+------+-----+
| 35|Frisco| Jonathan West|   TX|   200|75034|
| 28|Dallas|Andrea Foreman|   TX|   201|75001|
+---+------+--------------+-----+------+-----+
```

```
jsonDF.createOrReplaceTempView("jsonDF")

val uzDF = spark.sql("SELECT userid, zip FROM jsonDF")

uzDF.show
+------+-----+
|userid|  zip|
+------+-----+
|   200|75034|
|   201|75001|
|   202|75025|
|   203|75002|
+------+-----+
```

> **Note** The DataFrame and Dataset APIs have been unified in Spark 2.0. The DataFrame is now just a type alias for a Dataset of Row, where a Row is a generic untyped object. In contrast, Dataset is a collection of strongly typed objects Dataset[T]. Scala supports strongly typed and untyped API, while in Java, Dataset[T] is the main abstraction. DataFrames is the main programming interface for R and Python due to its lack of support for compile-time type safety.

Spark Data Sources

Some of the most common tasks in data processing are reading from and writing to different data sources. I provide several examples in the next few pages. I cover Spark and Kudu integration in Chapter 6.

CSV

Spark provides you with different ways to read data from CSV files. You can read the data into an RDD first and then convert it to DataFrame.

```
val dataRDD = sc.textFile("/sparkdata/customerdata.csv")
val parsedRDD = dataRDD.map{_.split(",")}
```

```
case class CustomerData(customerid: Int, name: String, city: String, state:
String, zip: String)
val dataDF = parsedRDD.map{ a => CustomerData (a(0).toInt, a(1).toString,
a(2).toString,a(3).toString,a(4).toString) }.toDF
```

You can also use the Databricks CSV package. This method reads data directly in a DataFrame.

```
spark-shell --packages com.databricks:spark-csv_2.10:1.5.0

val dataDF = sqlContext.read.format("csv")
  .option("header", "true")
  .option("inferSchema", "true")
  .load("/sparkdata/customerdata.csv")
```

Starting in Spark 2.0, the CSV connector is already built in so there's no need to use Databrick's third-party package.

```
val dataDF = spark.read
            .option("header", "true")
            .option("inferSchema", "true")
            .csv("/sparkdata/customerdata.csv")
```

XML

Databricks has a Spark xml package that makes it easy to reads XML data.

```
cat users.xml

<userid>100</userid><name>Wendell Ryan</name><city>San Diego</city><state>CA</state><zip>92102</zip>
<userid>101</userid><name>Alicia Thompson</name><city>Berkeley</city><state>CA</state><zip>94705</zip>
<userid>102</userid><name>Felipe Drummond</name><city>Palo Alto</city><state>CA</state><zip>94301</zip>
<userid>103</userid><name>Teresa Levine</name><city>Walnut Creek</city><state>CA</state><zip>94507</zip>
```

```
hadoop fs -mkdir /xmldata
hadoop fs -put users.xml /xmldata

spark-shell --packages  com.databricks:spark-xml:2.10:0.4.1
```

Create a DataFrame using Spark XML. In this example, we specify the row tag and the path in HDFS where the XML file is located.

```
val xmlDF = sqlContext.read.format("com.databricks.spark.xml").option
("rowTag", "user").load("/xmldata/");

xmlDF: org.apache.spark.sql.DataFrame = [city: string, name: string,
state: string, userid: bigint, zip: bigint]
```

Let's also take a look at the data.

```
xmlDF.show

+------------+---------------+-----+------+-----+
|        city|           name|state|userid|  zip|
+------------+---------------+-----+------+-----+
|   San Diego|   Wendell Ryan|   CA|   100|92102|
|    Berkeley|Alicia Thompson|   CA|   101|94705|
|   Palo Alto|Felipe Drummond|   CA|   102|94301|
|Walnut Creek|  Teresa Levine|   CA|   103|94507|
+------------+---------------+-----+------+-----+
```

JSON

We'll create a JSON file as sample data for this example. Make sure the file is in a folder in HDFS called /jsondata.

```
cat users.json

{"userid": 200, "name": "Jonathan West", "city":"Frisco", "state":"TX",
"zip": "75034", "age":35}
{"userid": 201, "name": "Andrea Foreman", "city":"Dallas", "state":"TX",
"zip": "75001", "age":28}
```

```
{"userid": 202, "name": "Kirsten Jung", "city":"Plano", "state":"TX",
"zip": "75025", "age":69}
{"userid": 203, "name": "Jessica Nguyen", "city":"Allen", "state":"TX",
"zip": "75002", "age":52}
```

Create a DataFrame from the JSON file.

```
val jsonDF = sqlContext.read.json("/jsondata")

jsonDF: org.apache.spark.sql.DataFrame = [age: bigint, city: string, name:
string, state: string, userid: bigint, zip: string]
```

Check the data.

```
jsonDF.show

+---+------+--------------+-----+------+-----+
|age|  city|          name|state|userid|  zip|
+---+------+--------------+-----+------+-----+
| 35|Frisco| Jonathan West|   TX|   200|75034|
| 28|Dallas|Andrea Foreman|   TX|   201|75001|
| 69| Plano|  Kirsten Jung|   TX|   202|75025|
| 52| Allen|Jessica Nguyen|   TX|   203|75002|
+---+------+--------------+-----+------+-----+
```

Relational Databases Using JDBC

We use MySQL in this example, but other relational databases such as Oracle, SQL Server, Teradata, and PostgreSQL, to mention a few, are also supported. As long as the relational database has a JDBC/ODBC driver, it should be accessible from Spark. Performance is dependent on your JDBC/ODBC driver's support for batch operations. Please check your JDBC driver's documentation for more details.

```
mysql -u root -pmypassword

create databases salesdb;

use salesdb;

create table customers (
```

```
customerid INT,
name VARCHAR(100),
city VARCHAR(100),
state CHAR(3),
zip  CHAR(5));

spark-shell --driver-class-path mysql-connector-java-5.1.40-bin.jar --jars
mysql-connector-java-5.1.40-bin.jar
```

Note In some versions of Spark --jars does not add the JAR in the driver's class path.[vii] It is recommended that you include the JDBC driver in your --jars and the Spark classpath.[viii]

Start the spark-shell. Take note that I had to include the MySQL driver as a parameter in both the –driver-class-path and –jars. You may not need to do this in newer versions of Spark.

Read the csv file into a DataFrame

```
val dataRDD = sc.textFile("/home/hadoop/test.csv")
val parsedRDD = dataRDD.map{_.split(",")}
case class CustomerData(customerid: Int, name: String, city: String, state:
String, zip: String)

val dataDF = parsedRDD.map{ a => CustomerData (a(0).toInt, a(1).toString,
a(2).toString,a(3).toString,a(4).toString)}.toDF
```

Register the data frame as a temp table so that we can run SQL queries against it. In Spark 2.x, use createOrReplaceTempView.

```
dataDF.registerTempTable("dataDF")
```

Let's set up the connection properties.

```
val jdbcUsername = "myuser"
val jdbcPassword = "mypass"
val jdbcHostname = "10.0.1.112"
val jdbcPort = 3306
val jdbcDatabase ="salesdb"
```

```
val jdbcrewriteBatchedStatements = "true"
val jdbcUrl = s"jdbc:mysql://${jdbcHostname}:${jdbcPort}/${jdbcDatabase}?user=${jdbcUsername}&password=${jdbcPassword}&rewriteBatchedStatements=${jdbcrewriteBatchedStatements}"

val connectionProperties = new java.util.Properties()
```

This will allow us to specify the correct save mode - Append, Overwrite, etc.

```
import org.apache.spark.sql.SaveMode
```

Insert the data returned by the SELECT statement to the customer table stored in the MySQL salesdb database.

```
sqlContext.sql("select * from dataDF").write.mode(SaveMode.Append).jdbc(jdbcUrl, "customers", connectionProperties)
```

Let's read a table using JDBC. Let's populate the users table in MySQL with some test data. Make sure the users table exists in the salesdb database.

```
mysql -u root -pmypassword

use salesdb;

describe users;
+--------+--------------+------+-----+---------+-------+
| Field  | Type         | Null | Key | Default | Extra |
+--------+--------------+------+-----+---------+-------+
| userid | bigint(20)   | YES  |     | NULL    |       |
| name   | varchar(100) | YES  |     | NULL    |       |
| city   | varchar(100) | YES  |     | NULL    |       |
| state  | char(3)      | YES  |     | NULL    |       |
| zip    | char(5)      | YES  |     | NULL    |       |
| age    | tinyint(4)   | YES  |     | NULL    |       |
+--------+--------------+------+-----+---------+-------+

select * from users;
Empty set (0.00 sec)

insert into users values (300,'Fred Stevens','Torrance','CA',90503,23);
```

CHAPTER 5 INTRODUCTION TO SPARK

```
insert into users values (301,'Nancy Gibbs','Valencia','CA',91354,49);
insert into users values (302,'Randy Park','Manhattan Beach','CA',90267,21);
insert into users values (303,'Victoria Loma','Rolling Hills','CA',90274,75);
select * from users;
+--------+---------------+-----------------+-------+-------+------+
| userid | name          | city            | state | zip   | age  |
+--------+---------------+-----------------+-------+-------+------+
|    300 | Fred Stevens  | Torrance        | CA    | 90503 |   23 |
|    301 | Nancy Gibbs   | Valencia        | CA    | 91354 |   49 |
|    302 | Randy Park    | Manhattan Beach | CA    | 90267 |   21 |
|    303 | Victoria Loma | Rolling Hills   | CA    | 90274 |   75 |
+--------+---------------+-----------------+-------+-------+------+

spark-shell --driver-class-path mysql-connector-java-5.1.40-bin.jar --jars mysql-connector-java-5.1.40-bin.jar
```

Let's set up the jdbc url and connection properties.

```
val jdbcURL = s"jdbc:mysql://10.0.1.101:3306/salesdb?user=myuser&password=mypass"

val connectionProperties = new java.util.Properties()
```

We can create a DataFrame from an entire table.

```
val mysqlDF = sqlContext.read.jdbc(jdbcURL, "users", connectionProperties)

mysqlDF.show
+------+-------------+---------------+-----+-----+---+
|userid|         name|           city|state|  zip|age|
+------+-------------+---------------+-----+-----+---+
|   300| Fred Stevens|       Torrance|   CA|90503| 23|
|   301|  Nancy Gibbs|       Valencia|   CA|91354| 49|
|   302|   Randy Park|Manhattan Beach|   CA|90267| 21|
|   303|Victoria Loma|  Rolling Hills|   CA|90274| 75|
+------+-------------+---------------+-----+-----+---+
```

Parquet

Reading and writing to Parquet is straightforward.

```
val employeeDF = spark.read.load("/sparkdata/employees.parquet")

employeeDF.select("id","firstname","lastname","salary").write.
format("parquet").save("/sparkdata/myData.parquet")
```

You can run SELECT statements on Parquet files directly.

```
val myDF = spark.sql("SELECT * FROM parquet.`/sparkdata/myData.parquet`")
```

HBase

There are different ways to access HBase from Spark. As discussed earlier, Scala has access to all Java libraries including the HBase client APIs. This is not the preferred way to access HBase from Spark, but some developers might find them handy. Another way to access HBase is via Impala, which I discuss in Chapter 6.

Note Spark on HBase is the preferred way of accessing HBase from Spark. However, it only works on Spark 1.x at the time of this writing. Spark on HBase is not supported on Spark 2.x.

You can use SaveAsHadoopDataset to write data to HBase.

Start the HBase shell. Create an HBase table and populate it with test data.

```
hbase shell

create 'users', 'cf1'
```

Start the spark-shell.

```
spark-shell
```

Import all necessary packages.

```
import org.apache.hadoop.fs.Path;
import org.apache.hadoop.hbase.{HBaseConfiguration, HTableDescriptor}
import org.apache.hadoop.hbase.client.HBaseAdmin
```

```
import org.apache.hadoop.hbase.mapreduce.TableInputFormat
import org.apache.hadoop.hbase.HColumnDescriptor
import org.apache.hadoop.hbase.client.Put;
import org.apache.hadoop.hbase.client.Get;
import org.apache.hadoop.hbase.client.HTable;
import org.apache.hadoop.conf.Configuration;
import org.apache.hadoop.hbase.client.Result;
import org.apache.hadoop.hbase.util.Bytes;
import java.io.IOException;

val hconf = HBaseConfiguration.create()
val jobConf = new JobConf(hconf, this.getClass)
jobConf.setOutputFormat(classOf[TableOutputFormat])
jobConf.set(TableOutputFormat.OUTPUT_TABLE,"users")

val num = sc.parallelize(List(1,2,3,4,5,6))

val theRDD = num.filter.map(x=>{

        val rowkey = "row" + x

        val put = new Put(Bytes.toBytes(rowkey))

        put.add(Bytes.toBytes("cf1"), Bytes.toBytes("fname"), Bytes.
        toBytes("my fname" + x))

    (new ImmutableBytesWritable, put)
})
theRDD.saveAsHadoopDataset(jobConf)
```

You can also use the HBase client API from Spark to read and write data to HBase. Start the HBase shell. Create another HBase table and populate it with test data.

```
hbase shell

create 'employees', 'cf1'

put 'employees','400','cf1:name', 'Patrick Montalban'
put 'employees','400','cf1:city', 'Los Angeles'
put 'employees','400','cf1:state', 'CA'
put 'employees','400','cf1:zip', '90010'
```

CHAPTER 5 INTRODUCTION TO SPARK

```
put 'employees','400','cf1:age', '71'
put 'employees','401','cf1:name', 'Jillian Collins'
put 'employees','401','cf1:city', 'Santa Monica'
put 'employees','401','cf1:state', 'CA'
put 'employees','401','cf1:zip', '90402'
put 'employees','401','cf1:age', '45'

put 'employees','402','cf1:name', 'Robert Sarkisian'
put 'employees','402','cf1:city', 'Glendale'
put 'employees','402','cf1:state', 'CA'
put 'employees','402','cf1:zip', '91204'
put 'employees','402','cf1:age', '29'

put 'employees','403','cf1:name', 'Warren Porcaro'
put 'employees','403','cf1:city', 'Burbank'
put 'employees','403','cf1:state', 'CA'
put 'employees','403','cf1:zip', '91523'
put 'employees','403','cf1:age', '62'
```

Let's verify if the data was successfully inserted into our HBase table.

```
scan 'employees'
```

ROW	COLUMN+CELL
400	column=cf1:age, timestamp=1493105325812, value=71
400	column=cf1:city, timestamp=1493105325691, value=Los Angeles
400	column=cf1:name, timestamp=1493105325644, value=Patrick Montalban
400	column=cf1:state, timestamp=1493105325738, value=CA
400	column=cf1:zip, timestamp=1493105325789, value=90010
401	column=cf1:age, timestamp=1493105334417, value=45
401	column=cf1:city, timestamp=1493105333126, value=Santa Monica
401	column=cf1:name, timestamp=1493105333050, value=Jillian Collins
401	column=cf1:state, timestamp=1493105333145, value=CA

401	column=cf1:zip, timestamp=1493105333165, value=90402
402	column=cf1:age, timestamp=1493105346254, value=29
402	column=cf1:city, timestamp=1493105345053, value=Glendale
402	column=cf1:name, timestamp=1493105344979, value=Robert Sarkisian
402	column=cf1:state, timestamp=1493105345074, value=CA
402	column=cf1:zip, timestamp=1493105345093, value=91204
403	column=cf1:age, timestamp=1493105353650, value=62
403	column=cf1:city, timestamp=1493105352467, value=Burbank
403	column=cf1:name, timestamp=1493105352445, value=Warren Porcaro
403	column=cf1:state, timestamp=1493105352513, value=CA
403	column=cf1:zip, timestamp=1493105352549, value=91523

Start the spark-shell.

```
spark-shell
```

Import all necessary packages.

```
val configuration = HBaseConfiguration.create()
```

Specify the HBase table and rowkey.

```
val table = new HTable(configuration, "employees");
val g = new Get(Bytes.toBytes("401"))
val result = table.get(g);
```

Extract the values from the table.

```
val val2 = result.getValue(Bytes.toBytes("cf1"),Bytes.toBytes("name"));
val val3 = result.getValue(Bytes.toBytes("cf1"),Bytes.toBytes("city"));
val val4 = result.getValue(Bytes.toBytes("cf1"),Bytes.toBytes("state"));
val val5 = result.getValue(Bytes.toBytes("cf1"),Bytes.toBytes("zip"));
val val6 = result.getValue(Bytes.toBytes("cf1"),Bytes.toBytes("age"));
```

Convert the values to the appropriate data types.

```
val id = Bytes.toString(result.getRow())
val name = Bytes.toString(val2);
val city = Bytes.toString(val3);
val state = Bytes.toString(val4);
val zip = Bytes.toString(val5);
val age = Bytes.toShort(val6);
```

Print the values.

```
println("employee id: " + id + "name: " + name + "city: " + city + 
"state: " + state + "zip: " + zip + "age: " + age);
```

```
employee id: 401 name: Jillian Collins city: Santa Monica state: CA zip: 
90402 age: 13365
```

Let's write to HBase using the HBase API.

```
val configuration = HBaseConfiguration.create()
val table = new HTable(configuration, "employees");
```

Specify a new rowkey.

```
val p = new Put(new String("404").getBytes());
```

Populate the cells with the new values.

```
p.add("cf1".getBytes(), "name".getBytes(), new String("Denise Shulman").
getBytes());
p.add("cf1".getBytes(), "city".getBytes(), new String("La Jolla").
getBytes());
p.add("cf1".getBytes(), "state".getBytes(), new String("CA").getBytes());
p.add("cf1".getBytes(), "zip".getBytes(), new String("92093").getBytes());
p.add("cf1".getBytes(), "age".getBytes(), new String("56").getBytes());
```

Write to the HBase table.

```
table.put(p);
table.close();
```

CHAPTER 5 ■ INTRODUCTION TO SPARK

Confirm that the values were successfully inserted into the HBase table.
Start the HBase shell.

```
hbase shell

scan 'employees'
```

ROW	COLUMN+CELL
400	column=cf1:age, timestamp=1493105325812, value=71
400	column=cf1:city, timestamp=1493105325691, value=Los Angeles
400	column=cf1:name, timestamp=1493105325644, value=Patrick Montalban
400	column=cf1:state, timestamp=1493105325738, value=CA
400	column=cf1:zip, timestamp=1493105325789, value=90010
401	column=cf1:age, timestamp=1493105334417, value=45
401	column=cf1:city, timestamp=1493105333126, value=Santa Monica
401	column=cf1:name, timestamp=1493105333050, value=Jillian Collins
401	column=cf1:state, timestamp=1493105333145, value=CA
401	column=cf1:zip, timestamp=1493105333165, value=90402
402	column=cf1:age, timestamp=1493105346254, value=29
402	column=cf1:city, timestamp=1493105345053, value=Glendale
402	column=cf1:name, timestamp=1493105344979, value=Robert Sarkisian
402	column=cf1:state, timestamp=1493105345074, value=CA
402	column=cf1:zip, timestamp=1493105345093, value=91204
403	column=cf1:age, timestamp=1493105353650, value=62
403	column=cf1:city, timestamp=1493105352467, value=Burbank
403	column=cf1:name, timestamp=1493105352445, value=Warren Porcaro
403	column=cf1:state, timestamp=1493105352513, value=CA
403	column=cf1:zip, timestamp=1493105352549, value=91523
404	column=cf1:age, timestamp=1493123890714, value=56
404	column=cf1:city, timestamp=1493123890714, value=La Jolla
404	column=cf1:name, timestamp=1493123890714, value=Denise Shulman
404	column=cf1:state, timestamp=1493123890714, value=CA
404	column=cf1:zip, timestamp=1493123890714, value=92093

Amazon S3

Amazon S3 is a popular object store frequently used as data store for transient clusters. It's also a cost-effective storage for backups and cold data. Reading data from S3 is just like reading data from HDFS or any other file system.

Read a CSV file from Amazon S3. Make sure you've configured your S3 credentials.

```
val myCSV = sc.textFile("s3a://mydata/customers.csv")
```

Map CSV data to an RDD.

```
import org.apache.spark.sql.Row

val myRDD = myCSV.map(_.split(',')).map(e => Row(r(0).trim.toInt, r(1), r(2).trim.toInt, r(3)))
```

Create a schema.

```
import org.apache.spark.sql.types.{StructType, StructField, StringType, IntegerType};

val mySchema = StructType(Array(
StructField("customerid",IntegerType,false),
StructField("customername",StringType,false),
StructField("age",IntegerType,false),
StructField("city",StringType,false)))

val myDF = sqlContext.createDataFrame(myRDD, mySchema)
```

Solr

Solr is a popular search platform that provides full-text search and real-time indexing capabilities. You can interact with Solr from Spark using SolrJ.[ix]

```
import java.net.MalformedURLException;
import org.apache.solr.client.solrj.SolrServerException;
import org.apache.solr.client.solrj.impl.HttpSolrServer;
import org.apache.solr.client.solrj.SolrQuery;
import org.apache.solr.client.solrj.response.QueryResponse;
import org.apache.solr.common.SolrDocumentList;
```

```
val solr = new HttpSolrServer("http://master02:8983/solr/mycollection");

val query = new SolrQuery();

query.setQuery("*:*");
query.addFilterQuery("userid:3");
query.setFields("userid","name","age","city");
query.setStart(0);
query.set("defType", "edismax");

val response = solr.query(query);
val results = response.getResults();

println(results);
```

A much better way to access Solr collections from Spark is by using the spark-solr package. Lucidworks started the spark-solr project to provide Spark-Solr integration.[x] Using spark-solr is so much easier and powerful compared to SolrJ, allowing you to create DataFrames from Solr collections and using SparkSQL to interact with them.

Start by importing the jar file from spark-shell. You can download the jar file from Lucidworks's website.

```
spark-shell --jars spark-solr-3.0.1-shaded.jar
```

Specify the collection and connection information.

```
val myOptions = Map( "collection" -> "mycollection","zkhost" -> "{ master02:8983/solr}")
```

Create a DataFrame.

```
val solrDF = spark.read.format("solr")
  .options(myOptions)
  .load
```

Microsoft Excel

I've encountered several requests on how to access Excel worksheets from Spark. While this is not something that I would normally do, working with Excel is a reality in almost every corporate IT environment.

A company called Crealytics developed a Spark plug-in for interacting with Excel. The library requires Spark 2.x. The package can be added using the --packages command-line option.[xi]

```
spark-shell --packages com.crealytics:spark-excel_2.11:0.9.12
```

Create a DataFrame from an Excel worksheet.

```
val ExcelDF = spark.read
   .format("com.crealytics.spark.excel")
   .option("sheetName", "sheet1")
   .option("useHeader", "true")
   .option("inferSchema", "true")
   .option("treatEmptyValuesAsNulls", "true")
   .load("budget.xlsx")
```

Write a DataFrame to an Excel worksheet.

```
ExcelDF2.write
  .format("com.crealytics.spark.excel")
  .option("sheetName", "sheet1")
  .option("useHeader", "true")
  .mode("overwrite")
  .save("budget2.xlsx")
```

You can find more details from their github page: github.com/crealytics.

Secure FTP

Downloading files from SFTP and writing DataFrames to an SFTP server is also a popular request. SpringML provides a Spark SFTP connector library. The library requires Spark 2.x and utilizes jsch, a Java implementation of SSH2. Reading from and writing to SFTP servers will be executed as a single process.

The package can be added using the --packages command-line option.[xii]

```
spark-shell --packages com.springml:spark-sftp_2.11:1.1.1
```

Create a DataFrame from the file in SFTP server.

```
val SftpDF = spark.read.
            format("com.springml.spark.sftp").
            option("host", "sftpserver.com").
            option("username", "myusername").
            option("password", "mypassword").
            option("inferSchema", "true").
            option("fileType", "csv").
            option("delimiter", ",").
            load("/myftp/myfile.csv")
```

Write DataFrame as CSV file to FTP server.

```
SftpDF2.write.
       format("com.springml.spark.sftp").
       option("host", "sftpserver.com").
       option("username", "myusername").
       option("password", "mypassword").
       option("fileType", "csv").
       option("delimiter", ",").
       save("/myftp/myfile.csv")
```

You can find more details from their github page: github.com/springml/spark-sftp.

Spark MLlib (DataFrame-Based API)

Machine Learning is one of Spark's main applications. The DataFrame-based API (Spark ML or Spark ML Pipelines) is now the primary API for Spark. The RDD-based API (Spark MLlib) is entering maintenance mode. We won't cover the old RDD-based API. Previously, the DataFrame-based API was informally referred to as Spark ML and Spark ML Pipelines (spark.ml package) to differentiate it from the RDD-based API, which was named based on the original spark.mllib package. The RDD-based API will be deprecated in Spark 2.3 once the DataFrames-based API reaches feature parity.[xiii] The RDD-based API will be removed in Spark 3.0. For now, Spark MLlib includes both APIs.

CHAPTER 5 INTRODUCTION TO SPARK

The DataFrames based API is faster and easier to use than the RDD-based API, allowing users to use SQL and take advantage of Catalyst and Tungsten optimizations. The DataFrames-based API makes it easy to transform features by providing a higher-level abstraction for representing tabular data similar to a relational database table, making it a natural choice for implementing pipelines.

A majority of the operations performed in a typical machine learning pipeline is feature transformation. As shown in Figure 5-3, DataFrames makes it easy to perform feature transformation.[xiv] The tokenizer breaks the text into a bag of words, appending the words into the output second DataFrame. TF-IDF takes the second DataFrame as input, converts the bag of words into a feature vector, and adds them to the third DataFrame.

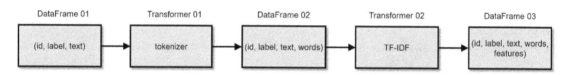

Figure 5-3. *Example Feature Transformation in a typical Spark ML pipeline*

Pipeline

A pipeline is just a sequence of connected stages to create a machine learning workflow. A stage can be either a Transformer or Estimator.

Transformer

A transformer takes a DataFrame as input and outputs a new DataFrame with additional columns appended to the new DataFrame. The new DataFrame includes the columns from the input DataFrame and the additional columns.

Estimator

An estimation is a machine learning algorithm that fits a model on training data. An estimator Estimators accept training data and produces a machine learning model.

ParamGridBuilder

A ParamGridBuilder is used to build a parameter grid. The CrossValidator performs a grid search and trains models with a combination of user-specified hyperparameters in the parameter grid.

CrossValidator

A CrossValidator cross evaluates fitted machine learning models and outputs the best one by trying to fit the underlying estimator with user-specified combinations of hyperparameters.

Evaluator

An evaluator calculates the performance of your machine learning models. An evaluator outputs a metric such as precision or recall to measure how well a fitted model performs.

Example

Let's work on an example. We'll use the Heart Disease Data Set[xv] from the UCI Machine Learning Repository to predict the presence of heart disease. The data was collected by Robert Detrano, MD, PhD, from the VA Medical Center, Long Beach and Cleveland Clinic Foundation. Historically, the Cleveland data set has been the subject of numerous research so we'll use that data set. The original data set has 76 attributes, but only 14 of them are traditionally used by ML researchers (Table 5-2). We will simply perform binomial classification and determine if the patient has heart disease or not.

CHAPTER 5 INTRODUCTION TO SPARK

Table 5-2. *Cleveland Heart Disease Data Set Attribute Information*

Attribute	Description
age	Age
sex	Sex
cp	Chest pain type
trestbps	Resting blood pressure
chol	Serum cholesterol in mg/dl
fbs	Fasting blood sugar > 120 mg/dl
restecg	Resting electrocardiographic results
thalach	Maximum heart rate achieved
exang	Exercise induced angina
oldpeak	ST depression induced by exercise relative to rest
slope	The slope of the peak exercise ST segment
ca	Number of major vessels (0–3) colored by flourosopy
thal	Thalium stress test result
num	The predicted attribute – diagnosis of heart disease

Let's start. Add the column names to the CSV file before starting. We'll need to download the file and copy it to HDFS.

wget http://archive.ics.uci.edu/ml/machine-learning-databases/heart-disease/cleveland.data

head -n 10 processed.cleveland.data

63.0,1.0,1.0,145.0,233.0,1.0,2.0,150.0,0.0,2.3,3.0,0.0,6.0,0
67.0,1.0,4.0,160.0,286.0,0.0,2.0,108.0,1.0,1.5,2.0,3.0,3.0,2
67.0,1.0,4.0,120.0,229.0,0.0,2.0,129.0,1.0,2.6,2.0,2.0,7.0,1
37.0,1.0,3.0,130.0,250.0,0.0,0.0,187.0,0.0,3.5,3.0,0.0,3.0,0
41.0,0.0,2.0,130.0,204.0,0.0,2.0,172.0,0.0,1.4,1.0,0.0,3.0,0
56.0,1.0,2.0,120.0,236.0,0.0,0.0,178.0,0.0,0.8,1.0,0.0,3.0,0

```
62.0,0.0,4.0,140.0,268.0,0.0,2.0,160.0,0.0,3.6,3.0,2.0,3.0,3
57.0,0.0,4.0,120.0,354.0,0.0,0.0,163.0,1.0,0.6,1.0,0.0,3.0,0
63.0,1.0,4.0,130.0,254.0,0.0,2.0,147.0,0.0,1.4,2.0,1.0,7.0,2
53.0,1.0,4.0,140.0,203.0,1.0,2.0,155.0,1.0,3.1,3.0,0.0,7.0,1
```

```
hadoop fs -put processed.cleveland.data /tmp/data
```

Then use the spark-shell to interactively create our model using Spark MLlib as shown in Listing 5-2.

Listing 5-2. Performing binary classification using Random Forest

```
spark-shell --packages com.databricks:spark-csv_2.10:1.5.0

import org.apache.spark._
import org.apache.spark.rdd.RDD
import org.apache.spark.sql.SQLContext
import org.apache.spark.sql.functions._
import org.apache.spark.sql.types._
import org.apache.spark.sql._
import org.apache.spark.ml.classification.RandomForestClassifier
import org.apache.spark.ml.evaluation.BinaryClassificationEvaluator
import org.apache.spark.ml.feature.StringIndexer
import org.apache.spark.ml.feature.VectorAssembler
import org.apache.spark.ml.tuning.{ParamGridBuilder, CrossValidator}
import org.apache.spark.ml.{Pipeline, PipelineStage}
import org.apache.spark.mllib.evaluation.RegressionMetrics
import org.apache.spark.ml.param.ParamMap
import org.apache.kudu.spark.kudu._

val dataDF = sqlContext.read.format("csv")
  .option("header", "true")
  .option("inferSchema", "true")
  .load("/tmp/data/processed.cleveland.data")
```

CHAPTER 5 INTRODUCTION TO SPARK

```
dataDF.printSchema
root
 |-- id: string (nullable = false)
 |-- age: float (nullable = true)
 |-- sex: float (nullable = true)
 |-- cp: float (nullable = true)
 |-- trestbps: float (nullable = true)
 |-- chol: float (nullable = true)
 |-- fbs: float (nullable = true)
 |-- restecg: float (nullable = true)
 |-- thalach: float (nullable = true)
 |-- exang: float (nullable = true)
 |-- oldpeak: float (nullable = true)
 |-- slope: float (nullable = true)
 |-- ca: float (nullable = true)
 |-- thal: float (nullable = true)
 |-- num: float (nullable = true)
```

```
val myFeatures = Array("age", "sex", "cp", "trestbps", "chol", "fbs",
      "restecg", "thalach", "exang", "oldpeak", "slope",
      "ca", "thal", "num")

val myAssembler = new VectorAssembler().setInputCols(myFeatures).
setOutputCol("features")

val dataDF2 = myAssembler.transform(dataDF)

val myLabelIndexer = new StringIndexer().setInputCol("num").
setOutputCol("label")

val dataDF3 = mylabelIndexer.fit(dataDF2).transform(dataDF2)

val dataDF4 = dataDF3.where(dataDF3("ca").isNotNull).where(dataDF3("thal").
isNotNull).where(dataDF3("num").isNotNull)

val Array(trainingData, testData) = dataDF4.randomSplit(Array(0.8, 0.2), 101)

val myRFclassifier = new RandomForestClassifier().setFeatureSubsetStrategy
("auto").setSeed(101)
```

```
val myEvaluator = new BinaryClassificationEvaluator().setLabelCol("label")

val myParamGrid = new ParamGridBuilder()
    .addGrid(myRFclassifier.maxBins, Array(10, 20, 30))
    .addGrid(myRFclassifier.maxDepth, Array(5, 10, 15))
    .addGrid(myRFclassifier.numTrees, Array(20, 30, 40))
    .addGrid(myRGclassifier.impurity, Array("gini", "entropy"))
    .build()

val myPipeline = new Pipeline().setStages(Array(myRFclassifier))

val myCV = new CrosValidator()
    .setEstimator(myPipeline)
    .setEvaluator(myEvaluator)
    .setEstimatorParamMaps(myParamGrid)
    .setNumFolds(3)
```

We can now fit the model

```
val myCrossValidatorModel = myCV.fit(trainingData)
```

Let's evaluate the model.

```
val myEvaluatorParamMap = ParamMap(myEvaluator.metricName -> "areaUnderROC")

val aucTrainingData = myEvaluator.evaluate(CrossValidatorPrediction, myEvaluatorParamMap)
```

You can now make some predictions on our data.

```
val myCrossValidatorPrediction = myCrossValidatorModel.transform(testData)
```

Spark MLlib provides features for building pipelines, featurization, and popular machine learning algorithms for regression, classification, clustering, and collaborative filtering. *Advanced Analytics with Spark, 2nd edition,* by Sandy Ryza, Uri Laserson, Sean Owen, and Josh Wills (O'Reilly, 2017) provides a more in-depth treatment of machine learning with Spark. We'll use Kudu as a feature store in Chapter 6.

GraphX

Spark includes a graph processing framework called GraphX. There is a separate package called GraphFrames that is based on DataFrames. GraphFrames is currently not part of core Apache Spark. GraphX and GraphFrames are still considered immature and are not supported by Cloudera Enterprise at the time of this writing.[xvi] I won't cover them in this book, but feel free to visit Spark's online documentation for more details.

Spark Streaming

I cover Spark Streaming in Chapter 6. Spark 2.0 includes a new stream processing framework called Structured Streaming, a high-level streaming API built on top of Spark SQL. Structured Streaming is not supported by Cloudera at the time of this writing.[xvii]

Hive on Spark

Cloudera supports Hive on Spark for faster batch processing. Early benchmarks show an average of 3x faster performance than Hive on MapReduce.[xviii] Note that Hive on Spark is still mainly for batch processing and does not replace Impala for low-latency SQL queries. Hive on Spark is useful for organizations who want to take advantage of Spark's performance without having to learn Scala or Python. Some may find it non-trivial to refactor data processing pipelines due to the amount and complexity of HiveQL queries. Hive for Spark is ideal for those scenarios.

Spark 1.x vs Spark 2.x

Although plenty of code base out there still runs on Spark 1.x, most of your development should now be on Spark 2.x. Most of the Spark 2.x API is similar to 1.x, but there are some changes in 2.x that break API compatibility. Spark 2 is not compatible with Scala 2.10; only Scala 2.11 is supported. JDK 8 is also a requirement for Spark 2.2. Refer to Spark's online documentation for more details.

CHAPTER 5 INTRODUCTION TO SPARK

Monitoring and Configuration

There are several tools that you can use to monitor and configure Apache Spark. Cloudera Manager is the de facto administration tool for Cloudera Enterprise. Spark also includes system administration and monitoring capabilities.

Cloudera Manager

Cloudera Manager is the cluster administration tool that comes with Cloudera Enterprise. You can use Cloudera Manager to perform all sorts of administration tasks such as updating a configuration (Figure 5-4) and monitoring its performance (Figure 5-5).

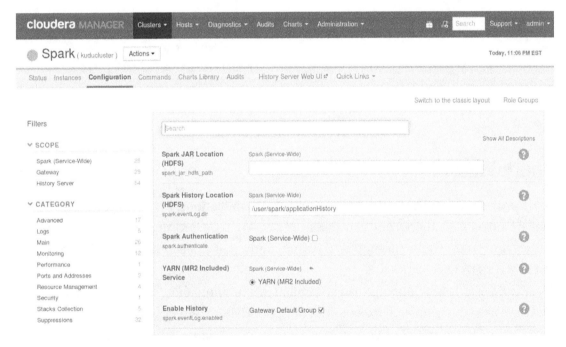

Figure 5-4. Using Cloudera Manager to configure Spark

153

CHAPTER 5   INTRODUCTION TO SPARK

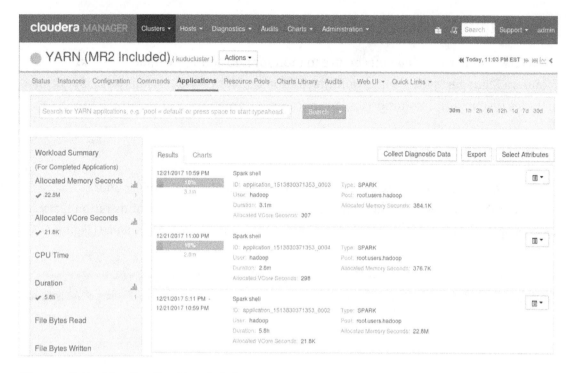

Figure 5-5. Monitoring Spark Jobs

Spark Web UI

Spark provides a couple of ways to monitor Spark applications. For Spark applications that are currently running, you can access its performance information on port 4040. If there are multiple jobs running on the same node, you can access the web UI on port 4041, 4042, and so on.

The Spark History Server provides detailed information about Spark applications that have already completed execution. The Spark History Server can be accessed on port 18088 (Figure 5-6). Detailed performance metrics (Figure 5-7 and Figure 5-8) and information about the Spark environment (Figure 5-9) can be viewed to aid in monitoring and troubleshooting.

CHAPTER 5 INTRODUCTION TO SPARK

Spark 1.6.0 History Server

Event log directory: hdfs://kuducluster:8020/user/spark/applicationHistory

Showing 1-20 of 77

App ID	App Name	Started	Completed	Duration	Spark User	Last Updated
application_1513754335415_0003	Spark shell	2017/12/20 18:27:55	2017/12/20 23:02:05	4.6 h	hadoop	2017/12/20 23:02:05
application_1513754335415_0002	Spark shell	2017/12/20 18:26:38	2017/12/20 22:15:36	3.8 h	hadoop	2017/12/20 22:15:36
application_1513754335415_0001	Spark shell	2017/12/20 18:23:48	2017/12/20 18:26:05	2.3 min	hadoop	2017/12/20 18:26:05
application_1512961464611_0004	Spark shell	2017/12/11 23:02:01	2017/12/12 00:03:23	1.0 h	hadoop	2017/12/12 00:03:23
application_1512961464611_0003	Spark shell	2017/12/11 22:58:25	2017/12/11 23:01:37	3.2 min	hadoop	2017/12/11 23:01:37
application_1512961464611_0002	Spark shell	2017/12/11 22:50:38	2017/12/11 22:56:25	5.8 min	hadoop	2017/12/11 22:56:25
application_1512961464611_0001	Spark shell	2017/12/11 22:47:02	2017/12/11 22:49:53	2.8 min	hadoop	2017/12/11 22:49:53
application_1494112645086_0002	Spark shell	2017/05/07 22:24:51	2017/05/08 00:10:54	1.8 h	hadoop	2017/05/08 00:10:54
application_1494112645086_0001	Spark shell	2017/05/07 16:37:54	2017/05/07 19:38:24	3.0 h	hadoop	2017/05/07 19:38:25
application_1494079501183_0004	Spark shell	2017/05/07 00:59:11	2017/05/07 02:00:56	1.0 h	hadoop	2017/05/07 02:00:56
application_1494079501183_0003	Spark shell	2017/05/07 00:53:39	2017/05/07 00:54:49	1.2 min	hadoop	2017/05/07 00:54:50
application_1494079501183_0002	Spark shell	2017/05/07 00:52:46	2017/05/07 00:53:25	38 s	hadoop	2017/05/07 00:53:25
application_1494079501183_0001	Spark shell	2017/05/07 00:50:52	2017/05/07 00:52:33	1.7 min	hadoop	2017/05/07 00:52:34
application_1494051656793_0006	Spark shell	2017/05/06 21:48:29	2017/05/06 23:19:58	1.5 h	hadoop	2017/05/06 23:19:58
application_1494051656793_0005	Spark shell	2017/05/06 21:47:18	2017/05/06 21:48:17	59 s	hadoop	2017/05/06 21:48:17
application_1494051656793_0004	Spark shell	2017/05/06 21:38:04	2017/05/06 21:47:06	9.0 min	hadoop	2017/05/06 21:47:06
application_1494051656793_0003	Spark shell	2017/05/06 20:24:51	2017/05/06 21:37:50	1.2 h	hadoop	2017/05/06 21:37:50
application_1494051656793_0002	Spark shell	2017/05/06 20:11:27	2017/05/06 20:24:29	13 min	hadoop	2017/05/06 20:24:30
application_1494051656793_0001	Spark shell	2017/05/06 16:47:05	2017/05/06 20:11:18	3.4 h	hadoop	2017/05/06 20:11:18
application_1493885505680_0001	Spark shell	2017/05/04 19:29:52	2017/05/04 19:40:50	11 min	hadoop	2017/05/04 19:40:50

Show incomplete applications

Figure 5-6. *Spark History Server*

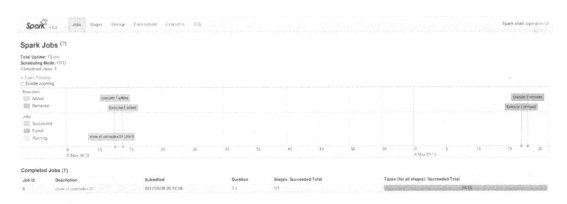

Figure 5-7. *Detailed performance information about a particular Spark job*

CHAPTER 5 INTRODUCTION TO SPARK

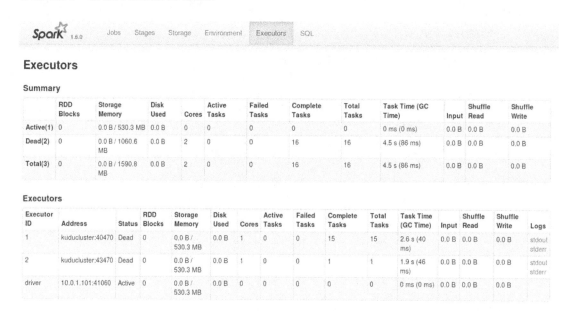

Figure 5-8. Performance metrics on Spark executors

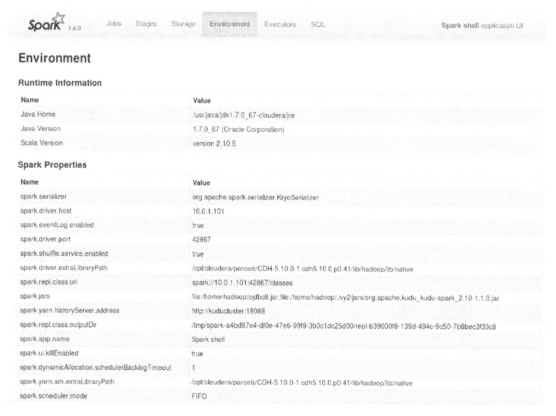

Figure 5-9. Information about the current Spark environment

Summary

Apache Spark has superseded MapReduce as the de facto big data processing framework. Data engineers and scientists appreciate Spark's simple and easy-to-use API, its ability to handle multiple workloads, and fast performance. Its focus on SparkSQL, Dataset, and DataFrame APIs are welcome improvements in making Spark more accessible and easier to use. Spark is the ideal data processing engine for Kudu. I discuss Spark and Kudu integration in Chapter 6.

References

i. Apache Spark; "Spark Overview," Apache Spark, 2018, https://spark.apache.org/docs/2.2.0/

ii. "Apache Software Foundation; "The Apache Software Foundation Announces Apache Spark as a Top-Level Project," Apache Software Foundation, 2018, https://blogs.apache.org/foundation/entry/the_apache_software_foundation_announces50

iii. Apache Spark; "Apache Spark News," Apache Spark, 2018, https://spark.apache.org/news/

iv. Matei Zaharia; "I'm Matei Zaharia, creator of Spark and CTO at Databricks. AMA!," Reddit, 2018, https://www.reddit.com/r/IAmA/comments/31bkue/im_matei_zaharia_creator_of_spark_and_cto_at/?st=j1svbrx9&sh=a8b9698e

v. Databricks; "What is Apache Spark?," Databricks, 2018, https://databricks.com/spark/about

vi. Jules Damji; "How to use SparkSession in Apache Spark 2.0," Databricks, 2018, https://databricks.com/blog/2016/08/15/how-to-use-sparksession-in-apache-spark-2-0.html

vii. Holden Karau, Rachel Warren; "High Performance Spark," O'Reilly, June 2017, https://www.safaribooksonline.com/library/view/high-performance-spark/9781491943199/

viii. Apache Spark; "JDBC To Other Databases," Apache Spark, 2018, http://spark.apache.org/docs/latest/sql-programming-guide.html#jdbc-to-other-databases

ix. Apache Lucene; "Using SolrJ," Apache Lucene, 2018, https://lucene.apache.org/solr/guide/6_6/using-solrj.html

x. Lucidworks; "Tools for reading data from Solr as a Spark RDD and indexing objects from Spark into Solr using SolrJ," Lucidworks, 2018, https://github.com/lucidworks/spark-solr

xi. Crealytics; "A Spark plugin for reading Excel files via Apache POI," Crealytics, 2018, https://github.com/crealytics/spark-excel

xii. SpringML; "Spark SFTP Connector Library," SpringML, 2018, https://github.com/springml/spark-sftp

xiii. Apache Spark; "Machine Learning Library (MLlib) Guide", Apache Spark, 2018, http://spark.apache.org/docs/latest/ml-guide.html

xiv. Xiangrui Meng, Joseph Bradley, Evan Sparks and Shivaram Venkataraman; "ML Pipelines: A New High-Level API for MLlib," Databricks, 2018, https://databricks.com/blog/2015/01/07/ml-pipelines-a-new-high-level-api-for-mllib.html

xv. David Aha; "Heart Disease Data Set," UCI Machine Learning Repository, 1988, http://archive.ics.uci.edu/ml/datasets/heart+Disease

xvi. Cloudera; "GraphX is not Supported," Cloudera, 2018, https://www.cloudera.com/documentation/spark2/latest/topics/spark2_known_issues.html#ki_graphx

xvii. Cloudera; "Structured Streaming is not Supported", Cloudera, 2018, https://www.cloudera.com/documentation/spark2/latest/topics/spark2_known_issues.html#ki_structured_streaming

xviii. Santosh Kumar; "Faster Batch Processing with Hive-on-Spark," Cloudera, 2016, https://vision.cloudera.com/faster-batch-processing-with-hive-on-spark/

CHAPTER 6

High Performance Data Processing with Spark and Kudu

Kudu is just a storage engine. You need a way to get data into it and out. As Cloudera's default big data processing framework, Spark is the ideal data processing and ingestion tool for Kudu. Not only does Spark provide excellent scalability and performance, Spark SQL and the DataFrame API make it easy to interact with Kudu.

If you are coming from a data warehousing background or if you are familiar with a relational database such as Oracle and SQL Server, you can consider Spark a more powerful and versatile equivalent to procedural extensions to SQL such as PL/SQL and T-SQL.

Spark and Kudu

You use Spark with Kudu using the Data Source API. You can use the —packages option in spark-shell or spark-submit to include kudu-spark dependency. You can also manually download the jar file from central.maven.org and include it in your —jars option. Consult the Apache Spark online documentation for more details on how to use sbt and Maven as your project build tool.

Spark 1.6.x

Use the kudu-spark_2.10 artifact if you are using Spark with Scala 2.10. For example:

```
spark-shell -packages org.apache.kudu:kudu-spark_2.10:1.1.0
spark-shell -jars kudu-spark_2.10-1.1.0.jar
```

CHAPTER 6   HIGH PERFORMANCE DATA PROCESSING WITH SPARK AND KUDU

Spark 2.x

Use the kudu-spark2_2.11 artifact if you are using Spark 2 with Scala 2.11. For example:

```
spark-shell --packages org.apache.kudu:kudu-spark2_2.11:1.1.0
spark-shell -jars kudu-spark2_2.11-1.1.0.jar
```

Kudu Context

You use a Kudu context in order to execute a DML statement against a Kudu table.[i] You have to specify the Kudu master servers and the port. We only have one Kudu context in our example below. There are usually multiple Kudu masters in a production environment; in that case you have to specify a comma-separated list of all the hostnames of the Kudu masters. I present examples on how to use Spark with Kudu. I also show examples on how to integrate Kudu with other data sources using Spark.

```
import org.apache.kudu.spark.kudu._
val kuduContext = new KuduContext("kudumaster01:7051")
```

We need to create our table before we can start.

```
impala-shell

CREATE TABLE customers
(
 id BIGINT PRIMARY KEY,
 name STRING,
 age SMALLINT
)
PARTITION BY HASH PARTITIONS 4
STORED AS KUDU;
```

Create a case class that will provide a schema to our sample data.

```
case class CustomerData(id: Long, name: String, age: Short)
```

Inserting Data

Create some sample data.

```
val data = Array(CustomerData(101,"Lisa Kim",60), CustomerData(102,"Casey Fernandez",45))

val insertRDD = sc.parallelize(data)
val insertDF = sqlContext.createDataFrame(insertRDD)

insertDF.show
```

```
+----------+---------------+---+
|customerid|           name|age|
+----------+---------------+---+
|       101|       Lisa Kim| 60|
|       102|Casey Fernandez| 45|
+----------+---------------+---+
```

Insert the DataFrame. Note the name of the table. This format is needed if the table was created in Impala. In this example, default is the name of the database and customers is the name of the table. This convention was adopted to prevent table name collision between tables created in Impala and tables created natively using the Kudu API or Spark.

```
kuduContext.insertRows(insertDF, "impala::default.customers")
```

Confirm that the data was successfully inserted.

```
val df = sqlContext.read.options(Map("kudu.master" -> "kudumaster01:7051","kudu.table" -> "impala::default.customers")).kudu
df.select("id","name","age").show()
```

```
+---+---------------+---+
| id|           name|age|
+---+---------------+---+
|102|Casey Fernandez| 45|
|101|       Lisa Kim| 60|
+---+---------------+---+
```

CHAPTER 6　HIGH PERFORMANCE DATA PROCESSING WITH SPARK AND KUDU

Updating a Kudu Table

Create an updated data set. Note that we modified the last name and the age.

```
val data = Array(CustomerData(101,"Lisa Kim",120), CustomerData(102,"Casey Jones",90))

val updateRDD = sc.parallelize(data)
val updateDF = sqlContext.createDataFrame(updateRDD)

updateDF.show

+---+----------+---+
| id|      name|age|
+---+----------+---+
|101|  Lisa Kim|120|
|102|Casey Jones| 90|
+---+----------+---+
```

Update the table.

```
kuduContext.updateRows(updateDF, "impala::default.customers");
```

Confirm that the table was successfully updated.

```
val df = sqlContext.read.options(Map("kudu.master" -> "kudumaster01:7051","kudu.table" -> "impala::default.customers")).kudu

df.select("id","name","age").show()

+---+----------+---+
| id|      name|age|
+---+----------+---+
|102|Casey Jones| 90|
|101|  Lisa Kim|120|
+---+----------+---+
```

Upserting Data

Create some sample data.

```
val data = Array(CustomerData(101,"Lisa Kim",240), CustomerData(102,"Casey Cullen",90),CustomerData(103,"Byron Miller",25))

val upsertRDD = sc.parallelize(data)
val upsertDF = sqlContext.createDataFrame(upsertRDD)

upsertDF.show
```

```
+---+------------+---+
| id|        name|age|
+---+------------+---+
|101|    Lisa Kim|240|
|102|Casey Cullen| 90|
|103|Byron Miller| 25|
+---+------------+---+
```

Upsert data – update all columns if the primary key exists, and insert rows if it doesn't.

```
kuduContext.upsertRows(upsertDF, "impala::default.customers")
```

Confirm if the data was successfully upserted.

```
val df = sqlContext.read.options(Map("kudu.master" -> "kudumaster01:7051","kudu.table" -> "impala::default.customers")).kudu

df.select("id","name","age").show()
```

```
+---+------------+---+
| id|        name|age|
+---+------------+---+
|102|Casey Cullen| 90|
|103|Byron Miller| 25|
|101|    Lisa Kim|240|
+---+------------+---+
```

CHAPTER 6 HIGH PERFORMANCE DATA PROCESSING WITH SPARK AND KUDU

Deleting Data

Inspect the data in the table.

```
val df = sqlContext.read.options(Map("kudu.master" -> "kudumaster01:7051","kudu.table" -> "impala::default.customers")).kudu

df.select("id","name","age").show()

+---+------+----------+
| id|  name|age|
+---+------+----------+
|102|Casey Cullen| 90|
|103|Byron Miller| 25|
|101|    Lisa Kim|240|
+---+------+----------+
```

Register the table so that we can use the table in our SQL query.

```
df.registerTempTable("customers")
```

Delete the data based on our query.

```
kuduContext.deleteRows(sqlContext.sql("select id from customers where name like 'Casey%'"), "impala::default.customers")
```

Confirm that the data was successfully deleted.

```
val df = sqlContext.read.options(Map("kudu.master" -> "kudumaster01:7051","kudu.table" -> "impala::default.customers")).kudu

df.select("id","name","age").show()

+---+------+----------+
| id|  name|age|
+---+------+----------+
|103|Byron Miller| 25|
|101|    Lisa Kim|240|
+---+------+----------+
```

Selecting Data

Select the data in the table.

```
val df = sqlContext.read.options(Map("kudu.master" ->
"kudumaster01:7051","kudu.table" -> "impala::default.customers")).kudu

df.select("id","name","age").show()

+---+------------+---+
| id|        name|age|
+---+------------+---+
|103|Byron Miller| 25|
|101|    Lisa Kim|240|
+---+------------+---+
```

You can also run your query by registering the table and using it in a SQL query. Note that you should use createOrReplaceTempView if you are using Spark 2.x.

```
df.registerTempTable("customers")

val df2 = sqlContext.sql("select * from customers where age=240")
df2.show

+---+--------+---+
| id|    name|age|
+---+--------+---+
|101|Lisa Kim|240|
+---+--------+---+
```

Creating a Kudu Table

This table will not be visible to Impala since this table was created in Spark. You have to create an external table in Impala and refer to this table.

```
Import org.apache.kudu.client.CreateTableOptions;

kuduContext.createTable("customer2", df.schema, Seq("customerid"), new
CreateTableOptions().setRangePartitionColumns(List("customerid").asJava).
setNumReplicas(1))
```

Inserting CSV into Kudu

Let's insert multiple rows from a CSV file.

```
val dataRDD = sc.textFile("/sparkdata/customerdata.csv")

val parsedRDD = dataRDD.map{_.split(",")}

case class CustomerData(customerid: Long, name: String, age: Short)

val dataDF = parsedRDD.map{a => CustomerData (a(0).toLong, a(1), a(2).toShort) }.toDF

kuduContext.insertRows(dataDF, "customer2");
```

Because we created the customer2 table via Spark we only had to specify the table name (customer2) instead of impala::default.customer2.

Inserting CSV into Kudu Using the spark-csv Package

We can also use the spark-csv package to process the sample CSV data. The command below will automatically download the spark-csv dependencies so make sure you have an Internet connection in your cluster node. Note the use of the comma separating the list of packages.

```
spark-shell -packages com.databricks:spark-csv_2.10:1.5.0,org.apache.kudu:kudu-spark_2.10:1.1.0

val dataDF = sqlContext.read.format("csv")
  .option("header", "true")
  .option("inferSchema", "true")
  .load("/sparkdata/customerdata.csv ")

kuduContext.insertRows(dataDF, "customer2");
```

Because we created the customer2 table via Spark, we only had to specify the table name (customer2) instead of impala::default.customer2.

Insert CSV into Kudu by Programmatically Specifying the Schema

You can use StructType to define a schema for your data set. Programmatically specifying a schema is helpful when the schema cannot be determined ahead of time.

Read a CSV file from HDFS.

```
val myCSV = sc.textFile("/tmp/mydata/customers.csv")
```

Map CSV data to an RDD

```
import org.apache.spark.sql.Row

val myRDD = myCSV.map(_.split(',')).map(e ⇒ Row(r(0).trim.toInt, r(1), r(2).trim.toInt, r(3)))
```

Create a schema.

```
import org.apache.spark.sql.types.{StructType, StructField, StringType, IntegerType};

val mySchema = StructType(Array(
StructField("customerid",IntegerType,false),
StructField("customername",StringType,false),
StructField("age",IntegerType,false),
StructField("city",StringType,false)))

val myDF = sqlContext.createDataFrame(myRDD, mySchema)
```

Insert DataFrame to Kudu.

```
kuduContext.insertRows(myDF, "impala::default.customers")
```

Remember that we created the customers table via Impala. Therefore, we need to use the format impala::database.table when referring to the customers table.

CHAPTER 6 HIGH PERFORMANCE DATA PROCESSING WITH SPARK AND KUDU

Inserting XML into Kudu Using the spark-xml Package

We'll create an XML file as a sample data for this example. We need to copy the file to HDFS.

```
cat users.xml
```

```
<userid>100</userid><name>Wendell Ryan</name><city>San Diego</city><state>CA</state><zip>92102</zip>
<userid>101</userid><name>Alicia Thompson</name><city>Berkeley</city><state>CA</state><zip>94705</zip>
<userid>102</userid><name>Felipe Drummond</name><city>Palo Alto</city><state>CA</state><zip>94301</zip>
<userid>103</userid><name>Teresa Levine</name><city>Walnut Creek</city><state>CA</state><zip>94507</zip>
```

```
hadoop fs -mkdir /xmldata
hadoop fs -put users.xml /xmldata
```

We'll use the spark-xml package to process the sample XML data. This package works similarly to the spark-csv package. The command below will automatically download the spark-xml package so make sure you have an Internet connection in your cluster node. Include the kudu-spark dependency as well.

```
spark-shell -packages com.databricks:spark-xml:2.10:0.4.1,org.apache.kudu:kudu-spark_2.10:1.1.0
```

Create a DataFrame using Spark XML. In this example, we specify the row tag and the path in HDFS where the XML file is located.

```
val xmlDF = sqlContext.read.format("com.databricks.spark.xml").option("rowTag", "user").load("/xmldata/");

xmlDF: org.apache.spark.sql.DataFrame = [city: string, name: string, state: string, userid: bigint, zip: bigint]
```

Let's also take a look at the data.

```
xmlDF.show
+------------+---------------+-----+------+-----+
|        city|           name|state|userid|  zip|
+------------+---------------+-----+------+-----+
|   San Diego|   Wendell Ryan|   CA|   100|92102|
|    Berkeley|Alicia Thompson|   CA|   101|94705|
|   Palo Alto|Felipe Drummond|   CA|   102|94301|
|Walnut Creek|   Teresa Levine|  CA|   103|94507|
+------------+---------------+-----+------+-----+
```

Let's inspect the schema.

```
xmlDF.printSchema
root
 |- age: long (nullable = true)
 |- city: string (nullable = true)
 |- name: string (nullable = true)
 |- state: string (nullable = true)
 |- userid: long (nullable = true)
 |- zip: long (nullable = true)
```

Now let's go back to the impala-shell and compare the schema with the structure of the users table. As you can see, the data type of the age and zip columns in the users table is different from the corresponding columns in the data frame. We'll get an error message if we try to insert this DataFrame into the Kudu table.

```
describe users;
+--------+---------+---------+-------------+
| name   | type    | comment | primary_key |
+--------+---------+---------+-------------+
| userid | bigint  |         | true        |
| name   | string  |         | false       |
| city   | string  |         | false       |
| state  | string  |         | false       |
| zip    | string  |         | false       |
| age    | tinyint |         | false       |
+--------+---------+---------+-------------+
```

CHAPTER 6 HIGH PERFORMANCE DATA PROCESSING WITH SPARK AND KUDU

We'll need to cast the data types before inserting the DataFrame to the Kudu table. We're introducing the use of selectExpr method here to convert data types, but another option is to programmatically specify the schema using StructType.

```
val convertedDF = xmlDF.selectExpr("userid","name","city","state","cast(zip as string) zip","cast(age as tinyint) age");

convertedDF: org.apache.spark.sql.DataFrame = [usersid: bigint, name: string, city: string, state: string, zip: string, age: tinyint]
```

Create the kudu context and insert the DataFrame to the destination table.

```
import org.apache.kudu.spark.kudu._

val kuduContext = new KuduContext("kudumaster01:7051")

kuduContext.insertRows(convertedDF, "impala::default.users")
```

It looks like the DataFrame was successfully inserted into the Kudu table. Using the impala-shell, check the data in the table to confirm.

```
select * from users;
+--------+------------------+--------------+-------+-------+-----+
| userid | name             | city         | state | zip   | age |
+--------+------------------+--------------+-------+-------+-----+
| 102    | Felipe Drummond  | Palo Alto    | CA    | 94301 | 33  |
| 103    | Teresa Levine    | Walnut Creek | CA    | 94507 | 47  |
| 100    | Wendell Ryan     | San Diego    | CA    | 92102 | 24  |
| 101    | Alicia Thompson  | Berkeley     | CA    | 94705 | 52  |
+--------+------------------+--------------+-------+-------+-----+
```

Inserting JSON into Kudu

We'll create a JSON file as sample data for this example. Make sure the file is in a folder in HDFS called /jsondata.

cat users.json

{"userid": 200, "name": "Jonathan West", "city":"Frisco", "state":"TX", "zip": "75034", "age":35}
{"userid": 201, "name": "Andrea Foreman", "city":"Dallas", "state":"TX", "zip": "75001", "age":28}
{"userid": 202, "name": "Kirsten Jung", "city":"Plano", "state":"TX", "zip": "75025", "age":69}
{"userid": 203, "name": "Jessica Nguyen", "city":"Allen", "state":"TX", "zip": "75002", "age":52}

Create a DataFrame from the JSON file.

val jsonDF = sqlContext.read.json("/jsondata")

jsonDF: org.apache.spark.sql.DataFrame = [age: bigint, city: string, name: string, state: string, userid: bigint, zip: string]

Check the data.

jsonDF.show

```
+---+------+--------------+-----+------+-----+
|age|  city|          name|state|userid|  zip|
+---+------+--------------+-----+------+-----+
| 35|Frisco| Jonathan West|   TX|   200|75034|
| 28|Dallas|Andrea Foreman|   TX|   201|75001|
| 69| Plano|  Kirsten Jung|   TX|   202|75025|
| 52| Allen|Jessica Nguyen|   TX|   203|75002|
+---+------+--------------+-----+------+-----+
```

Check the schema.

```
jsonDF.printSchema
root
 |- age: long (nullable = true)
 |- city: string (nullable = true)
 |- name: string (nullable = true)
 |- state: string (nullable = true)
 |- userid: long (nullable = true)
 |- zip: string (nullable = true)
```

Convert the data type of the age column to tinyint to match the table's data type.

```
val convertedDF = jsonDF.selectExpr("userid","name","city","state","zip","cast(age as tinyint) age");

convertedDF: org.apache.spark.sql.DataFrame = [userid: bigint, name: string, city: string, state: string, zip: string, age: tinyint]
```

Create the kudu context and insert the DataFrame to the destination table.

```
import org.apache.kudu.spark.kudu._

val kuduContext = new KuduContext("kudumaster01:7051")

kuduContext.insertRows(convertedDF, "impala::default.users")
```

Using the impala-shell, check if the rows were successfully inserted.

```
select * from users order by userid;
```

userid	name	city	state	zip	age
100	Wendell Ryan	San Diego	CA	92102	24
101	Alicia Thompson	Berkeley	CA	94705	52
102	Felipe Drummond	Palo Alto	CA	94301	33
103	Teresa Levine	Walnut Creek	CA	94507	47
200	Jonathan West	Frisco	TX	75034	35
201	Andrea Foreman	Dallas	TX	75001	28
202	Kirsten Jung	Plano	TX	75025	69
203	Jessica Nguyen	Allen	TX	75002	52

Inserting from MySQL into Kudu

Let's populate the users table in MySQL with some test data. Make sure the users table exists in the salesdb database. We will insert this data to a table in Kudu.

```
mysql -u root -p mypassword

use salesdb;

describe users;
```

```
+--------+--------------+------+-----+---------+-------+
| Field  | Type         | Null | Key | Default | Extra |
+--------+--------------+------+-----+---------+-------+
| userid | bigint(20)   | YES  |     | NULL    |       |
| name   | varchar(100) | YES  |     | NULL    |       |
| city   | varchar(100) | YES  |     | NULL    |       |
| state  | char(3)      | YES  |     | NULL    |       |
| zip    | char(5)      | YES  |     | NULL    |       |
| age    | tinyint(4)   | YES  |     | NULL    |       |
+--------+--------------+------+-----+---------+-------+
```

```
select * from users;
Empty set (0.00 sec)
```

insert into users values (300,'Fred Stevens','Torrance','CA',90503,23);

insert into users values (301,'Nancy Gibbs','Valencia','CA',91354,49);

insert into users values (302,'Randy Park','Manhattan Beach','CA',90267,21);

insert into users values (303,'Victoria Loma','Rolling Hills','CA',90274,75);

CHAPTER 6　HIGH PERFORMANCE DATA PROCESSING WITH SPARK AND KUDU

```
select * from users;
+--------+---------------+------------------+-------+-------+------+
| userid | name          | city             | state | zip   | age  |
+--------+---------------+------------------+-------+-------+------+
|    300 | Fred Stevens  | Torrance         | CA    | 90503 |   23 |
|    301 | Nancy Gibbs   | Valencia         | CA    | 91354 |   49 |
|    302 | Randy Park    | Manhattan Beach  | CA    | 90267 |   21 |
|    303 | Victoria Loma | Rolling Hills    | CA    | 90274 |   75 |
+--------+---------------+------------------+-------+-------+------+
```

Note In some versions of Spark -jars does not add the JAR in the driver's class path.[ii] It is recommended that you include the JDBC driver in your -jars and the Spark classpath.[iii]

Start the spark-shell. Take note that I had to include the MySQL driver as a parameter in both the –driver-class-path and –jars.

spark-shell -packages org.apache.kudu:kudu-spark_2.10:1.1.0 -driver-class-path mysql-connector-java-5.1.40-bin.jar -jars mysql-connector-java-5.1.40-bin.jar

Let's set up the jdbc url and connection properties.

val jdbcURL = s"jdbc:mysql://10.0.1.101:3306/salesdb?user=root&password=cloudera"

val connectionProperties = new java.util.Properties()

We can create a DataFrame from an entire table.

val mysqlDF = sqlContext.read.jdbc(jdbcURL, "users", connectionProperties)

```
mysqlDF.show
+------+-------------+---------------+-----+-----+---+
|userid|         name|           city|state|  zip|age|
+------+-------------+---------------+-----+-----+---+
|   300| Fred Stevens|       Torrance|   CA|90503| 23|
|   301|  Nancy Gibbs|       Valencia|   CA|91354| 49|
|   302|   Randy Park|Manhattan Beach|   CA|90267| 21|
|   303|Victoria Loma|   Rolling Hills|  CA|90274| 75|
+------+-------------+---------------+-----+-----+---+
```

Let's take advantage of pushdown optimization, running the query in the database and only returning the required results.

```
val mysqlDF = sqlContext.read.jdbc(jdbcURL, "users", connectionProperties).select("userid", "city", "state","age").where("age < 25")
```

```
mysqlDF.show
+------+---------------+-----+---+
|userid|           city|state|age|
+------+---------------+-----+---+
|   300|       Torrance|   CA| 23|
|   302|Manhattan Beach|   CA| 21|
+------+---------------+-----+---+
```

Let's specify an entire query. This is a more convenient and flexible method. Also, unlike the previous method, we are not required to specify the columns in our SELECT list if the column is specified in the WHERE clause.

```
val query = "(SELECT userid,name FROM users WHERE city IN ('Torrance','Rolling Hills')) as users"
```

```
val mysqlDF = sqlContext.read.jdbc(jdbcURL, query, connectionProperties)
```

```
mysqlDF.show
+------+-------------+
|userid|         name|
+------+-------------+
|   300| Fred Stevens|
|   303|Victoria Loma|
+------+-------------+
```

CHAPTER 6 HIGH PERFORMANCE DATA PROCESSING WITH SPARK AND KUDU

We've just tried different ways on how to select data from a MySQL table. Let' just go ahead and insert the whole table to Kudu.

```
val mysqlDF = sqlContext.read.jdbc(jdbcURL, "users", connectionProperties)
```

mysqlDF: org.apache.spark.sql.DataFrame = [userid: bigint, name: string, city: string, state: string, zip: string, age: int]

mysqlDF.show

```
+------+-------------+---------------+-----+-----+---+
|userid|         name|           city|state|  zip|age|
+------+-------------+---------------+-----+-----+---+
|   300| Fred Stevens|       Torrance|   CA|90503| 23|
|   301|  Nancy Gibbs|       Valencia|   CA|91354| 49|
|   302|   Randy Park|Manhattan Beach|   CA|90267| 21|
|   303|Victoria Loma|  Rolling Hills|   CA|90274| 75|
+------+-------------+---------------+-----+-----+---+
```

Verify the schema.

mysqlDF.printSchema

```
root
 |- userid: long (nullable = true)
 |- name: string (nullable = true)
 |- city: string (nullable = true)
 |- state: string (nullable = true)
 |- zip: string (nullable = true)
 |- age: integer (nullable = true)
```

But first let's convert age from integer to TINYINT. Otherwise you won't be able to insert this DataFrame into Kudu. Again, we could've just defined a schema using StructType.

```
val convertedDF = mysqlDF.selectExpr("userid","name","city","state","zip","cast(age as tinyint) age");
```

convertedDF: org.apache.spark.sql.DataFrame = [userid: bigint, name: string, city: string, state: string, zip: string, age: tinyint]

CHAPTER 6　HIGH PERFORMANCE DATA PROCESSING WITH SPARK AND KUDU

As you can see, the data type of the age column is now TINYINT. Let's go ahead and insert the data to Kudu.

```
import org.apache.kudu.spark.kudu._

val kuduContext = new KuduContext("kudumaster01:7051")

kuduContext.insertRows(convertedDF, "impala::default.users")
```

Now go to impala-shell and check if the data was successfully inserted.

```
select * from users order by userid
```

userid	name	city	state	zip	age
100	Wendell Ryan	San Diego	CA	92102	24
101	Alicia Thompson	Berkeley	CA	94705	52
102	Felipe Drummond	Palo Alto	CA	94301	33
103	Teresa Levine	Walnut Creek	CA	94507	47
200	Jonathan West	Frisco	TX	75034	35
201	Andrea Foreman	Dallas	TX	75001	28
202	Kirsten Jung	Plano	TX	75025	69
203	Jessica Nguyen	Allen	TX	75002	52
300	Fred Stevens	Torrance	CA	90503	23
301	Nancy Gibbs	Valencia	CA	91354	49
302	Randy Park	Manhattan Beach	CA	90267	21
303	Victoria Loma	Rolling Hills	CA	90274	75

Alternatively, you can also check using Spark.

```
import org.apache.kudu.spark.kudu._

val kuduDF = sqlContext.read.options(Map("kudu.master" -> "kudumaster01:7051","kudu.table" -> "impala::default.users")).kudu

kuduDF.select("userid","name","city","state","zip","age").sort($"userid".asc).show()
```

```
+------+---------------+---------------+-----+-----+---+
|userid|           name|           city|state|  zip|age|
+------+---------------+---------------+-----+-----+---+
|   100|   Wendell Ryan|      San Diego|   CA|92102| 24|
|   101|Alicia Thompson|       Berkeley|   CA|94705| 52|
|   102|Felipe Drummond|      Palo Alto|   CA|94301| 33|
|   103|  Teresa Levine|   Walnut Creek|   CA|94507| 47|
|   200|  Jonathan West|         Frisco|   TX|75034| 35|
|   201| Andrea Foreman|         Dallas|   TX|75001| 28|
|   202|   Kirsten Jung|          Plano|   TX|75025| 69|
|   203| Jessica Nguyen|          Allen|   TX|75002| 52|
|   300|   Fred Stevens|       Torrance|   CA|90503| 23|
|   301|    Nancy Gibbs|       Valencia|   CA|91354| 49|
|   302|     Randy Park|Manhattan Beach|   CA|90267| 21|
|   303|  Victoria Loma|   Rolling Hills|  CA|90274| 75|
+------+---------------+---------------+-----+-----+---+
```

Inserting from SQL Server into Kudu

The first thing you need to do is download the Microsoft JDBC driver for SQL Server. You can download the JDBC driver here: https://docs.microsoft.com/en-us/sql/connect/jdbc/microsoft-jdbc-driver-for-sql-server.

CHAPTER 6 HIGH PERFORMANCE DATA PROCESSING WITH SPARK AND KUDU

You should see a page similar to the one in Figure 6-1. Click the "Download JDBC Driver" link.

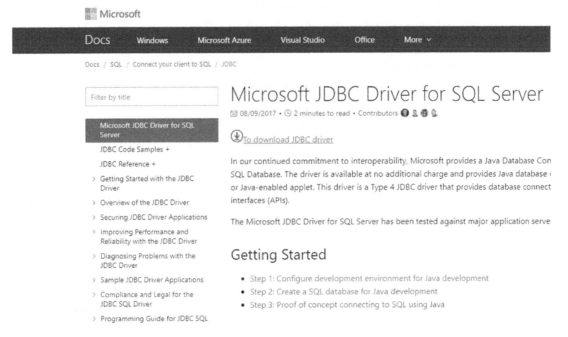

Figure 6-1. *Microsoft JDBC Driver for SQL Server*

Choose the language and click "Download."
Untar and unzip the Tarball. Choose the driver based on the version of your JRE.

```
tar zxvf sqljdbc_6.0.8112.100_enu.tar.gz

sqljdbc_6.0/enu/auth/x64/sqljdbc_auth.dll
sqljdbc_6.0/enu/auth/x86/sqljdbc_auth.dll
sqljdbc_6.0/enu/install.txt
sqljdbc_6.0/enu/jre7/sqljdbc41.jar
sqljdbc_6.0/enu/jre8/sqljdbc42.jar
sqljdbc_6.0/enu/license.txt
sqljdbc_6.0/enu/release.txt
sqljdbc_6.0/enu/samples/adaptive/executeStoredProcedure.java
sqljdbc_6.0/enu/samples/adaptive/readLargeData.java
sqljdbc_6.0/enu/samples/adaptive/updateLargeData.java
```

CHAPTER 6 HIGH PERFORMANCE DATA PROCESSING WITH SPARK AND KUDU

```
sqljdbc_6.0/enu/samples/alwaysencrypted/AlwaysEncrypted.java
sqljdbc_6.0/enu/samples/connections/connectDS.java
sqljdbc_6.0/enu/samples/connections/connectURL.java
sqljdbc_6.0/enu/samples/datatypes/basicDT.java
sqljdbc_6.0/enu/samples/datatypes/sqlxmlExample.java
sqljdbc_6.0/enu/samples/resultsets/cacheRS.java
sqljdbc_6.0/enu/samples/resultsets/retrieveRS.java
sqljdbc_6.0/enu/samples/resultsets/updateRS.java
sqljdbc_6.0/enu/samples/sparse/SparseColumns.java
sqljdbc_6.0/enu/xa/x64/sqljdbc_xa.dll
sqljdbc_6.0/enu/xa/x86/sqljdbc_xa.dll
sqljdbc_6.0/enu/xa/xa_install.sql
```

I'm going to use SQL Server 2016 throughout this book. You will also need to install SQL Server Management Studio separately (Figure 6-2).

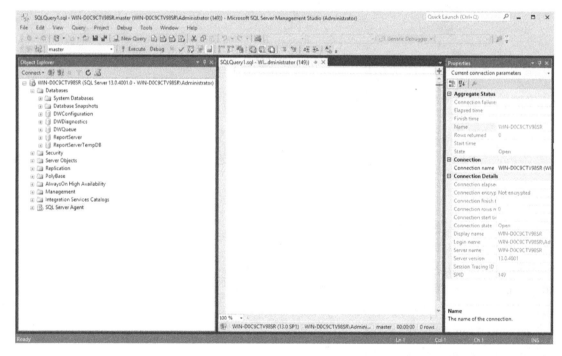

Figure 6-2. *SQL Server Management Studio*

CHAPTER 6 HIGH PERFORMANCE DATA PROCESSING WITH SPARK AND KUDU

We'll create the salesdb database and the users table. In the Object Explorer, right-click the Databases node and click "New Databases" (Figure 6-3). You will be shown a window where you can specify the database name and other database configuration options. Enter a database name "salesdb" and click OK. For testing purposes we'll leave the other options with the default values.

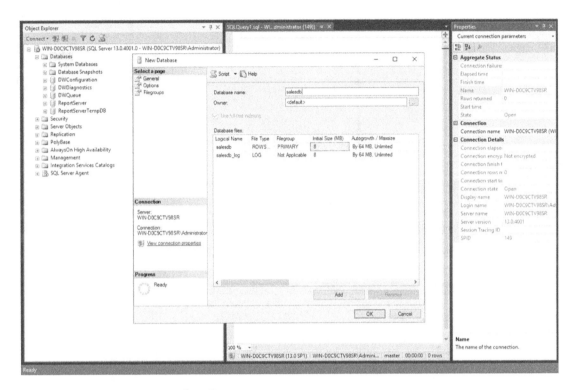

Figure 6-3. *Create new database*

Expand the salesdb node. Right-click "Tables," click "New," then "Table." Fill in the column names and data types. To make it easier for you to follow the examples in the book, make sure you keep the column names and data types the same as the MySQL and Kudu tables. Click the "Save" icon near the upper-right corner of the window, and then enter the table name "users" (Figure 6-4).

181

CHAPTER 6 HIGH PERFORMANCE DATA PROCESSING WITH SPARK AND KUDU

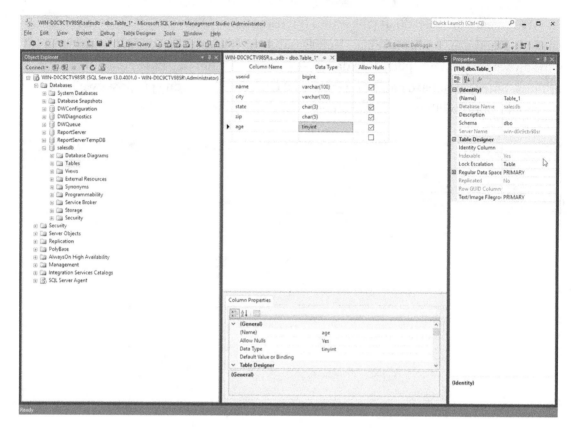

Figure 6-4. *Create new table*

CHAPTER 6   HIGH PERFORMANCE DATA PROCESSING WITH SPARK AND KUDU

Let's insert some test data into the users table we just created. Click "New Query" button on the standard toolbar to open a new editor window (Figure 6-5).

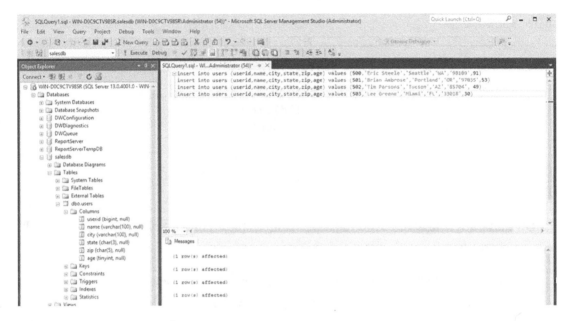

Figure 6-5.   *Insert test data*

We're going to copy these four rows from SQL Server to Kudu.

Start the spark-shell. Don't forget to pass the SQL Server driver as a parameter in both the -driver-class-path and -jars.

spark-shell -packages org.apache.kudu:kudu-spark_2.10:1.1.0 -driver-class-path sqljdbc41.jar -jars sqljdbc41.jar

Let's set up the jdbc url and connection properties.

val jdbcURL = "jdbc:sqlserver://192.168.56.102;databaseName=salesdb;user=salesuser;password=salespassword"
val connectionProperties = new java.util.Properties()

183

CHAPTER 6 HIGH PERFORMANCE DATA PROCESSING WITH SPARK AND KUDU

We can create a DataFrame from an entire table.

```
val sqlDF = sqlContext.read.jdbc(jdbcURL, "users", connectionProperties)

sqlDF.show
+------+-------------+--------+-----+-----+---+
|userid|         name|    city|state|  zip|age|
+------+-------------+--------+-----+-----+---+
|   500|  Eric Steele| Seattle|   WA|98109| 91|
|   501|Brian Ambrose|Portland|   OR|97035| 53|
|   502|  Tim Parsons|  Tucson|   AZ|85704| 49|
|   503|   Lee Greene|   Miami|   FL|33018| 30|
+------+-------------+--------+-----+-----+---+
```

Let's take advantage of pushdown optimization to run the query in the database and only return the results.

```
val sqlDF = sqlContext.read.jdbc(jdbcURL, "users", connectionProperties).
select("userid", "city", "state","age").where("age < 50")

sqlDF.show
+------+------+-----+---+
|userid|  city|state|age|
+------+------+-----+---+
|   502|Tucson|   AZ | 49|
|   503| Miami|   FL | 30|
+------+------+-----+---+
```

We can specify an entire query.

```
val query = "(SELECT userid,name FROM users WHERE city IN
('Seattle','Portland')) as users"

val sqlDF = sqlContext.read.jdbc(jdbcURL, query, connectionProperties)
```

```
sqlDF.show
+------+-------------+
|userid|         name|
+------+-------------+
|   500|  Eric Steele|
|   501|Brian Ambrose|
+------+-------------+
```

Let's just go ahead and insert the entire table to Kudu.

```
val sqlDF = sqlContext.read.jdbc(jdbcURL, "users", connectionProperties)
```

```
sqlDF.show
+------+-------------+--------+-----+-----+---+
|userid|         name|    city|state|  zip|age|
+------+-------------+--------+-----+-----+---+
|   500|  Eric Steele| Seattle|   WA|98109| 91|
|   501|Brian Ambrose|Portland|   OR|97035| 53|
|   502|  Tim Parsons|  Tucson|   AZ|85704| 49|
|   503|   Lee Greene|   Miami|   FL|33018| 30|
+------+-------------+--------+-----+-----+---+
```

Checking the schema, it looks like the age was converted into an integer.

```
sqlDF.printSchema
root
 |- userid: long (nullable = true)
 |- name: string (nullable = true)
 |- city: string (nullable = true)
 |- state: string (nullable = true)
 |- zip: string (nullable = true)
 |- age: integer (nullable = true)
```

We need to convert age from integer to TINYINT. Otherwise you won't be able to insert this DataFrame into kudu.

```
val convertedDF = sqlDF.selectExpr("userid","name","city","state","zip","cast(age as tinyint) age");
```

CHAPTER 6 HIGH PERFORMANCE DATA PROCESSING WITH SPARK AND KUDU

convertedDF: org.apache.spark.sql.DataFrame = [userid: bigint, name: string, city: string, state: string, zip: string, age: tinyint]

As you can see, the data type of the age column is now TINYINT. Let's go ahead and insert the DataFrame to Kudu.

import org.apache.kudu.spark.kudu._

val kuduContext = new KuduContext("kudumaster01:7051")

kuduContext.insertRows(convertedDF, "impala::default.users")

Let's now go to impala-shell and confirm if the data was successfully inserted.

select * from users order by userid

userid	name	city	state	zip	age
100	Wendell Ryan	San Diego	CA	92102	24
101	Alicia Thompson	Berkeley	CA	94705	52
102	Felipe Drummond	Palo Alto	CA	94301	33
103	Teresa Levine	Walnut Creek	CA	94507	47
200	Jonathan West	Frisco	TX	75034	35
201	Andrea Foreman	Dallas	TX	75001	28
202	Kirsten Jung	Plano	TX	75025	69
203	Jessica Nguyen	Allen	TX	75002	52
300	Fred Stevens	Torrance	CA	90503	23
301	Nancy Gibbs	Valencia	CA	91354	49
302	Randy Park	Manhattan Beach	CA	90267	21
303	Victoria Loma	Rolling Hills	CA	90274	75
400	Patrick Montalban	Los Angeles	CA	90010	71
401	Jillian Collins	Santa Monica	CA	90402	45
402	Robert Sarkisian	Glendale	CA	91204	29
403	Warren Porcaro	Burbank	CA	91523	62
500	Eric Steele	Seattle	WA	98109	91
501	Brian Ambrose	Portland	OR	97035	53
502	Tim Parsons	Tucson	AZ	85704	49
503	Lee Greene	Miami	FL	33018	30

Alternatively, you can also check using Spark.

import org.apache.kudu.spark.kudu._

val kuduDF = sqlContext.read.options(Map("kudu.master" -> "kudumaster01:7051","kudu.table" -> "impala::default.users")).kudu

kuduDF.select("userid","name","city","state","zip","age").sort($"userid".asc).show()

```
+------+-----------------+---------------+-----+-----+---+
|userid|             name|           city|state|  zip|age|
+------+-----------------+---------------+-----+-----+---+
|   100|     Wendell Ryan|      San Diego|   CA|92102| 24|
|   101|  Alicia Thompson|       Berkeley|   CA|94705| 52|
|   102|  Felipe Drummond|      Palo Alto|   CA|94301| 33|
|   103|    Teresa Levine|   Walnut Creek|   CA|94507| 47|
|   200|    Jonathan West|         Frisco|   TX|75034| 35|
|   201|   Andrea Foreman|         Dallas|   TX|75001| 28|
|   202|     Kirsten Jung|          Plano|   TX|75025| 69|
|   203|   Jessica Nguyen|          Allen|   TX|75002| 52|
|   300|    Fred Stevens |       Torrance|   CA|90503| 23|
|   301|      Nancy Gibbs|       Valencia|   CA|91354| 49|
|   302|      Randy Park |Manhattan Beach|   CA|90267| 21|
|   303|    Victoria Loma|  Rolling Hills|   CA|90274| 75|
|   400|Patrick Montalban|    Los Angeles|   CA|90010| 71|
|   401|  Jillian Collins|   Santa Monica|   CA|90402| 45|
|   402| Robert Sarkisian|       Glendale|   CA|91204| 29|
|   403|   Warren Porcaro|        Burbank|   CA|91523| 62|
|   500|      Eric Steele|        Seattle|   WA|98109| 91|
|   501|    Brian Ambrose|       Portland|   OR |97035| 53|
|   502|      Tim Parsons|         Tucson|   AZ |85704| 49|
|   503|      Lee Greene |          Miami|   FL |33018| 30|
+------+-----------------+---------------+-----+-----+---+
```

CHAPTER 6 HIGH PERFORMANCE DATA PROCESSING WITH SPARK AND KUDU

Inserting from HBase into Kudu

There are several ways to transfer data from HBase to Kudu. We can use the HBase client API. There's the Spark-HBase connector developed by Hortonworks.[iv] Astro, developed by Huwawei, provides an SQL layer HBase using Spark SQL.[v] The SparkOnHBase project from Cloudera was integrated into HBase recently, but it will probably take time before it makes it into a release of HBase.[vi]

We'll use the most straightforward way by using JDBC. This might not be the fastest way to move data from HBase to Kudu, but it should be adequate for most tasks. We'll create a Hive table on top of the HBase table, and then we'll create a Spark DataFrame using JDBC via Impala. Once we have the DataFrame, we can easily insert it to Kudu.

The first thing we need to do is download the Impala JDBC driver. Point your browser to `https://www.cloudera.com/downloads.html` (Figure 6-6).

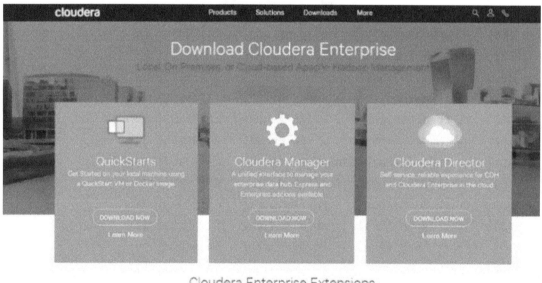

Figure 6-6. *Cloudera Enterprise download page*

CHAPTER 6 HIGH PERFORMANCE DATA PROCESSING WITH SPARK AND KUDU

Clicking the Impala JDBC Driver Downloads link near the bottom right of the page will bring you to the download page of the latest version of the JDBC driver for Impala. Download and unzip the files.

```
ls -l
```

```
-rw-r-r- 1 hadoop hadoop    693530 Mar  9 10:04 Cloudera-JDBC-Driver-for-Impala-Install-Guide.pdf
-rw-r-r- 1 hadoop hadoop     43600 Mar  8 11:11 Cloudera-JDBC-Driver-for-Impala-Release-Notes.pdf
-rw-r-r- 1 hadoop hadoop     46725 Mar  4 13:12 commons-codec-1.3.jar
-rw-r-r- 1 hadoop hadoop     60686 Mar  4 13:12 commons-logging-1.1.1.jar
-rw-r-r- 1 hadoop hadoop   7670596 Mar  4 13:16 hive_metastore.jar
-rw-r-r- 1 hadoop hadoop    596600 Mar  4 13:16 hive_service.jar
-rw-r-r- 1 hadoop hadoop    352585 Mar  4 13:12 httpclient-4.1.3.jar
-rw-r-r- 1 hadoop hadoop    181201 Mar  4 13:12 httpcore-4.1.3.jar
-rw-r-r- 1 hadoop hadoop   1562600 Mar  4 13:19 ImpalaJDBC41.jar
-rw-r-r- 1 hadoop hadoop    275186 Mar  4 13:12 libfb303-0.9.0.jar
-rw-r-r- 1 hadoop hadoop    347531 Mar  4 13:12 libthrift-0.9.0.jar
-rw-r-r- 1 hadoop hadoop    367444 Mar  4 13:12 log4j-1.2.14.jar
-rw-r-r- 1 hadoop hadoop    294796 Mar  4 13:16 ql.jar
-rw-r-r- 1 hadoop hadoop     23671 Mar  4 13:12 slf4j-api-1.5.11.jar
-rw-r-r- 1 hadoop hadoop      9693 Mar  4 13:12 slf4j-log4j12-1.5.11.jar
-rw-r-r- 1 hadoop hadoop   1307923 Mar  4 13:16 TCLIServiceClient.jar
-rw-r-r- 1 hadoop hadoop    792964 Mar  4 13:12 zookeeper-3.4.6.jar
```

The first thing we need to do is to start "hbase shell" and create the HBase table. We also need to add test data to our HBase table. Consult the Apache HBase Reference Guide online if you are unfamiliar with the HBase commands.

```
hbase shell

create 'hbase_users', 'cf1'

put 'hbase_users','400','cf1:name', 'Patrick Montalban'
put 'hbase_users','400','cf1:city', 'Los Angeles'
put 'hbase_users','400','cf1:state', 'CA'
put 'hbase_users','400','cf1:zip', '90010'
```

```
put 'hbase_users','400','cf1:age', '71'

put 'hbase_users','401','cf1:name', 'Jillian Collins'
put 'hbase_users','401','cf1:city', 'Santa Monica'
put 'hbase_users','401','cf1:state', 'CA'
put 'hbase_users','401','cf1:zip', '90402'
put 'hbase_users','401','cf1:age', '45'

put 'hbase_users','402','cf1:name', 'Robert Sarkisian'
put 'hbase_users','402','cf1:city', 'Glendale'
put 'hbase_users','402','cf1:state', 'CA'
put 'hbase_users','402','cf1:zip', '91204'
put 'hbase_users','402','cf1:age', '29'

put 'hbase_users','403','cf1:name', 'Warren Porcaro'
put 'hbase_users','403','cf1:city', 'Burbank'
put 'hbase_users','403','cf1:state', 'CA'
put 'hbase_users','403','cf1:zip', '91523'
put 'hbase_users','403','cf1:age', '62'
```

Create the Hive external table on top of the HBase table.

```
create external table hbase_users
(userid bigint,
name string,
city string,
state string,
zip string,
age tinyint)
stored by
'org.apache.hadoop.hive.hbase.HBaseStorageHandler'
with
SERDEPROPERTIES ('hbase.columns.mapping'=':key, cf1:name, cf1:city,
cf1:state, cf1:zip, cf1:age')
TBLPROPERTIES ('hbase.table.name'='hbase_users');
```

CHAPTER 6 HIGH PERFORMANCE DATA PROCESSING WITH SPARK AND KUDU

Using the impala-shell, verify that you can see the Hive external table.

```
show tables;
+------------+
| name       |
+------------+
| customers  |
| sample_07  |
| sample_08  |
| users      |
| web_logs   |
+------------+
```

It's not showing up. We need to invalidate the metadata to refresh Impala's memory.

```
invalidate metadata;

show tables;
+--------------+
| name         |
+--------------+
| customers    |
| hbase_users  |
| sample_07    |
| sample_08    |
| users        |
| web_logs     |
+--------------+

select * from hbase_users;
+--------+-----+--------------+-------------------+-------+-------+
| userid | age | city         | name              | state | zip   |
+--------+-----+--------------+-------------------+-------+-------+
| 400    | 71  | Los Angeles  | Patrick Montalban | CA    | 90010 |
| 401    | 45  | Santa Monica | Jillian Collins   | CA    | 90402 |
| 402    | 29  | Glendale     | Robert Sarkisian  | CA    | 91204 |
| 403    | 62  | Burbank      | Warren Porcaro    | CA    | 91523 |
+--------+-----+--------------+-------------------+-------+-------+
```

CHAPTER 6   HIGH PERFORMANCE DATA PROCESSING WITH SPARK AND KUDU

Start the spark-shell.

```
spark-shell -driver-class-path ImpalaJDBC41.jar -jars ImpalaJDBC41.jar
-packages org.apache.kudu:kudu-spark_2.10:1.1.0
```

Create a DataFrame from the HBase table

```
val jdbcURL = s"jdbc:impala://10.0.1.101:21050;AuthMech=0"

val connectionProperties = new java.util.Properties()

val hbaseDF = sqlContext.read.jdbc(jdbcURL, "hbase_users", connectionProperties)

hbaseDF: org.apache.spark.sql.DataFrame = [userid: bigint, age: int, city: string, name: string, state: string, zip: string]

hbaseDF.show
```

```
+------+---+------------+-----------------+-----+-----+
|userid|age|        city|             name|state|  zip|
+------+---+------------+-----------------+-----+-----+
|   400| 71| Los Angeles|Patrick Montalban|   CA|90010|
|   401| 45|Santa Monica|   Jillian Collins|   CA|90402|
|   402| 29|    Glendale|  Robert Sarkisian|   CA|91204|
|   403| 62|     Burbank|    Warren Porcaro|   CA|91523|
+------+---+------------+-----------------+-----+-----+
```

We still need to cast age to TINYINT before we can insert the data to the Kudu users table. Defining the schema using StructType is an option here.

```
val convertedDF = hbaseDF.selectExpr("userid","name","city","state","zip","cast(age as tinyint) age");

convertedDF: org.apache.spark.sql.DataFrame = [userid: bigint, name: string, city: string, state: string, zip: string, age: tinyint]
```

CHAPTER 6 HIGH PERFORMANCE DATA PROCESSING WITH SPARK AND KUDU

We can now insert the data to Kudu.

```
import org.apache.kudu.spark.kudu._

val kuduContext = new KuduContext("kudumaster01:7051")

kuduContext.insertRows(convertedDF, "impala::default.users")
```

Confirm if the data was successfully inserted.

```
val kuduDF = sqlContext.read.options(Map("kudu.master" -> "kudumaster01:7051","kudu.table" -> "impala::default.users")).kudu

kuduDF.select("userid","name","city","state","zip","age").sort($"userid".asc).show()
```

```
+------+-----------------+----------------+-----+-----+---+
|userid|             name|            city|state|  zip|age|
+------+-----------------+----------------+-----+-----+---+
|   100|     Wendell Ryan|       San Diego|   CA|92102| 24|
|   101|  Alicia Thompson|        Berkeley|   CA|94705| 52|
|   102|  Felipe Drummond|       Palo Alto|   CA|94301| 33|
|   103|    Teresa Levine|    Walnut Creek|   CA|94507| 47|
|   200|    Jonathan West|          Frisco|   TX|75034| 35|
|   201|   Andrea Foreman|          Dallas|   TX|75001| 28|
|   202|     Kirsten Jung|           Plano|   TX|75025| 69|
|   203|   Jessica Nguyen|           Allen|   TX|75002| 52|
|   300|    Fred Stevens|         Torrance|   CA|90503| 23|
|   301|      Nancy Gibbs|        Valencia|   CA|91354| 49|
|   302|      Randy Park|Manhattan Beach|   CA|90267| 21|
|   303|    Victoria Loma|   Rolling Hills|   CA|90274| 75|
|   400|Patrick Montalban|     Los Angeles|   CA|90010| 71|
|   401|  Jillian Collins|    Santa Monica|   CA|90402| 45|
|   402| Robert Sarkisian|        Glendale|   CA|91204| 29|
|   403|   Warren Porcaro|         Burbank|   CA|91523| 62|
+------+-----------------+----------------+-----+-----+---+
```

The rows were successfully inserted.

CHAPTER 6 HIGH PERFORMANCE DATA PROCESSING WITH SPARK AND KUDU

Inserting from Solr into Kudu

As discussed in Chapter 5, you can access Solr from Spark using SolrJ.[vii]

```
import java.net.MalformedURLException;
import org.apache.solr.client.solrj.SolrServerException;
import org.apache.solr.client.solrj.impl.HttpSolrServer;
import org.apache.solr.client.solrj.SolrQuery;
import org.apache.solr.client.solrj.response.QueryResponse;
import org.apache.solr.common.SolrDocumentList;

val solr = new HttpSolrServer("http://master02:8983/solr/mycollection");

val query = new SolrQuery();

query.setQuery("*:*");
query.addFilterQuery("userid:3");
query.setFields("userid","name","age","city");
query.setStart(0);
query.set("defType", "edismax");

val response = solr.query(query);
val results = response.getResults();

println(results);
```

A much better way to access Solr collections from Spark is by using the spark-solr package. Lucidworks started the spark-solr project to provide Spark-Solr integration.[viii] Using spark-solr is so much easier and powerful compared to SolrJ, allowing you to create DataFrames from Solr collections and using SparkSQL to interact with them. You can download the jar file from Lucidworks's website.

Start by importing the jar file from spark-shell.

```
spark-shell -jars spark-solr-3.0.1-shaded.jar
```

Specify the collection and connection information.

```
val myOptions = Map("collection" -> "mycollection","zkhost" -> "{master02:8983/solr}")
```

Create a DataFrame.

```
val solrDF = spark.read.format("solr")
  .options(myOptions)
  .load
```

Insert the data to Kudu.

```
kuduContext.insertRows(solrDF, "impala::default.users")
```

Insert from Amazon S3 into Kudu

Amazon S3 is a popular object store frequently used as data store for transient clusters. It's also a cost-effective storage for backups and cold data. Reading data from S3 is just like reading data from HDFS or any other file system.

Read a CSV file from Amazon S3. Make sure you've configured your S3 credentials.

```
val myCSV = sc.textFile("s3a://mydata/customers.csv")
```

Map CSV data to an RDD.

```
import org.apache.spark.sql.Row
val myRDD = myCSV.map(_.split(',')).map(e ⇒ Row(r(0).trim.toInt, r(1), r(2).trim.toInt, r(3)))
```

Create a schema.

```
import org.apache.spark.sql.types.{StructType, StructField, StringType, IntegerType};

val mySchema = StructType(Array(
StructField("customerid",IntegerType,false),
StructField("customername",StringType,false),
StructField("age",IntegerType,false),
StructField("city",StringType,false)))

val myDF = sqlContext.createDataFrame(myRDD, mySchema)
```

Insert DataFrame to Kudu.

```
kuduContext.insertRows(myDF, "impala::default.customers")
```

CHAPTER 6 HIGH PERFORMANCE DATA PROCESSING WITH SPARK AND KUDU

You've successfully inserted data from S3 into Kudu.

We've inserted data from different data sources into Kudu. Let's now insert data from Kudu to different data sources.

Inserting from Kudu into MySQL

Start the spark-shell.

spark-shell -packages org.apache.kudu:kudu-spark_2.10:1.1.0 -driver-class-path mysql-connector-java-5.1.40-bin.jar -jars mysql-connector-java-5.1.40-bin.jar

Connect to the Kudu master and check the data in the users table. We're going to sync this Kudu table with a MySQL table.

import org.apache.kudu.spark.kudu._

val kuduDF = sqlContext.read.options(Map("kudu.master" -> "kudumaster01:7051","kudu.table" -> "impala::default.users")).kudu

kuduDF.select("userid","name","city","state","zip","age").sort($"userid".asc).show()

```
+------+---------------+---------------+-----+-----+---+
|userid|           name|           city|state|  zip|age|
+------+---------------+---------------+-----+-----+---+
|   100|   Wendell Ryan|      San Diego|   CA|92102| 24|
|   101|Alicia Thompson|       Berkeley|   CA|94705| 52|
|   102|Felipe Drummond|      Palo Alto|   CA|94301| 33|
|   103|  Teresa Levine|   Walnut Creek|   CA|94507| 47|
|   200|  Jonathan West|         Frisco|   TX|75034| 35|
|   201| Andrea Foreman|         Dallas|   TX|75001| 28|
|   202|   Kirsten Jung|          Plano|   TX|75025| 69|
|   203| Jessica Nguyen|          Allen|   TX|75002| 52|
|   300|   Fred Stevens|       Torrance|   CA|90503| 23|
|   301|    Nancy Gibbs|       Valencia|   CA|91354| 49|
|   302|     Randy Park|Manhattan Beach|   CA|90267| 21|
|   303|   Victoria Loma|  Rolling Hills|   CA|90274| 75|
+------+---------------+---------------+-----+-----+---+
```

Register the DataFrame so we can run SQL queries against it.

```
kuduDF.registerTempTable("kudu_users")
```

Set up the JDBC URL and connection properties of the MySQL database.

```
val jdbcURL = s"jdbc:mysql://10.0.1.101:3306/salesdb?user=root&password=cloudera"

val connectionProperties = new java.util.Properties()
import org.apache.spark.sql.SaveMode
```

Check the data in the MySQL table using the MySQL command-line tool.

```
select * from users;
+--------+---------------+-----------------+-------+-------+------+
| userid | name          | city            | state | zip   | age  |
+--------+---------------+-----------------+-------+-------+------+
|    300 | Fred Stevens  | Torrance        | CA    | 90503 |   23 |
|    301 | Nancy Gibbs   | Valencia        | CA    | 91354 |   49 |
|    302 | Randy Park    | Manhattan Beach | CA    | 90267 |   21 |
|    303 | Victoria Loma | Rolling Hills   | CA    | 90274 |   75 |
+--------+---------------+-----------------+-------+-------+------+
```

Let's sync both tables by inserting all the rows with userid < 300 from Kudu to MySQL.

```
sqlContext.sql("select * from kudu_users where userid < 300").write.mode(SaveMode.Append).jdbc(jdbcUrl, "users", connectionProperties)
```

Check the MySQL table again and verify that the rows were added.

```
select * from users order by userid;
+--------+------------------+------------------+-------+-------+------+
| userid | name             | city             | state | zip   | age  |
+--------+------------------+------------------+-------+-------+------+
|    100 | Wendell Ryan     | San Diego        | CA    | 92102 |   24 |
|    101 | Alicia Thompson  | Berkeley         | CA    | 94705 |   52 |
|    102 | Felipe Drummond  | Palo Alto        | CA    | 94301 |   33 |
|    103 | Teresa Levine    | Walnut Creek     | CA    | 94507 |   47 |
|    200 | Jonathan West    | Frisco           | TX    | 75034 |   35 |
|    201 | Andrea Foreman   | Dallas           | TX    | 75001 |   28 |
|    202 | Kirsten Jung     | Plano            | TX    | 75025 |   69 |
|    203 | Jessica Nguyen   | Allen            | TX    | 75002 |   52 |
|    300 | Fred Stevens     | Torrance         | CA    | 90503 |   23 |
|    301 | Nancy Gibbs      | Valencia         | CA    | 91354 |   49 |
|    302 | Randy Park       | Manhattan Beach  | CA    | 90267 |   21 |
|    303 | Victoria Loma    | Rolling Hills    | CA    | 90274 |   75 |
+--------+------------------+------------------+-------+-------+------+
```

It looks like the rows were successfully inserted.

Inserting from Kudu into SQL Server

Start spark-shell.

```
spark-shell -packages org.apache.kudu:kudu-spark_2.10:1.1.0 -driver-class-path sqljdbc41.jar -jars sqljdbc41.jar
```

Create a DataFrame from the users table in the default database.

```
import org.apache.kudu.spark.kudu._

val kuduDF = sqlContext.read.options(Map("kudu.master" -> "kudumaster01:7051","kudu.table" -> "impala::default.users")).kudu
```

Verify the contents of the DataFrame.

```
kuduDF.select("userid","name","city","state","zip","age").sort($"userid".
asc).show()

+------+---------------+------------+-----+-----+---+
|userid|           name|        city|state|  zip|age|
+------+---------------+------------+-----+-----+---+
|   100|   Wendell Ryan|   San Diego|   CA|92102| 24|
|   101|Alicia Thompson|    Berkeley|   CA|94705| 52|
|   102|Felipe Drummond|   Palo Alto|   CA|94301| 33|
|   103|  Teresa Levine|Walnut Creek|   CA|94507| 47|
+------+---------------+------------+-----+-----+---+
```

Register the DataFrame so we can run SQL queries against it.

```
kuduDF.registerTempTable("kudu_users")
```

Set up the JDBC URL and connection properties of the SQL Server database.

```
val jdbcURL = "jdbc:sqlserver://192.168.56.103;databaseName=salesdb;user=sa;password=cloudera"

val connectionProperties = new java.util.Properties()

import org.apache.spark.sql.SaveMode
```

To make sure our test is consistent, make sure the users table in SQL Server is empty using SQL Server Management Studio (Figure 6-7).

CHAPTER 6   HIGH PERFORMANCE DATA PROCESSING WITH SPARK AND KUDU

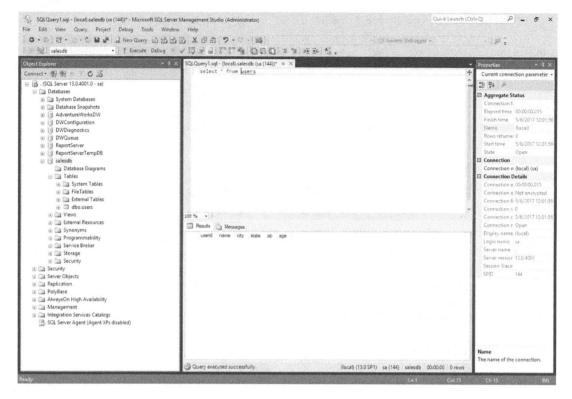

Figure 6-7. *Make sure the table is empty*

Insert the data from Kudu to SQL Server.

```
sqlContext.sql("select * from kudu_users").write.mode(SaveMode.Append).
jdbc(jdbcURL, "users", connectionProperties)
```

CHAPTER 6 HIGH PERFORMANCE DATA PROCESSING WITH SPARK AND KUDU

Check the SQL Server table again and verify that the rows were added (Figure 6-8).

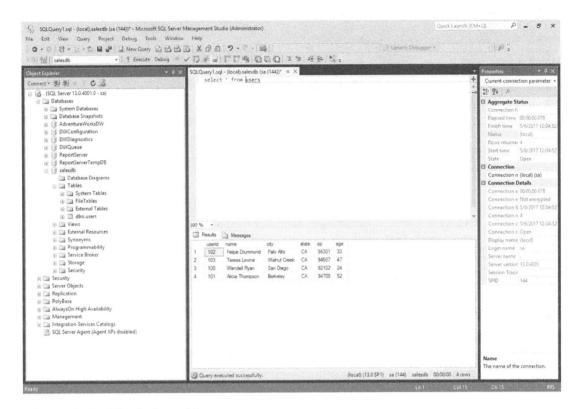

Figure 6-8. *Check the table*

Congratulations! The data was successfully inserted.

Inserting from Kudu into Oracle

The first thing we need to do is set up the Oracle environment. We'll create the user table in an existing pluggable database called EDWPDB. Log in as sysdba and start the instance. Consult the online Oracle documentation if you are unfamiliar with Oracle.

```
sqlplus / as sysdba

SQL*Plus: Release 12.2.0.1.0 Production on Sat May 6 18:12:45 2017

Copyright (c) 1982, 2016, Oracle.  All rights reserved.

Connected to an idle instance.
```

```
SQL> startup
ORACLE instance started.

Total System Global Area 1845493760 bytes
Fixed Size                  8793976 bytes
Variable Size             553648264 bytes
Database Buffers         1275068416 bytes
Redo Buffers                7983104 bytes
Database mounted.
Database opened.

SELECT name, open_mode FROM v$pdbs;

NAME                 OPEN_MODE
-------------------- ---------
PDB$SEED             READ ONLY
ORCLPDB              MOUNTED
EDWPDB               MOUNTED
```

Open the EDWPDB pluggable database and set it as the current container.

```
ALTER PLUGGABLE DATABASE EDWPDB OPEN;

SELECT name, open_mode FROM v$pdbs;

NAME                 OPEN_MODE
-------------------- -----
PDB$SEED             READ ONLY
ORCLPDB              MOUNTED
EDWPDB               READ WRITE

ALTER SESSION SET container = EDWPDB;
```

Create the Oracle table.

```
CREATE TABLE users (
userid NUMBER(19),
name VARCHAR(50),
city VARCHAR(50),
```

```
state VARCHAR (50),
zip VARCHAR(50),
age NUMBER(3));
```

Start the spark-shell. Don't forget to include the oracle driver. I'm using ojdbc6.jar in this example.

Note You might encounter the error "ORA-28040: No matching authentication protocol exception" when connecting to Oracle 12c R2 using the ojdbc6.jar driver. This is most likely caused by a bug in Oracle12c, Bug 14575666. The fix is to set SQLNET.ALLOWED_LOGON_VERSION=8 in the oracle/network/admin/sqlnet.ora file.

```
spark-shell -packages org.apache.kudu:kudu-spark_2.10:1.1.0 -driver-class-
path ojdbc6.jar -jars ojdbc6.jar
```

Create a DataFrame from the users table in the default database.

```
import org.apache.kudu.spark.kudu._

val kuduDF = sqlContext.read.options(Map("kudu.master" ->
"kudumaster01:7051","kudu.table" -> "impala::default.users")).kudu
```

Verify the contents of the DataFrame.

```
kuduDF.select("userid","name","city","state","zip","age").sort($"userid".
asc).show()

+------+---------------+------------+-----+-----+---+
|userid|           name|        city|state|  zip|age|
+------+---------------+------------+-----+-----+---+
|   100|   Wendell Ryan|   San Diego|   CA|92102| 24|
|   101|Alicia Thompson|    Berkeley|   CA|94705| 52|
|   102|Felipe Drummond|   Palo Alto|   CA|94301| 33|
|   103|  Teresa Levine|Walnut Creek|   CA|94507| 47|
+------+---------------+------------+-----+-----+---+
```

CHAPTER 6   HIGH PERFORMANCE DATA PROCESSING WITH SPARK AND KUDU

Register the DataFrame so we can run SQL queries against it.

```
kuduDF.registerTempTable("kudu_users")
```

Set up the JDBC URL and connection properties of the Oracle database.

```
val jdbcURL = "jdbc:oracle:thin:sales/cloudera@//192.168.56.30:1521/EDWPDB"
val connectionProperties = new java.util.Properties()
import org.apache.spark.sql.SaveMode
```

Make sure the users table in Oracle is empty using Oracle SQL Developer (Figure 6-9).

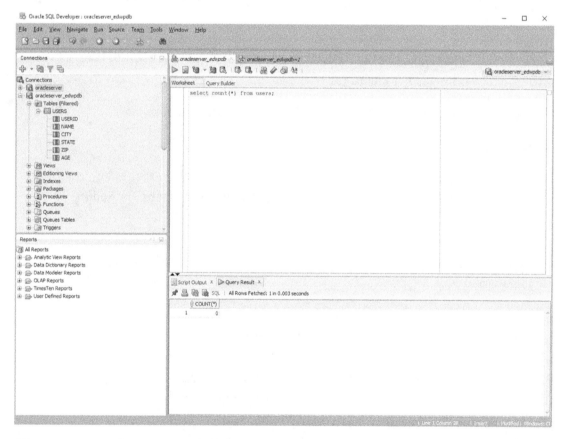

Figure 6-9. *Make sure Oracle table is empty*

CHAPTER 6 HIGH PERFORMANCE DATA PROCESSING WITH SPARK AND KUDU

Insert the data from Kudu to Oracle.

```
sqlContext.sql("select * from kudu_users").write.mode(SaveMode.Append).
jdbc(jdbcURL, "users", connectionProperties)
```

Check the Oracle table again and verify that the rows were added (Figure 6-10).

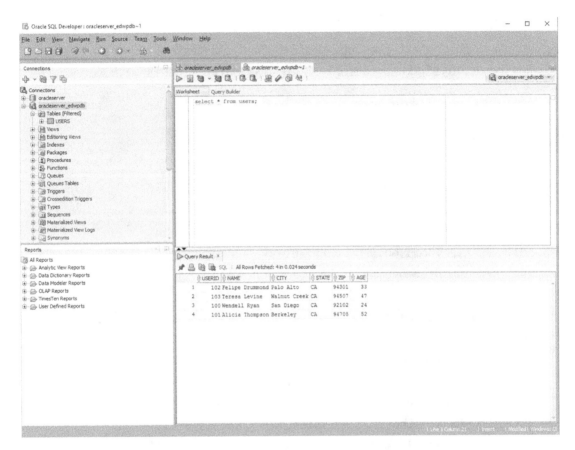

Figure 6-10. Check Oracle table

Congratulations! You've successfully copied rows from Kudu to Oracle.

Inserting from Kudu to HBase

We're going to insert data HBase via Impala so that we can use SQL. This is not the fastest method to write to HBase. If performance is critical, I suggest you use the saveAsHadoopDataset method or the HBase Java client API to write to HBase. There are various other ways of importing data into HBase.[ix]

Chapter 6 High Performance Data Processing with Spark and Kudu

Start the spark-shell and create a data frame from the kudu table.

```
spark-shell -driver-class-path ImpalaJDBC41.jar -jars ImpalaJDBC41.jar
-packages org.apache.kudu:kudu-spark_2.10:1.1.0

import org.apache.kudu.client.CreateTableOptions;
import org.apache.kudu.spark.kudu._

val kuduDF = sqlContext.read.options(Map("kudu.master" ->
"kudumaster01:7051","kudu.table" -> "impala::default.users")).kudu
```

Verify the contents of the table.

```
kuduDF.sort($"userid".asc).show()
```

```
+------+----------------+---------------+-----+-----+---+
|userid|            name|           city|state|  zip|age|
+------+----------------+---------------+-----+-----+---+
|   100|    Wendell Ryan|      San Diego|   CA|92102| 24|
|   101| Alicia Thompson|       Berkeley|   CA|94705| 52|
|   102| Felipe Drummond|      Palo Alto|   CA|94301| 33|
|   103|   Teresa Levine|   Walnut Creek|   CA|94507| 47|
|   200|   Jonathan West|         Frisco|   TX|75034| 35|
|   201|  Andrea Foreman|         Dallas|   TX|75001| 28|
|   202|    Kirsten Jung|          Plano|   TX|75025| 69|
|   203|  Jessica Nguyen|          Allen|   TX|75002| 52|
|   300|    Fred Stevens|       Torrance|   CA|90503| 23|
|   301|     Nancy Gibbs|       Valencia|   CA|91354| 49|
|   302|     Randy Park|Manhattan Beach|   CA|90267| 21|
|   303|   Victoria Loma|  Rolling Hills|   CA|90274| 75|
|   400|Patrick Montalban|    Los Angeles|   CA|90010| 71|
|   401| Jillian Collins|   Santa Monica|   CA|90402| 45|
|   402| Robert Sarkisian|       Glendale|   CA|91204| 29|
|   403|  Warren Porcaro|        Burbank|   CA|91523| 62|
+------+----------------+---------------+-----+-----+---+
```

Let's register the table so we can use it in a query.

```
kuduDF.registerTempTable("kudu_users")
```

Using impala-shell, verify the contents of the destination HBase table.

```
select * from hbase_users order by userid;
```

userid	age	city	name	state	zip
400	71	Los Angeles	Patrick Montalban	CA	90010
401	45	Santa Monica	Jillian Collins	CA	90402
402	29	Glendale	Robert Sarkisian	CA	91204
403	62	Burbank	Warren Porcaro	CA	91523

Go back to the spark-shell and set up the Impala connection.

```
val jdbcURL = s"jdbc:impala://10.0.1.101:21050;AuthMech=0"
```

```
val connectionProperties = new java.util.Properties()
```

Insert only the selected rows to the destination HBase table.

```
import org.apache.spark.sql.SaveMode
```

```
sqlContext.sql("select * from kudu_users where userid in (300,301,302,303)").write.mode(SaveMode.Append).jdbc(jdbcURL, "hbase_users", connectionProperties)
```

Go back to impala-shell and confirm that the rows were added to the destination HBase table.

CHAPTER 6 HIGH PERFORMANCE DATA PROCESSING WITH SPARK AND KUDU

```
select * from hbase_users order by userid;
```

```
+--------+-----+----------------+------------------+-------+-------+
| userid | age | city           | name             | state | zip   |
+--------+-----+----------------+------------------+-------+-------+
| 300    | 23  | Torrance       | Fred Stevens     | CA    | 90503 |
| 301    | 49  | Valencia       | Nancy Gibbs      | CA    | 91354 |
| 302    | 21  | Manhattan Beach| Randy Park       | CA    | 90267 |
| 303    | 75  | Rolling Hills  | Victoria Loma    | CA    | 90274 |
| 400    | 71  | Los Angeles    | Patrick Montalban| CA    | 90010 |
| 401    | 45  | Santa Monica   | Jillian Collins  | CA    | 90402 |
| 402    | 29  | Glendale       | Robert Sarkisian | CA    | 91204 |
| 403    | 62  | Burbank        | Warren Porcaro   | CA    | 91523 |
+--------+-----+----------------+------------------+-------+-------+
```

The data was successfully inserted into the HBase table.

Inserting Rows from Kudu to Parquet

Read the table.

```
spark-shell -packages org.apache.kudu:kudu-spark_2.10:1.1.0
```

```
import org.apache.kudu.client.CreateTableOptions;
import org.apache.kudu.spark.kudu._
```

```
val df = sqlContext.read.options(Map("kudu.master" -> "kudumaster01:7051","kudu.table" -> "impala::default.customers")).kudu
```

```
df.show
```

```
+---+------------+---+
| id|        name|age|
+---+------------+---+
|103|Byron Miller| 25|
|101|    Lisa Kim|240|
+---+------------+---+
```

CHAPTER 6 HIGH PERFORMANCE DATA PROCESSING WITH SPARK AND KUDU

Register the table and then create another DataFrame off of the result of a query.

```
df.registerTempTable("customers")

val df2 = sqlContext.sql("select * from customers where age=240")
```

Inspect the data.

```
df2.show

+---+--------+---+
| id|    name|age|
+---+--------+---+
|101|Lisa Kim|240|
+---+--------+---+
```

Append the DataFrame to the parquet table. You can also overwrite the destination by using the keyword "overwrite" instead of "append."

```
df2.write.mode("SaveMode.Append").parquet("/user/hive/warehouse/Mytable")
```

You will discover that Spark generates dozens or hundreds of small files when writing to HDFS. This is known as the "small file" problem.[x] This will eventually cause all sorts of performance problems on your cluster. If that happens you might want to use coalesce or repartition to specify how many files to write to HDFS. For example, you might want Spark to write 1 parquet file to HDFS.

```
df2.coalesce(1).write.mode("SaveMode.Append").parquet("/user/hive/warehouse/Mytable")
```

Using coalesce and repartition may cause performance issues since you are essentially reducing the degree of parallelism when writing data. Coalesce and repartition also trigger a shuffle that could cause performance issues depending on the amount of data that you're processing. You need to balance the number of generated files with processing performance. You may still have to perform regular compaction of your parquet tables after a period of time. This is a problem that you will not have with Kudu.

CHAPTER 6 HIGH PERFORMANCE DATA PROCESSING WITH SPARK AND KUDU

Insert SQL Server and Oracle DataFrames into Kudu

We will join data from SQL and Oracle and insert it into Kudu.

Start the spark-shell. Don't forget to include the necessary drivers and dependencies.

```
spark-shell -packages org.apache.kudu:kudu-spark_2.10:1.1.0 -driver-class-path ojdbc6.jar:sqljdbc41.jar -jars ojdbc6.jar,sqljdbc41.jar
```

Set up the Oracle connection.

```
val jdbcURL = "jdbc:oracle:thin:sales/cloudera@//192.168.56.30:1521/EDWPDB"
val connectionProperties = new java.util.Properties()
```

Create a DataFrame from the Oracle table.

```
val oraDF = sqlContext.read.jdbc(jdbcURL, "users", connectionProperties)

oraDF.show
```

```
+------+---------------+------------+-----+-----+---+
|USERID|           NAME|        CITY|STATE|  ZIP|AGE|
+------+---------------+------------+-----+-----+---+
|   102|Felipe Drummond|   Palo Alto|   CA|94301| 33|
|   103|  Teresa Levine|Walnut Creek|   CA|94507| 47|
|   100|   Wendell Ryan|   San Diego|   CA|92102| 24|
|   101|Alicia Thompson|    Berkeley|   CA|94705| 52|
+------+---------------+------------+-----+-----+---+
```

Register the table so we can run SQL against it.

```
oraDF.registerTempTable("ora_users")
```

Set up the SQL Server connection.

```
val jdbcURL = "jdbc:sqlserver://192.168.56.103;databaseName=salesdb;user=sa;password=cloudera"
```

```
val connectionProperties = new java.util.Properties()
```

Create a DataFrame from the SQL Server table.

```
val sqlDF = sqlContext.read.jdbc(jdbcURL, "userattributes", connectionProperties)
```

```
sqlDF.show

+------+------+------+------------------+
|userid|height|weight|        occupation|
+------+------+------+------------------+
|   100|   175|   170|       Electrician|
|   101|   180|   120|          Librarian|
|   102|   180|   215|    Data Scientist|
|   103|   178|   132|Software Developer|
+------+------+------+------------------+
```

Register the table so that we can join it to the Oracle table.

```
sqlDF.registerTempTable("sql_userattributes")
```

Join both tables. We'll insert the results to a Kudu table.

```
val joinDF = sqlContext.sql("select ora_users.userid,ora_users.name,ora_users.city,ora_users.state,ora_users.zip,ora_users.age,sql_userattributes.height,sql_userattributes.weight,sql_userattributes.occupation from ora_users  INNER JOIN sql_userattributes ON ora_users.userid=sql_userattributes.userid")

joinDF.show
```

```
+------+----------------+------------+-----+-----+---+------+------+----------+
|userid|            name|        city|state|  zip|age|height|weight|occupation|
+------+----------------+------------+-----+-----+---+------+------+----------+
|   100|    Wendell Ryan|   San Diego|   CA|92102| 24|   175|   170|Electrician|
|   101| Alicia Thompson|    Berkeley|   CA|94705| 52|   180|   120| Librarian|
|   102| Felipe Drummond|   Palo Alto|   CA|94301| 33|   180|   215|      Data
                                                                     Scientist|
|   103|   Teresa Levine|Walnut Creek|   CA|94507| 47|   178|   132|  Software
                                                                     Developer|
+------+----------------+------------+-----+-----+---+------+------+----------+
```

CHAPTER 6 HIGH PERFORMANCE DATA PROCESSING WITH SPARK AND KUDU

You can also join both DataFrames using this method.

```
val joinDF2 = oraDF.join(sqlDF,"userid")
```

```
joinDF2.show
```

```
+------+---------------+------------+-----+-----+---+------+------+------------+
|userid|           NAME|        CITY|STATE|  ZIP|AGE|height|weight|  occupation|
+------+---------------+------------+-----+-----+---+------+------+------------+
|   100|   Wendell Ryan|   San Diego|   CA|92102| 24|   175|   170|  Electrician|
|   101|Alicia Thompson|    Berkeley|   CA|94705| 52|   180|   120|    Librarian|
|   102|Felipe Drummond|   Palo Alto|   CA|94301| 33|   180|   215|        Data
                                                                    Scientist|
|   103|  Teresa Levine|Walnut Creek|   CA|94507| 47|   178|   132|    Software
                                                                    Developer|
+------+---------------+------------+-----+-----+---+------+------+------------+
```

Create the destination Kudu table in Impala.

```
impala-shell

create table users2 (
userid BIGINT PRIMARY KEY,
name STRING,
city STRING,
state STRING,
zip STRING,
age STRING,
height STRING,
weight STRING,
occupation STRING
)
PARTITION BY HASH PARTITIONS 16
STORED AS KUDU;
```

Go back to the spark-shell and set up the Kudu connection

```
import org.apache.kudu.spark.kudu._
val kuduContext = new KuduContext("kudumaster01:7051")
```

Insert the data to Kudu.

```
kuduContext.insertRows(JoinDF, "impala::default.users2")
```

Confirm that the data was successfully inserted into the Kudu table.

```
impala-shell

select * from users2;

+------+---------------+------------+-----+------+---+------+------+----------+
|userid|name           |city        |state|zip   |age|height|weight|occupation|
+------+---------------+------------+-----+------+---+------+------+----------+
|102   |Felipe Drummond|Palo Alto   |CA   |94301 |33 |180   |215   | Data
                                                                    Scientist|
|103   |Teresa Levine  |Walnut Creek|CA   |94507 |47 |178   |132   | Software
                                                                    Developer|
|100   |Wendell Ryan   |San Diego   |CA   |92102 |24 |175   |170   |Electrician|
|101   |Alicia Thompson|Berkeley    |CA   |94705 |52 |180   |120   |Librarian |
+------+---------------+------------+-----+------+---+------+------+----------+
```

Looks good to me.

CHAPTER 6   HIGH PERFORMANCE DATA PROCESSING WITH SPARK AND KUDU

Insert Kudu and SQL Server DataFrames into Oracle

Create the destination table in Oracle using Oracle SQL Developer (Figure 6-11).

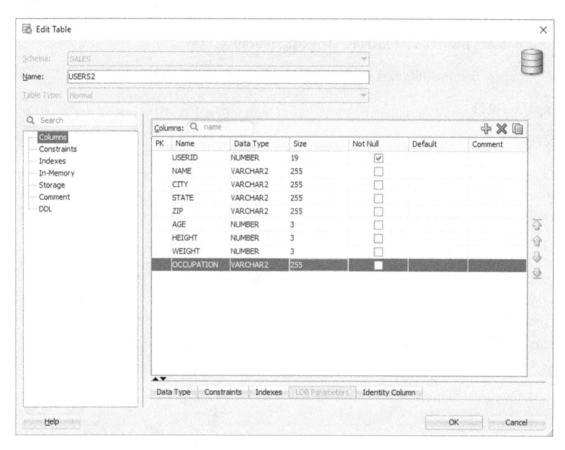

Figure 6-11. *Create an Oracle table*

Start the spark-shell. Don't forget to include the necessary drivers and dependencies.

```
spark-shell -packages org.apache.kudu:kudu-spark_2.10:1.1.0 -driver-class-
path ojdbc6.jar:sqljdbc41.jar -jars ojdbc6.jar,sqljdbc41.jar
```

Create a DataFrame from the Kudu users table in the default database.

```
import org.apache.kudu.spark.kudu._

val kuduDF = sqlContext.read.options(Map("kudu.master" ->
"kudumaster01:7051","kudu.table" -> "impala::default.users")).kudu
```

Verify the contents of the DataFrame.

```
kuduDF.select("userid","name","city","state","zip","age").sort($"userid".
asc).show()
```

```
+------+---------------+------------+-----+-----+---+
|userid|           name|        city|state|  zip|age|
+------+---------------+------------+-----+-----+---+
|   100|   Wendell Ryan|   San Diego|   CA|92102| 24|
|   101|Alicia Thompson|    Berkeley|   CA|94705| 52|
|   102|Felipe Drummond|   Palo Alto|   CA|94301| 33|
|   103|  Teresa Levine|Walnut Creek|   CA|94507| 47|
+------+---------------+------------+-----+-----+---+
```

Register the DataFrame so we can run SQL queries against it.

```
kuduDF.registerTempTable("kudu_users")
```

```
val jdbcURL = "jdbc:sqlserver://192.168.56.103;databaseName=salesdb;user=sa
;password=cloudera"
```

```
val connectionProperties = new java.util.Properties()
```

Create a DataFrame from the SQL Server table.

```
val sqlDF = sqlContext.read.jdbc(jdbcURL, "userattributes",
connectionProperties)
```

```
sqlDF.show
```

```
+------+------+------+------------------+
|userid|height|weight|        occupation|
+------+------+------+------------------+
|   100|   175|   170|       Electrician|
|   101|   180|   120|         Librarian|
|   102|   180|   215|    Data Scientist|
|   103|   178|   132|Software Developer|
+------+------+------+------------------+
```

Register the DataFrame as a temp table.

```
sqlDF.registerTempTable("sql_userattributes")
```

Join both tables. We'll insert the results to an Oracle database.

```
val joinDF = sqlContext.sql("select
kudu_users.userid,kudu_users.name,kudu_users.city,kudu_users.state,kudu_
users.zip,kudu_users.age,sql_userattributes.height,sql_userattributes.
weight,sql_userattributes.occupation from kudu_users  INNER JOIN sql_
userattributes ON kudu_users.userid=sql_userattributes.userid")
joinDF.show
```

```
+------+---------------+------------+-----+-----+---+------+------+-----------+
|userid|           name|        city|state|  zip|age|height|weight|occupation |
+------+---------------+------------+-----+-----+---+------+------+-----------+
|   100|   Wendell Ryan|   San Diego|   CA|92102| 24|   175|   170|Electrician|
|   101|Alicia Thompson|    Berkeley|   CA|94705| 52|   180|   120|Librarian  |
|   102|Felipe Drummond|   Palo Alto|   CA|94301| 33|   180|   215|Data
                                                                    Scientist|
|   103|  Teresa Levine|Walnut Creek|   CA|94507| 47|   178|   132|Software
                                                                    Developer|
+------+---------------+------------+-----+-----+---+------+------+-----------+
```

You can achieve the same result using this method.

```
val joinDF2 = kuduDF.join(sqlDF,"userid")
```

```
joinDF2.show
```

```
+------+---------------+------------+-----+-----+---+------+------+-----------+
|userid|           name|        city|state|  zip|age|height|weight| occupation|
+------+---------------+------------+-----+-----+---+------+------+-----------+
|   100|   Wendell Ryan|   San Diego|   CA|92102| 24|   175|   170|Electrician|
|   101|Alicia Thompson|    Berkeley|   CA|94705| 52|   180|   120|  Librarian|
|   102|Felipe Drummond|   Palo Alto|   CA|94301| 33|   180|   215|Data
                                                                    Scientist|
|   103|  Teresa Levine|Walnut Creek|   CA|94507| 47|   178|   132|Software
                                                                    Developer|
+------+---------------+------------+-----+-----+---+------+------+-----------+
```

CHAPTER 6 HIGH PERFORMANCE DATA PROCESSING WITH SPARK AND KUDU

Set up the JDBC URL and connection properties of the Oracle database.

```
val jdbcURL = "jdbc:oracle:thin:sales/cloudera@//192.168.56.30:1521/EDWPDB"
val connectionProperties = new java.util.Properties()

import org.apache.spark.sql.SaveMode
```

Insert the DataFrame to Oracle.

```
joinDF.write.mode(SaveMode.Append).jdbc(jdbcURL, "users2",
connectionProperties)
```

Verify that the rows were successfully added to the Oracle database (Figure 6-12).

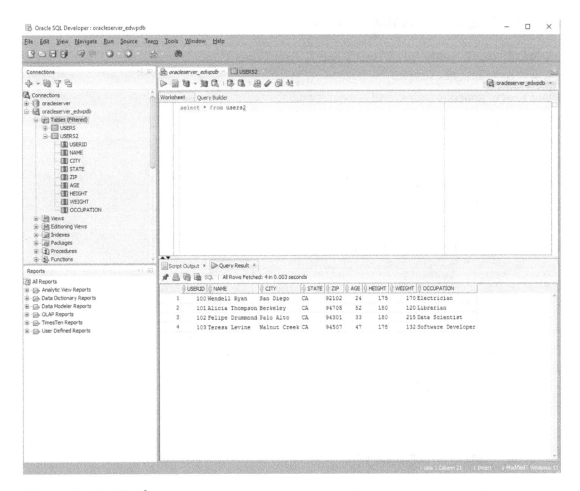

Figure 6-12. Verify rows

217

CHAPTER 6 HIGH PERFORMANCE DATA PROCESSING WITH SPARK AND KUDU

Spark Streaming and Kudu

There are use cases where you need to ingest data into Kudu in near real time. This is a requirement for Internet of Things (IoT) use cases, for example. In Listing 6-1, we show a sample Spark Streaming application that ingests sensor data from Flume. We perform basic event stream processing to tag the status of each event as NORMAL, WARNING, or CRITICAL depending on the temperature returned by the sensor. The status is then saved in a Kudu table together with the rest of the data.

Users can query the Kudu table as data is being inserted into it. In Chapter 9, we discuss a real-time data visualization tool called Zoomdata. You can use Zoomdata to visualize data stored in Kudu in real time.

Listing 6-1. Spark Streaming with Kudu

```
import java.io.IOException;
import org.apache.spark._
import org.apache.spark.rdd.NewHadoopRDD
import org.apache.spark.SparkConf
import org.apache.spark.storage.StorageLevel
import org.apache.spark.streaming.flume._
import org.apache.spark.streaming.Seconds
import org.apache.spark.streaming.StreamingContext
import org.apache.spark.util.IntParam
import org.apache.spark.sql.SQLContext

object FlumeStreaming {

  case class SensorData(tableid: String, deviceid: String, thedate:
  String, thetime: String, temp: Short, status: String)

   def parseSensorData(str: String): SensorData = {
     val myData = str.split(",")

     val myTableid = myData(0)
     val myDeviceid = myData(1)
     val myDate = myData(2)
     val myTime = myData(3)
     val myTemp = myData(4)
```

```
    val myStatus = myData(5)

    SensorData(myTableid, myDeviceid, myDate, myTime, myTemp.toShort,
    myStatus)
  }

  def main(args: Array[String]) {

    val sparkConf = new SparkConf().setMaster("local[2]").
    setAppName("FlumeStreaming")
    val sc = new SparkContext(sparkConf)
    val ssc = new StreamingContext(sc, Seconds(1))

    val flumeStream = FlumeUtils.createPollingStream(ssc,args(0),args(1).toInt)

    val sensorDStream = flumeStream.map (x => new String(x.event.getBody.
    array)).map(parseSensorData)

    sensorDStream.foreachRDD { rdd =>

        val sqlContext = new SQLContext(sc)
        import sqlContext.implicits._

           val kuduContext = new KuduContext("kudumaster01:7051")
// convert the RDD into a DataFrame and insert it into the Kudu table
           val DataDF = rdd.toDF
           kuduContext.insertRows(DataDF, "impala::default.sensortable")

           DataDF.registerTempTable("currentDF")

           // Update the table based on the thresholds

           val WarningFilteredDF = sqlContext.sql("select * from
           currentDF where temp > 50 and temp <= 60")
           WarningFilteredDF.registerTempTable("warningtable")

           val UpdatedWarningDF = sqlContext.sql("select tableid,deviceid
           ,thedate,thetime,temp,'WARNING' as status from warningtable")

           kuduContext.updateRows(UpdatedWarningDF, "impala::default.
           sensortable")
```

CHAPTER 6 HIGH PERFORMANCE DATA PROCESSING WITH SPARK AND KUDU

```
            val CriticalFilteredDF = sqlContext.sql("select * from
            currentDF where temp > 61")
            CriticalFilteredDF.registerTempTable("criticaltable")

            val UpdatedCriticalDF = sqlContext.sql("select tableid,deviceid,
            thedate,thetime,temp,'CRITICAL' as status from criticaltable")

            kuduContext.updateRows(UpdatedCriticalDF, "impala::default.
            sensortable")
      }
    ssc.start()
    ssc.awaitTermination()
  }
}
```

This is my SBT file. Consult Chapter 5 for more information on building applications. Consult Maven's online documentation if you are using Maven.

```
name := "My Test App"
version := "1.0"

scalaVersion := "2.10.5"

resolvers ++= Seq(
  "Apache Repository" at "https://repository.apache.org/content/
repositories/releases/",
  "Cloudera repo" at "https://repository.cloudera.com/artifactory/
cloudera-repos/"
)

libraryDependencies ++= Seq (
        "org.apache.spark" % "spark-core_2.10" % "1.5.0",
        "org.apache.spark" % "spark-streaming_2.10" % "1.5.0",
        "org.apache.spark" % "spark-streaming-flume_2.10" % "1.5.0",
        "org.apache.spark" % "spark-sql_2.10" % "1.5.0"
)
```

After executing sbt package to package the spark application, you can now launch the application using the spark-submit tool that comes with Spark. The parameters are for testing purposes only. Change the parameters based on your data ingestion requirements. Note that I'm using the jar file to include kudu-spark dependency. The parameters localhost and 9999 will be used as the Flume sink destination.

```
spark-submit \
-class FlumeStreaming \
-jars kudu-spark_2.10-0.10.0.jar \
-master yarn-client \
-driver-memory=512m \
-executor-memory=512m \
-executor-cores 4  \
/mydir/spark/flume_streaming_kudu/target/scala-2.10/butch-app_2.10-1.0.jar \ localhost 9999
```

This is a sample flume.conf file using spooldir as a data source. The settings are adequate for testing purposes.

```
agent1.sources  = source1
agent1.channels = channel1
agent1.sinks = spark

agent1.sources.source1.type = spooldir
agent1.sources.source1.spoolDir = /tmp/streaming
agent1.sources.source1.channels = channel1

agent1.channels.channel1.type = memory
agent1.channels.channel1.capacity = 10000
agent1.channels.channel1.transactionCapacity = 1000

agent1.sinks.spark.type = org.apache.spark.streaming.flume.sink.SparkSink
agent1.sinks.spark.hostname = 127.0.0.1
agent1.sinks.spark.port =  9999
agent1.sinks.spark.channel = channel1
agent1.sinks.spark.batchSize=5
```

CHAPTER 6   HIGH PERFORMANCE DATA PROCESSING WITH SPARK AND KUDU

Kudu as a Feature Store for Spark MLlib

Kudu makes it easier for data scientists to prepare and clean data. Kudu can serve as a fast, highly scalable and mutable feature store for machine learning applications. It also smoothly integrates with Spark SQL and DataFrame API.

Let's work on an example. We'll use the Heart Disease Data Set[xi] from the UCI Machine Learning Repository to predict the presence of heart disease. The data was collected by Robert Detrano, MD, PhD, from the VA Medical Center, Long Beach and Cleveland Clinic Foundation. Historically, the Cleveland data set has been the subject of numerous research so we'll use that data set. The original data set has 76 attributes, but only 14 of them are traditionally used by ML researchers (Table 6-1). We will simply perform binomial classification and determine if the patient has heart disease or not (Listing 6-2). This is the same example we used in Chapter 5, but this time we'll use Kudu to store our features.

Table 6-1. *Cleveland Heart Disease Data Set Attribute Information*

Attribute	Description
age	Age
sex	Sex
cp	Chest pain type
trestbps	Resting blood pressure
chol	Serum cholesterol in mg/dl
fbs	Fasting blood sugar > 120 mg/dl
restecg	Resting electrocardiographic results
thalach	Maximum heart rate achieved
exang	Exercise induced angina
oldpeak	ST depression induced by exercise relative to rest
slope	The slope of the peak exercise ST segment
ca	Number of major vessels (0–3) colored by flourosopy
thal	Thalium stress test result
num	The predicted attribute – diagnosis of heart disease

CHAPTER 6   HIGH PERFORMANCE DATA PROCESSING WITH SPARK AND KUDU

Let's start. We'll need to download the file, copy it to HDFS, and create an external table on top of it. We will then copy the data to a Kudu table.

```
wget http://archive.ics.uci.edu/ml/machine-learning-databases/heart-disease/cleveland.data
```

```
head -n 10 processed.cleveland.data
```

```
63.0,1.0,1.0,145.0,233.0,1.0,2.0,150.0,0.0,2.3,3.0,0.0,6.0,0
67.0,1.0,4.0,160.0,286.0,0.0,2.0,108.0,1.0,1.5,2.0,3.0,3.0,2
67.0,1.0,4.0,120.0,229.0,0.0,2.0,129.0,1.0,2.6,2.0,2.0,7.0,1
37.0,1.0,3.0,130.0,250.0,0.0,0.0,187.0,0.0,3.5,3.0,0.0,3.0,0
41.0,0.0,2.0,130.0,204.0,0.0,2.0,172.0,0.0,1.4,1.0,0.0,3.0,0
56.0,1.0,2.0,120.0,236.0,0.0,0.0,178.0,0.0,0.8,1.0,0.0,3.0,0
62.0,0.0,4.0,140.0,268.0,0.0,2.0,160.0,0.0,3.6,3.0,2.0,3.0,3
57.0,0.0,4.0,120.0,354.0,0.0,0.0,163.0,1.0,0.6,1.0,0.0,3.0,0
63.0,1.0,4.0,130.0,254.0,0.0,2.0,147.0,0.0,1.4,2.0,1.0,7.0,2
53.0,1.0,4.0,140.0,203.0,1.0,2.0,155.0,1.0,3.1,3.0,0.0,7.0,1
```

```
hadoop fs -put processed.cleveland.data /tmp/data
impala-shell
```

```
CREATE EXTERNAL TABLE cleveland_csv (
            age float,
            sex float,
            cp float,
            trestbps float,
            chol float,
            fbs float,
            restecg float,
            thalach float,
            exang float,
            oldpeak float,
            slope float ,
            ca float,
            thal float,
            num float
            )
```

```
            ROW FORMAT
      DELIMITED FIELDS TERMINATED BY ','
      LINES TERMINATED BY '\n' STORED AS TEXTFILE
      LOCATION '/tmp/data';
CREATE TABLE cleveland_kudu (
            id string,
            age float,
            sex float,
            cp float,
            trestbps float,
            chol float,
            fbs float,
            restecg float,
            thalach float,
            exang float,
            oldpeak float,
            slope float ,
            ca float,
            thal float,
            num float,
            primary key(id)
            )
      PARTITION BY HASH PARTITIONS 4
      STORED AS KUDU
      TBLPROPERTIES ('kudu.num_tablet_replicas' = '1');
INSERT INTO cleveland_kudu
SELECT
      uuid(),
            age,
            sex
            cp,
            trestbps,
            chol,
            fbs,
```

```
            restecg,
            thalach,
            exang,
            oldpeak,
            slope,
            ca,
            thal,
            num
FROM
cleveland_csv;
```

Note that we used the Impala function uuid() to generate a unique primary key for our Kudu table. The Kudu table should now have some data in it. Let's use the spark-shell to fit our model using Spark MLlib.

Listing 6-2. Performing binary classifcation using Kudu as a feature store

```
spark-shell -packages org.apache.kudu:kudu-spark_2.10:1.1.0

import org.apache.spark._
import org.apache.spark.rdd.RDD
import org.apache.spark.sql.SQLContext
import org.apache.spark.sql.functions._
import org.apache.spark.sql.types._
import org.apache.spark.sql._
import org.apache.spark.ml.classification.RandomForestClassifier
import org.apache.spark.ml.evaluation.BinaryClassificationEvaluator
import org.apache.spark.ml.feature.StringIndexer
import org.apache.spark.ml.feature.VectorAssembler
import org.apache.spark.ml.tuning.{ParamGridBuilder, CrossValidator}
import org.apache.spark.ml.{Pipeline, PipelineStage}
import org.apache.spark.mllib.evaluation.RegressionMetrics
import org.apache.spark.ml.param.ParamMap
import org.apache.kudu.spark.kudu._

val kuduDF = sqlContext.read.options(Map("kudu.master" ->
"kudumaster01:7051","kudu.table" -> "impala::default.cleveland_kudu")).kudu
```

```
kuduDF.printSchema
root
 |- id: string (nullable = false)
 |- age: float (nullable = true)
 |- sex: float (nullable = true)
 |- cp: float (nullable = true)
 |- trestbps: float (nullable = true)
 |- chol: float (nullable = true)
 |- fbs: float (nullable = true)
 |- restecg: float (nullable = true)
 |- thalach: float (nullable = true)
 |- exang: float (nullable = true)
 |- oldpeak: float (nullable = true)
 |- slope: float (nullable = true)
 |- ca: float (nullable = true)
 |- thal: float (nullable = true)
 |- num: float (nullable = true)
```

```
val myFeatures = Array("age", "sex", "cp", "trestbps", "chol", "fbs",
    "restecg", "thalach", "exang", "oldpeak", "slope",
    "ca", "thal", "num")
```

```
val myAssembler = new VectorAssembler().setInputCols(myFeatures).
setOutputCol("features")
```

```
val kuduDF2 = myAssembler.transform(kuduDF)
```

```
val myLabelIndexer = new StringIndexer().setInputCol("num").
setOutputCol("label")
```

```
val kuduDF3 = mylabelIndexer.fit(kuduDF2).transform(kuduDF2)
```

```
val kuduDF4 = kuduDF3.where(kuduDF3("ca").isNotNull).where(kuduDF3("thal").
isNotNull).where(kuduDF3("num").isNotNull)
```

```
val Array(trainingData, testData) = kuduDF4.randomSplit(Array(0.8, 0.2), 101)
```

```
val myRFclassifier = new RandomForestClassifier().setFeatureSubsetStrategy(
"auto").setSeed(101)
```

```
val myEvaluator = new BinaryClassificationEvaluator().setLabelCol("label")
```

```
val myParamGrid = new ParamGridBuilder()
    .addGrid(myRFclassifier.maxBins, Array(10, 20, 30))
    .addGrid(myRFclassifier.maxDepth, Array(5, 10, 15))
    .addGrid(myRFclassifier.numTrees, Array(20, 30, 40))
    .addGrid(myRGclassifier.impurity, Array("gini", "entropy"))
    .build()

val myPipeline = new Pipeline().setStages(Array(myRFclassifier))

val myCV = new CrosValidator()
    .setEstimator(myPipeline)
    .setEvaluator(myEvaluator)
    .setEstimatorParamMaps(myParamGrid)
    .setNumFolds(3)
```

We can now fit the model.

```
val myCrossValidatorModel = myCV.fit(trainingData)
```

Let's evaluate the model.

```
val myEvaluatorParamMap = ParamMap(myEvaluator.metricName ->
"areaUnderROC")

val aucTrainingData = myEvaluator.evaluate(CrossValidatorPrediction,
myEvaluatorParamMap)
```

You can now make some predictions on our data.

```
val myCrossValidatorPrediction = myCrossValidatorModel.transform(testData)
```

We used a very small data set (300+ observations) for our example, but imagine if the data set contains billions of rows. If you need to add or update your training data, you simply run the DML statement against the Kudu table. The ability to update a highly scalable storage engine such as Kudu greatly simplifies data preparation and feature engineering.

CHAPTER 6 HIGH PERFORMANCE DATA PROCESSING WITH SPARK AND KUDU

Note Kudu allows up to a maximum of 300 columns per table. HBase is a more appropriate storage engine if you needs to store more than 300 features. HBase tables can contain thousands or millions of columns. The downside in using HBase is that it is not as efficient in handling full-table scans compared to Kudu. There is discussion within the Apache Kudu community to address the 300-column limitation in future versions of Kudu.

Strictly speaking, you can bypass Kudu's 300-column limit by setting an unsafe flag. For example, if you need the ability to create a Kudu table with 1000 columns, you can start the Kudu master with the following flags: -unlock-unsafe-flags -max-num-columns=1000. This has not been thoroughly tested by the Kudu development team and is therefore not recommended for production use.

Summary

Spark, Impala, and Kudu are perfect together. Spark provides a highly scalable data processing framework, while Kudu provides a highly scalable storage engine. With Impala providing a fast SQL interface, you have everything you need to implement a cost-effective enterprise data management and analytic platform. Of course, you are not limited to these three Apache open source projects. In subsequent chapters, I will cover other open source and commercial applications that can further enhance your big data platform.

References

i. Apache Kudu; "Kudu Integration with Spark," Apache Kudu, 2018, https://kudu.apache.org/docs/developing.html#_kudu_integration_with_spark

ii. Holden Karau, Rachel Warren; "High Performance Spark," O'Reilly, June 2017 https://www.safaribooksonline.com/library/view/high-performance-spark/9781491943199/

iii. Apache Spark; "JDBC To Other Databases," Apache Spark, 2018, http://spark.apache.org/docs/latest/sql-programming-guide.html#jdbc-to-other-databases

iv. Zhan Zhang; "SPARK-ON-HBASE: DATAFRAME BASED HBASE CONNECTOR," Hortonworks, 2016, https://hortonworks.com/blog/spark-hbase-dataframe-based-hbase-connector/

v. Huawei; "Astro: High performance SQL layer over HBase by using Spark SQL framework," Huawei, 2018, http://huaweibigdata.github.io/astro/

vi. Ted Malaska; "Apache Spark Comes to Apache HBase with HBase-Spark Module," Cloudera, 2018 https://blog.cloudera.com/blog/2015/08/apache-spark-comes-to-apache-hbase-with-hbase-spark-module/

vii. Apache Lucene; "Using SolrJ," Apache Lucene, 2018, https://lucene.apache.org/solr/guide/6_6/using-solrj.html

viii. Lucidworks; "Lucidworks Spark/Solr Integration,", Lucidworks, 2018, https://github.com/lucidworks/spark-solr

ix. Cloudera; "Importing Data Into HBase," Cloudera, 2018, https://www.cloudera.com/documentation/enterprise/latest/topics/admin_hbase_import.html

x. Tom White; "The Small Files Problem," Cloudera, 2009, https://blog.cloudera.com/blog/2009/02/the-small-files-problem/

xi. David W. Aha; "Heart Disease Data Set," University of California, Irvine, 1988, http://archive.ics.uci.edu/ml/datasets/heart+Disease

CHAPTER 7

Batch and Real-Time Data Ingestion and Processing

Data ingestion is the process of transferring, loading, and processing data into a data management or storage platform. This chapter discusses various tools and methods on how to ingest data into Kudu in batch and real time. I'll cover native tools that come with popular Hadoop distributions. I'll show examples on how to use Spark to ingest data to Kudu using the Data Source API, as well as the Kudu client APIs in Java, Python, and C++. There is a group of next-generation commercial data ingestion tools that provide native Kudu support. Internet of Things (IoT) is also a hot topic. I'll discuss all of them in detail in this chapter starting with StreamSets.

StreamSets Data Collector

StreamSets Data Collector is a powerful, enterprise-grade streaming platform that you can use to ingest, route, and process real-time streaming and batch data from a large variety of sources. StreamSets was founded by Girish Pancha, Informatica's ex-chief product officer; and Arvind Prabhakar, an early employee of Cloudera, where he led the development of Apache Flume and Apache Sqoop.[i] StreamSets is used by companies such as CBS Interactive, Cox Automotive, and Vodafone.[ii]

Data Collector can perform all kinds of data enrichment, transformation, and cleansing in-stream, then write your data to a large number of destinations such as HDFS, Solr, Kafka, or Kudu, all without writing a single line of code. For more complex data processing, you can write code in one of the following supported languages and frameworks: Java, JavaScript, Jython (Python), Groovy, Java Expression Language (EL), and Spark. Data Collector can run in stand-alone or cluster mode to support the largest environments.

© Butch Quinto 2018
B. Quinto, *Next-Generation Big Data*, https://doi.org/10.1007/978-1-4842-3147-0_7

Pipelines

To ingest data, Data Collector requires that you design a pipeline. A pipeline consists of multiple stages that you configure to define your data sources (origins), any data transformation or routing needed (processors), and where you want to write your data (destinations).

After designing your pipeline, you can start it, and it will immediately start ingesting data. Data Collector will sit on standby and quietly wait for data to arrive at all times until you stop the pipeline. You can monitor Data Collector by inspecting data as it is being ingested or by viewing real-time metrics about your pipelines.

Origins

In a StreamSets pipeline, data sources are called Origins. They're configurable components that you can add to your canvas without any coding. StreamSets includes several origins to save you development time and efforts. Some of the available origins include MQTT Subscriber, Directory, File tail, HDFS, S3, MongoDB, Kafka, RabbitMQ, MySQL binary blog, and SQL Server and Oracle CDC client to name a few. Visit StreamSets.com for a complete list of supported origins.

Processors

Processors allows you to perform data transformation on your data. Some of the available processors are Field Hasher, Field Masker, Expression Evaluator, Record Deduplicator, JSON Parser, XML Parser, and JDBC Lookup to name a few. Some processors, such as the Stream Selector, let you easily route data based on conditions. In addition, you can use evaluators that can process data based on custom code. Supported languages and frameworks include JavaScript, Groovy, Jython, and Spark. Visit StreamSets.com for a complete list of supported processors.

Destinations

Destinations are the target destination of your pipeline. Available destinations include Kudu, S3, Azure Data Lake Store, Solr, Kafka, JDBC Producer, Elasticsearch, Cassandra, HBase, MQTT Publisher, and HDFS to name a few. Visit StreamSets.com for a complete list of supported destinations.

CHAPTER 7 BATCH AND REAL-TIME DATA INGESTION AND PROCESSING

Executors

Executors allow you to run tasks such as a MapReduce job, Hive query, Shell script, or Spark application when it receives an event.

Data Collector Console

The console is Data Collector's main user interface (see Figure 7-1). This is where all the action happens. This is where you design your pipeline, configure your stages, run and troubleshoot pipelines, and more.

Figure 7-1. *StreamSets Data Collector console*

StreamSets Data Collector can ingest data in real time or batch (see Figure 7-2). Real-time data sources or origins include MQTT, Kafka, and Kinesis to name a few. Batch data sources includes HDFS, S3, Oracle, SQL Server, MongoDB, and so on. Data eventually lands on one or more destinations. In between the data sources and destinations are processors that transform and process data in-stream.

Chapter 7 Batch and Real-Time Data Ingestion and Processing

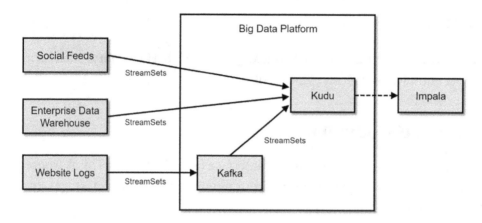

Figure 7-2. *Typical StreamSets Architecture*

There are several pre-built processors for some of the most common data transformations and enrichments. Examples of processors include column splitters and hashers, data type converters, XML parsers, JDBC lookups, and Stream processors to name a few. These processors can be configured without writing a single line of code. However, there are times when you want to write code, mainly, because none of the existing processors can handle the specific type of transformation that you require. StreamSets provide evaluators that support JavaScript, Java, Java EL, Spark, Jython, and Groovy. StreamSets connects all these stages (origins, processors, data sources) to form a pipeline. StreamSets executes the pipeline in-memory, providing maximum performance and scalability.

Real-Time Streaming

As discussed in previous chapters, Kudu is especially suitable for real-time ingestion and processing because of its support for efficient inserts and updates as well as fast columnar data scans.[iii] For real-time workloads, users need to select the right type of destination. There are storage engines that are not designed for real-time streaming. Examples of these storage engine includes HDFS and S3.

HDFS was designed for batch-oriented workloads. It splits and writes data in 128MB block sizes (configurable) to enable high throughput parallel processing. The problem arises when real-time data ingestion tools such as StreamSets or Flume start writing data to HDFS as fast as it receives them in order to make the data available to users in real time or near real time. This will cause HDFS to produce tons of small files. As mentioned earlier, HDFS was not designed to handle small files.[iv] Each file, directory, or block in

HDFS has metadata stored in the name node's memory. Sooner or later, the amount of memory consumed by these small files will overwhelm the name node. Another problem with small files is that they are physically inefficient to access. Applications need to perform more IO than necessary to read small files. A common workaround is to use coalesce or repartition in Spark to limit the number of files written to HDFS or compact the files regularly using Impala, but that is not always feasible.

S3 has its own IO characteristics and works differently than HDFS. S3 is an eventually consistent object store. With eventually consistent filesystems, users may not always see their updates immediately.[v] This might cause inconsistencies when reading and writing data to and from S3. Often, data appears after just a few milliseconds or seconds of writing to S3, but there are cases where it can take as long as 12 hours for some of the data to appear and become consistent. In rare cases, some objects can take almost 24 hours to appear.[vi] Another limitation of S3 is latency. Accessing data stored on S3 is significantly slower compared to HDFS and other storage systems.

Each S3 operation is an API call that might take tens to hundreds of milliseconds. The latency will add up and can become a performance bottleneck if you are processing millions of objects from StreamSets.[vii] S3 is great for storing backups or historical data but inappropriate for streaming data. In addition to Kudu, other destinations that are suitable for real-time data ingestion include HBase and MemSQL to name a few.

Batch-Oriented Data Ingestion

As mentioned earlier, StreamSets also supports batch workloads. As an origin and destination, I suggest Kudu for most batch-oriented jobs. Kudu provides most of the benefits of HDFS without the administration overheads. Performance benchmarks performed by the Kudu development team suggests that Kudu is still slightly slower than HDFS (Parquet) in some operations. If the difference in performance matters to your application (or if you're processing unstructured data), then I suggest you use HDFS. For most applications, the difference in performance may not be important. In that case, I would suggest you stick with Kudu (see Figure 7-3). With StreamSets, batch pipelines are built in the same way as real-time pipelines. The only difference is when you use batch-oriented origins such as S3, HDFS, RDBMS, or file directory, StreamSets detects the type of origin and automatically reads the data in batch mode.

CHAPTER 7 BATCH AND REAL-TIME DATA INGESTION AND PROCESSING

Figure 7-3. *Batch data ingestion with StreamSets and Kudu*

Internet of Things (IoT)

StreamSets is the Swiss army knife of data ingestion. Adding to its numerous features and capabilities is its support for IoT or Internet of Things[viii] (see Figure 7-4). StreamSets includes an MQTT Subscriber origin and MQTT Publisher destination that allows it to be used as an Internet of Things gateway. For reading data from SCADA networks and OPC historian, StreamSets includes an OPC UA Client origin. Finally, StreamSets supports CoAP (Constrained Application Protocol).[ix] CoAP is a protocol for low-power and low-bandwidth environments designed for machine-to-machine device communication. In Chapter 9, we implement a complete IoT data ingestion and visualization application using StreamSets, Kudu, and Zoomdata.

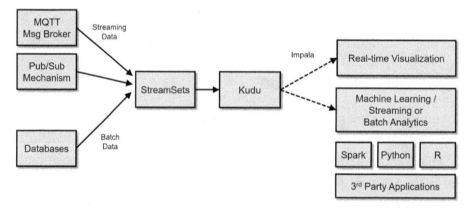

Figure 7-4. *StreamSets IoT Architecture*

CHAPTER 7 BATCH AND REAL-TIME DATA INGESTION AND PROCESSING

Deployment Options

StreamSets supports different deployment options, stand-alone or in-cluster modes. In stand-alone mode, StreamSets Data Collector is installed on one or more edge nodes. These edge nodes can be located in the same data center or in geographically disparate sites, as long as the network can support the latency and throughput requirements of your pipelines. In cluster mode, StreamSets pipelines are deployed and executed as Spark Streaming applications, utilizing YARN or Mesos as its cluster manager to take full advantage of the Hadoop cluster's scalability (see Figure 7-5).

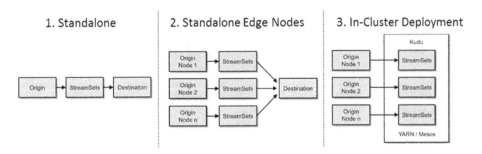

Figure 7-5. *StreamSets Deployment Options*

Consult the Data Collector User Guide for information on installing StreamSets from a Tar ball (Service Start), an RPM Package (Service Start), and Cloudera Manager (Cluster Mode).

Using StreamSets Data Collector

Let's start by creating our first pipeline. Navigate to the StreamSets Data Collector URL, and you will be presented with a login page. The default username is "admin." The default password is "admin." Log in, then click "Create New Pipeline." When prompted, enter a title and description for your new pipeline (see Figure 7-6).

CHAPTER 7 BATCH AND REAL-TIME DATA INGESTION AND PROCESSING

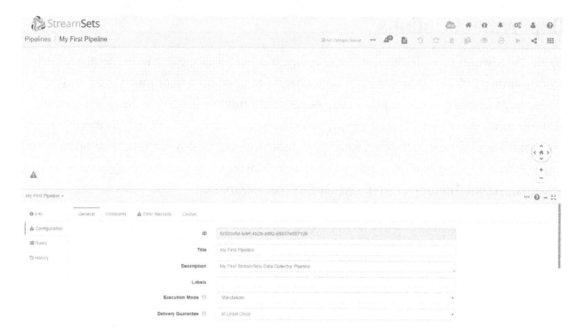

Figure 7-6. *StreamSets Console*

You can now design your first pipeline.

Ingesting XML to Kudu

In this example we will ingest sensor data in XML format into a Kudu table. Log in to StreamSets. On the Pipelines page, click the "Create New Pipeline" button (see Figure 7-7).

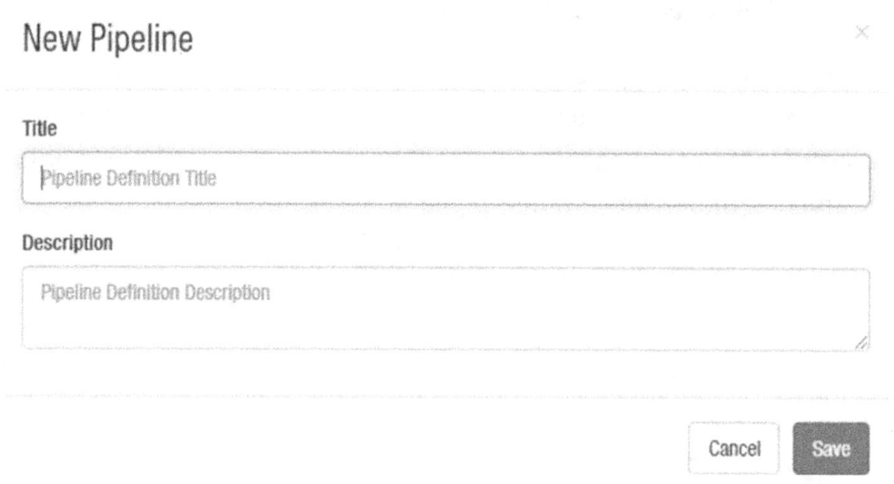

Figure 7-7. *New Pipeline*

CHAPTER 7 BATCH AND REAL-TIME DATA INGESTION AND PROCESSING

In the Data Collector Console, Click the "Stage Library" icon located near the "Help" icon (see Figure 7-8). You will be presented with a list of available stages you can use. Choose "Origins" in the list of stages. We will use Directory origins in this example. With the Directory origins, files that are copied to a designated directory will be ingested by StreamSets. Drag the Directory origins to the right side of the canvas.

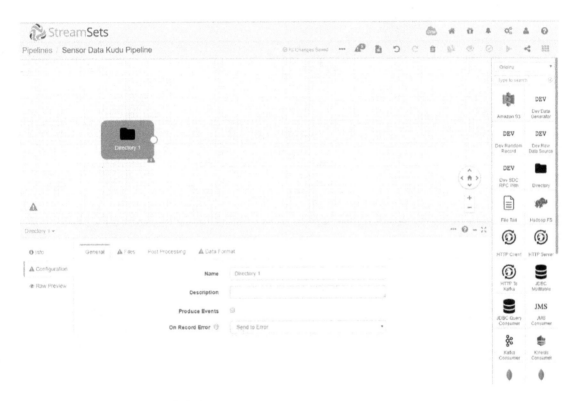

Figure 7-8. *Directory Origin*

Next, we need to add a processor. Choose "Processors" in the stage library. We will use an XML Parser to parse our XML data and convert it into a format that can be inserted into Kudu. Scroll down until you see the XML Parser icon. Drag the XML Parser processor to the canvas, near the Directory origins. Connect the Directory origins to the XML Parser processor in the canvas as shown in Figure 7-9. You will notice a yellow Pipeline Creation Help Bar. You can use the help bar to assist you in choosing stages. For now, we'll ignore the help bar.

CHAPTER 7 BATCH AND REAL-TIME DATA INGESTION AND PROCESSING

Figure 7-9. *XML Parser*

We need a Destination to complete our pipeline. Choose "Destinations" in the stage library. We will use a Kudu as our destination. Drag the Kudu destination to the canvas, near the XML Parser processor. Connect the XML Parser to the Kudu destination in the canvas as shown in Figure 7-10.

CHAPTER 7   BATCH AND REAL-TIME DATA INGESTION AND PROCESSING

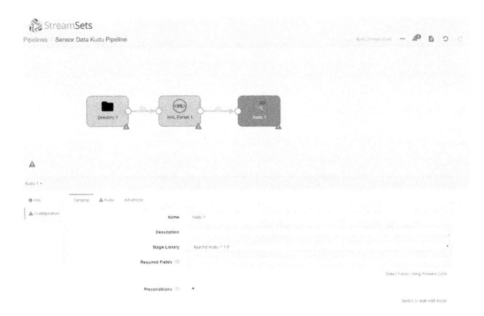

Figure 7-10. Kudu destination

Before we can start configuring our stages, we need to create the directory that will serve as the origin of our data source and the destination Kudu table. We'll also prepare some test data.

Open impala-shell and create a table.

```
CREATE TABLE sensordata
(
      rowid BIGINT,
      sensorid SMALLINT,
      sensortimestamp STRING,
      temperature TINYINT,
      pressure TINYINT,
      PRIMARY KEY(rowid)
)
PARTITION BY HASH PARTITIONS 16
STORED AS KUDU;
```

CHAPTER 7   BATCH AND REAL-TIME DATA INGESTION AND PROCESSING

Log in to the Cloudera cluster and create a directory that will contain our sensor data.

```
mkdir /sensordata
chown hadoop:hadoop /sensordata
```

Let's create some test data that we will use later in the exercise.

```
<sensordata><rowid>1000</rowid><sensorid>12</sensorid><sensortimestamp>20170409001904</sensortimestamp><temperature>23</temperature><pressure>30</pressure></sensordata>
<sensordata><rowid>1001</rowid><sensorid>39</sensorid><sensortimestamp>20170409001927</sensortimestamp><temperature>25</temperature><pressure>28</pressure></sensordata>
```

Configure Pipeline

You can set configuration options for the entire pipeline by clicking on the name of the pipeline; in our case click on the "Sensor Data Kudu Pipeline" link near the upper-left side of the panel (see Figure 7-11).

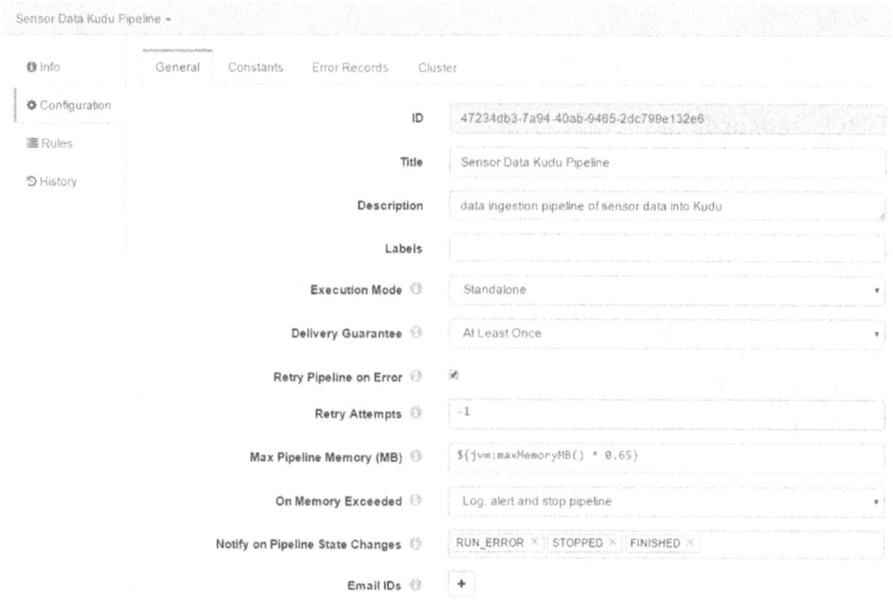

Figure 7-11. Configure Pipeline

In the "Error Records" tab, change "Error Records" to "Discard (Library: Basic)" (see Figure 7-12). For now, leave the rest of the default parameters.

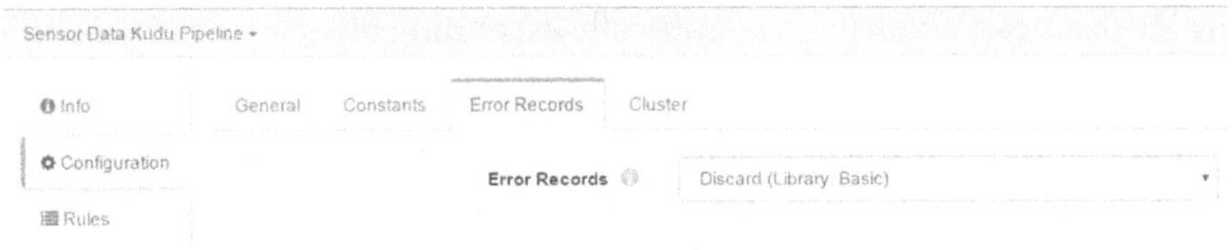

Figure 7-12. Error Records

Configure the Directory Origin

Click the Directory origin stage in the pipeline canvas. In the properties panel, navigate to the "Configurations" tab. Within the "Configurations" panel, you'll see another set of tabs (see Figure 7-13). We'll go through each tab.

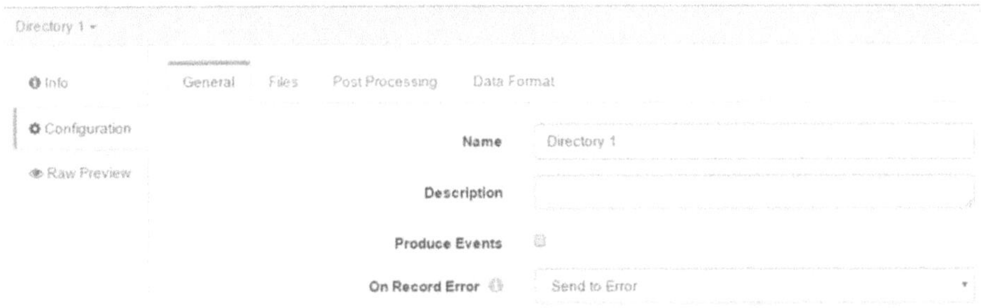

Figure 7-13. Configure Directory origin

The "General" allows users to specify a name and description for the origin. There's an option to enable the origin to generate events. If you enable "Produce Events," the directory origin produces event records every time the directory origin begins or ends reading a file. You can ignore this configuration item for now. "On Record Error" allows you to choose an action to take on records sent to error. The default option is "Send to Error." The other options are "Discard" and "Stop Pipeline." Leave the default option for now.

The "Files" tab contains various configuration options related to file ingestion. The most interesting option for us in our examples are "Files Directory," which is the local directory where you want to place files you want to ingest. "File Name Pattern Mode" lets you select whether the file name pattern specified uses regex syntax or glob syntax. Leave it set to "Glob" for now. "File Name Pattern" is an expression that defines the pattern

CHAPTER 7   BATCH AND REAL-TIME DATA INGESTION AND PROCESSING

of the file names in the directory. Since we're ingesting XML files, let's set the pattern to "*.xml". "Read Order" controls the order of how files are read and processed by Data Collector. There are two options, "Last Modified Timestamp" and "Lexicographically Ascending File Names." When lexicographically ascending file names are used, files are read in lexicographically ascending order based on file names.[x] It's important to note that this ordering will order numbers 1, 2, 3, 4, 5, 6, 7, 8, 9, 10, 11, 12, 13, 14, 15 as 1, 10, 11, 12, 13, 14, 15, 2, 3, 4, 5, 6, 7, 8, 9.If you must read files in lexicographically ascending order, consider adding leading zeros to the file names such as: file_001.xml, file_002.xml, file_003.xml, file_004.xml, file_010.xml, file_011.xml, file_012.xml.

When the last modified timestamp is used, files are read using the timestamp. Files with the same timestamp are ordered based on file names. For this exercise, let's set the read order to "Last Modified Timestamp" (see Figure 7-14).

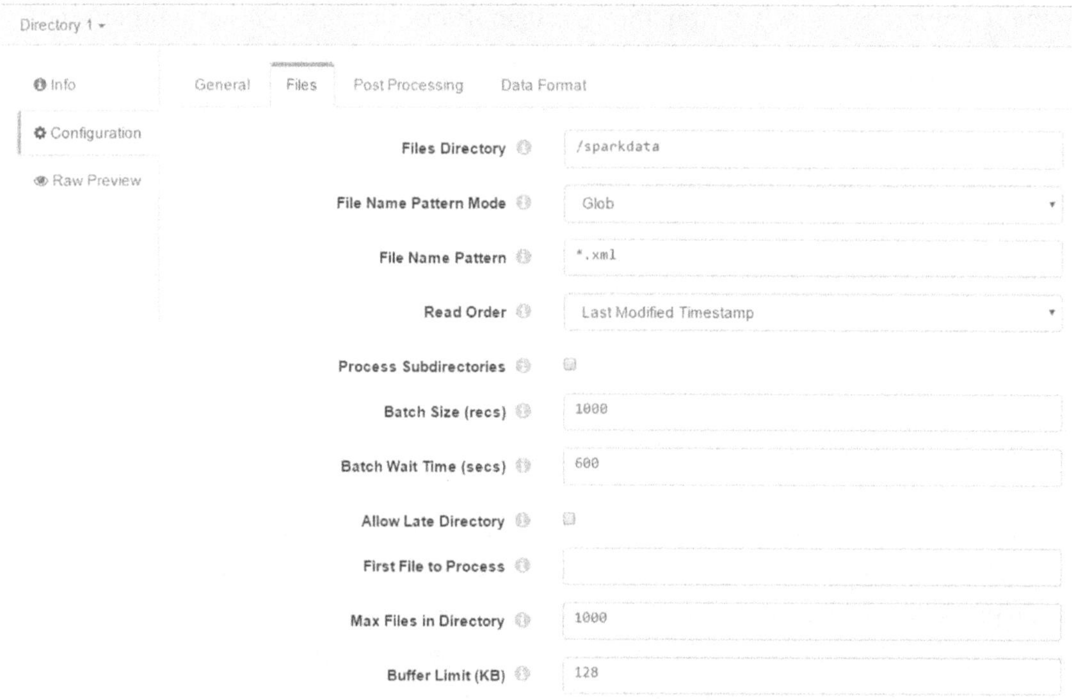

Figure 7-14.   Last Modified Timestamp

The "Post Processing" tab provides configuration options for files after they've been read and processed. "Error Directory" can be configured as the destination for files that can't be read. "File Post Processing" provides options to take after a file has been processed. You can archive or delete a file after processing, or you can do nothing. If you

CHAPTER 7 BATCH AND REAL-TIME DATA INGESTION AND PROCESSING

choose "Archive," additional options will be presented, such as the "Archive Directory" and "Archive Retention Time (mins)" (see Figure 7-15). Let's set "File Post Processing" to nothing for this example.

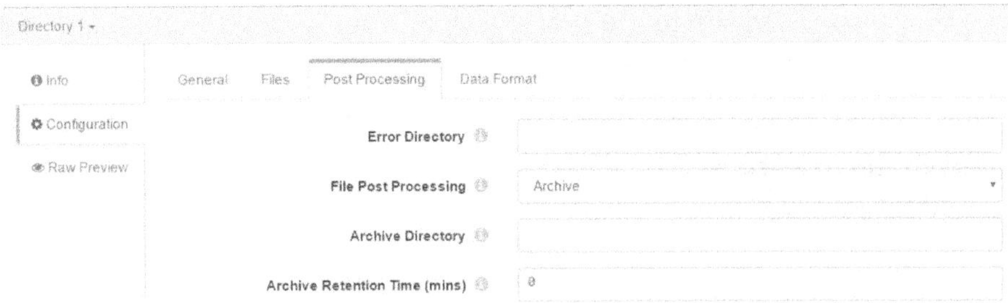

Figure 7-15. Archive

The "Data Format" tab allows you to set configuration parameters related to the format of your data (see Figure 7-16). The most interesting parameters for us are "Data Format." Note that even though we're ingesting XML data, we'll set this parameter to "Text." The XML data that we're ingesting is not strictly valid XML since it does not include a root element or XML prolog. So we will have to process our XML data using custom delimiters.[xi] Let's check "Use Customer Delimiters" and specify a customer delimiter; in our case let's set it to "</sensordata>." Finally, let's check "Include Customer Delimiters." We'll leave the other parameters to its default values for now.

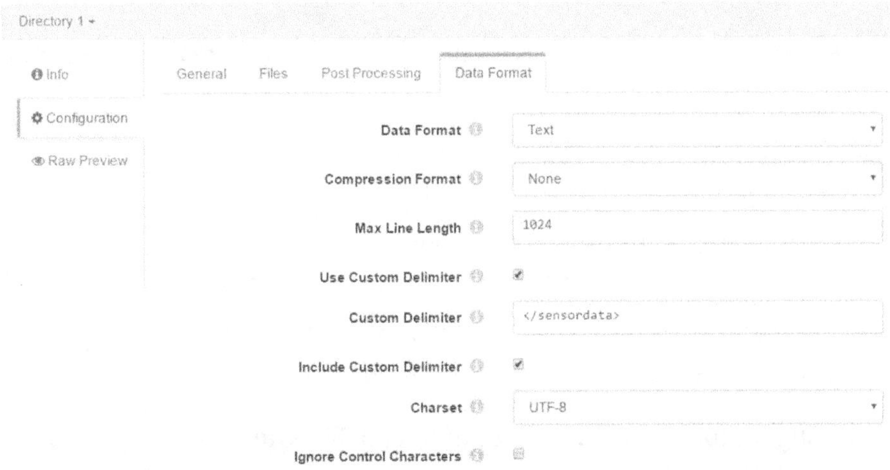

Figure 7-16. Data Format

245

CHAPTER 7 BATCH AND REAL-TIME DATA INGESTION AND PROCESSING

Configure the XML Parser Processor

Click the XML parser processor in the canvas and navigate to the "Configuration" section of the Properties Panel. The "General" tab will let you set a name and description for the XML Parser processor. Navigate to the "Parse" tab and set "Field to Parse" and "Target Field" to "/text" (see Figure 7-17). As mentioned earlier, the directory origins were configured to process text data. It writes records into a single text field named "text."[xii] This is a convention used by SDC.

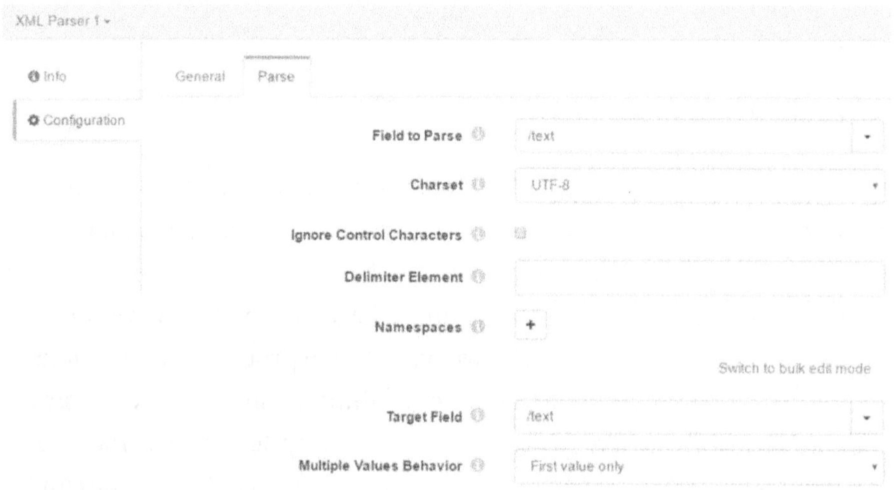

Figure 7-17. Parse tab

When the Directory origins processes our XML data, it will create two records, delimited by "</sensordata>."

```
<sensordata><rowid>1000</rowid><sensorid>12</sensorid><senso
rtimestamp>20170409001904</sensortimestamp><temperature>23</
temperature><pressure>30</pressure></sensordata>
<sensordata><rowid>1001</rowid><sensorid>39</sensorid><senso
rtimestamp>20170409001927</sensortimestamp><temperature>25</
temperature><pressure>28</pressure></sensordata>
```

In the "Configuration" tab, navigate to Kudu and fill in the required parameters. The hostnames and port of the "Kudu Masters." Note that because we created the "sensordata" table in Impala, it's considered an internally managed table. Kudu tables created in Impala follow the naming convention "impala::database.table_name." Set the

CHAPTER 7   BATCH AND REAL-TIME DATA INGESTION AND PROCESSING

"Table Name" to "impala::default.sensordata." You also need to set the field to column mapping as seen in Figure 7-18. The format of the SDC field looks like "/text/column[0]/value." Finally, you set the "Default Operation" to "INSERT." Note, that you have the option to perform other operations such as UPSERT, UPDATE, and DELETE.

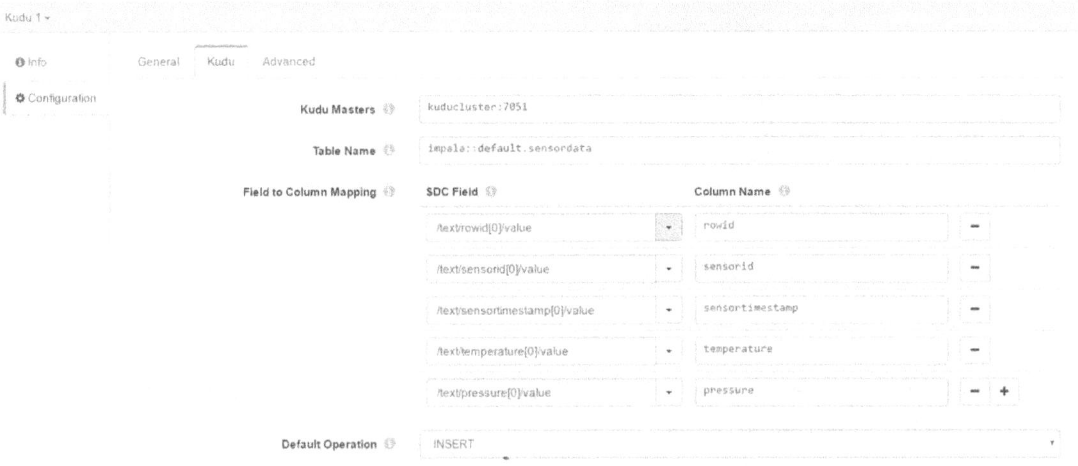

Figure 7-18. Kudu tab

Validate and Preview Pipeline

Now that you're done designing your pipeline, you need to check and make sure all the configurations are valid. Click the validate icon on the upper-right corner of the canvas to validate the pipeline. Once you've completed the validation, you can preview your pipeline to test it without actually running it. Previewing your pipeline is an easy way to debug and step through your pipeline, allowing you to inspect your data at each stage.[xiii] Click the preview icon. A "Preview Configuration" window will appear (see Figure 7-19). Accept the default values and click the "Run Preview" button.

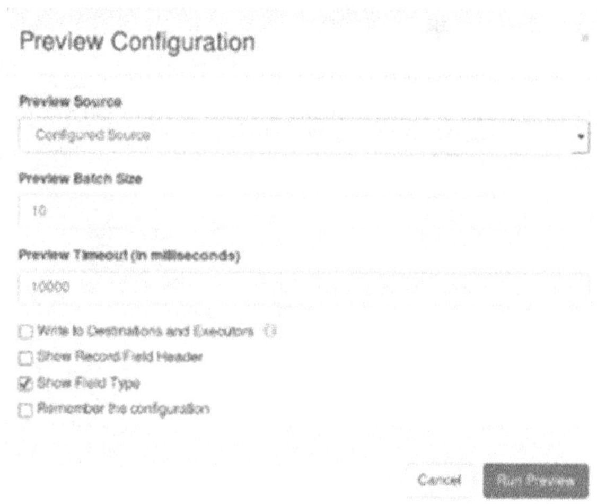

CHAPTER 7   BATCH AND REAL-TIME DATA INGESTION AND PROCESSING

Figure 7-19. Preview Configuration

You're now in preview mode. Click the Directory origin. In the preview stage pane, you can see a list of test records (see Figure 7-20). Collapsing each record shows the actual data. Note that the third record is empty. This is going to cause errors in the XML Parser processor.

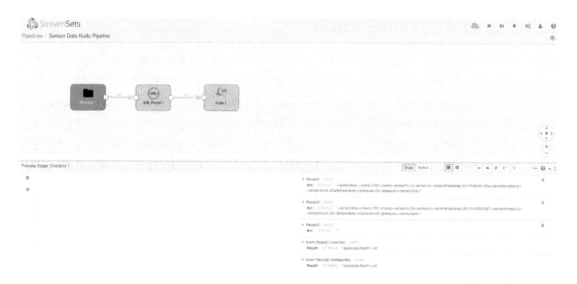

Figure 7-20. Preview – inspecting records

Click the XML Parser processor. Here you can see how the input data was processed and what the resulting output data looks like. As you can see, the XML data was processed correctly but as expected, there was an error processing the empty record (see Figure 7-21).

Figure 7-21. *Preview – error processing empty record*

CHAPTER 7 BATCH AND REAL-TIME DATA INGESTION AND PROCESSING

As expected, the Kudu destination's input data is the XML Parser's output (see Figure 7-22).

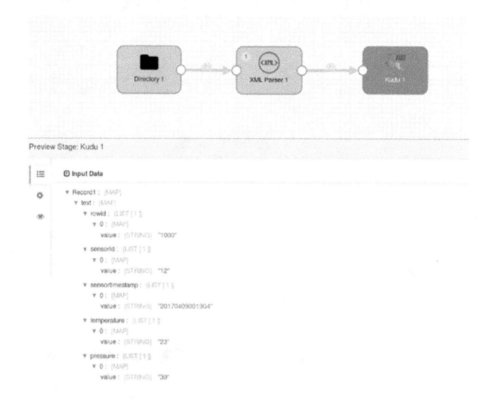

Figure 7-22. Preview – Kudu destination

CHAPTER 7 BATCH AND REAL-TIME DATA INGESTION AND PROCESSING

You can also preview multiple stages. Click "multiple" on top of the preview stage panel. The "Preview Multiple Stages" panel allows you to inspect the flow of the records from one stage to another (see Figure 7-23).

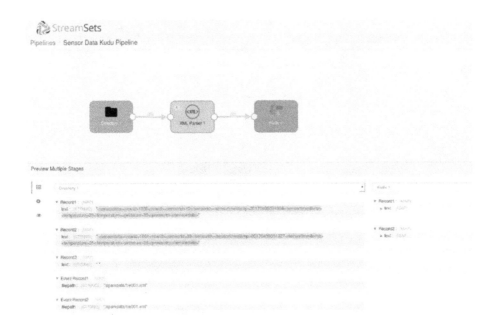

Figure 7-23. *Preview – Multiple Stages*

Start the Pipeline

Now you're ready to start the pipeline. Click the start icon, and then copy our test data (file001.xml) to /sparkdata.

```
cat file001.xml
<sensordata><rowid>1000</rowid><sensorid>12</sensorid><senso
rtimestamp>20170409001904</sensortimestamp><temperature>23</
temperature><pressure>30</pressure></sensordata>
<sensordata><rowid>1001</rowid><sensorid>39</sensorid><senso
rtimestamp>20170409001927</sensortimestamp><temperature>25</
temperature><pressure>28</pressure></sensordata>
cp file001.xml /sparkdata
```

251

CHAPTER 7 BATCH AND REAL-TIME DATA INGESTION AND PROCESSING

After a couple of seconds, you should see some runtime statistics on the monitoring panel such as record count, record throughput, and batch throughput to mention a few (see Figure 7-24).

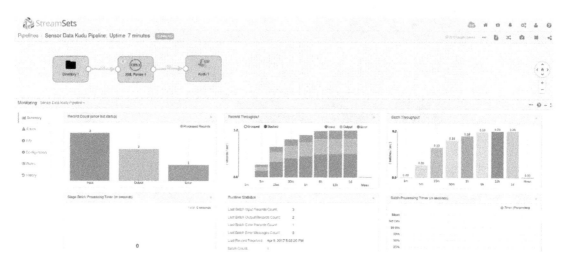

Figure 7-24. Monitoring pipeline

Click on each stage in the canvas to see runtime statistics specific for that stage. Let's click on the Directory origin located in the canvas (see Figure 7-25).

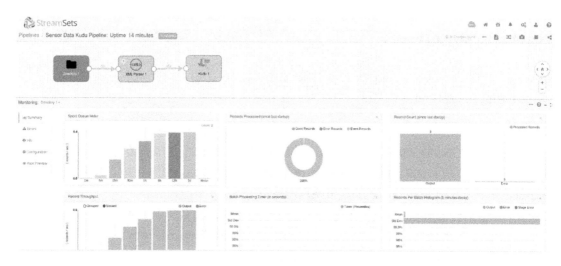

Figure 7-25. Monitoring– Directory Origin

CHAPTER 7 BATCH AND REAL-TIME DATA INGESTION AND PROCESSING

Clicking on the XML Parser processor gives you a different set of metrics (see Figure 7-26).

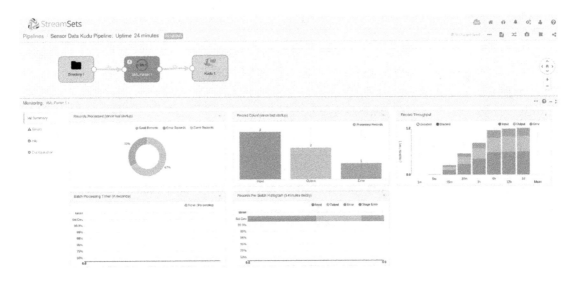

Figure 7-26. *Monitoring – XML Parser*

The Kudu destination has a different set of metrics as well. According to the screenshot in Figure 7-27, two rows were inserted to the sensordata Kudu table.

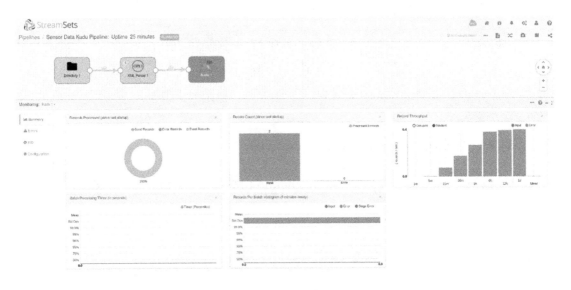

Figure 7-27. *Monitoring – Kudu destination*

CHAPTER 7 BATCH AND REAL-TIME DATA INGESTION AND PROCESSING

Let's verify if the rows were indeed successfully inserted in Kudu. Start impala-shell and query the sensordata table.

```
select * from sensordata;
```

```
+-------+---------+------------------+-------------+----------+
| rowid | sensorid | sensortimestamp | temperature | pressure |
+-------+---------+------------------+-------------+----------+
| 1001  | 39      | 20170409001927   | 25          | 28       |
| 1000  | 12      | 20170409001904   | 23          | 30       |
+-------+---------+------------------+-------------+----------+
```

The rows were successfully added to the table. Try adding more data (see Figure 7-28). You'll see the charts and other metrics get updated after adding more data.

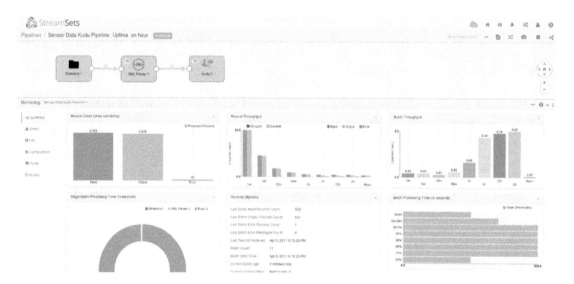

Figure 7-28. *Monitoring pipeline after one hour*

Congratulations! You've successfully built and run your first StreamSets data pipeline.

Stream Selector

There are times when you have a requirement to route data to certain streams based on conditions. For example, you can insert data to certain tables based on "Country." There is a default stream that handles records that do not match any of the user-defined conditions.[xiv]

```
Stream 1: ${record:value("/Country")=='USA'}
Stream 2: ${record:value("/Country")=='CANADA'}
Stream 3: ${record:value("/Country")=='BRAZIL'}
Stream 4: default
```

In this pipeline, all records where "Country" matches USA goes to Stream 1, Canada goes to Stream 2, and Brazil goes to Stream 3. Everything else goes to Stream 4. You'll be able to specify these conditions using the Stream Selector.

Let's design another pipeline. Instead of starting from scratch, let's make a copy of the pipeline we just created and modify it to use a Stream Selector. In this example, we'll write to three different versions of the XML data to three different Kudu tables. Here's what our new XML data looks like. The Stream Selector will examine the value of firmwareversion.

```
cat file002.xml

<sensordata>
        <rowid>1000</rowid>
        <sensorid>12</sensorid>
        <sensortimestamp>20170409001904</sensortimestamp>
        <temperature>23</temperature>
        <pressure>30</pressure>
        <firmwareversion>1</firmwareversion>
</sensordata>
<sensordata>
        <rowid>1001</rowid>
        <sensorid>39</sensorid>
        <sensortimestamp>20170409001927</sensortimestamp>
        <temperature>25</temperature>
        <pressure>28</pressure>
        <firmwareversion>2</firmwareversion>
```

```xml
        <humidity>110</humidity>
        <ozone>31</ozone>
</sensordata>
<sensordata>
        <rowid>1001</rowid>
        <sensorid>39</sensorid>
        <sensortimestamp>20170409001927</sensortimestamp>
        <temperature>25</temperature>
        <pressure>28</pressure>
        <firmwareversion>3</firmwareversion>
        <humidity>115</humidity>
        <ozone>12</ozone>
        <location>
                <altitude>121</altitude>
                <lat>37.8136</lat>
                <long>144.9631</long>
        </location>
</sensordata>
```

We'll create two more Kudu tables: sensordata2 and sensordata3. We'll also add a firmwareversion column in the existing sensordata table. We'll have a total of three sensordata tables, each with a different set of columns.

```
CREATE TABLE sensordata2
(
        rowid BIGINT,
        sensorid SMALLINT,
        sensortimestamp STRING,
        temperature TINYINT,
        pressure TINYINT,
        firmwareversion TINYINT,
        humidity TINYINT,
        ozone TINYINT,
        PRIMARY KEY(rowid)
)
```

```
PARTITION BY HASH PARTITIONS 16
STORED AS KUDU;

CREATE TABLE sensordata3
(
        rowid BIGINT,
        sensorid SMALLINT,
        sensortimestamp STRING,
        temperature TINYINT,
        pressure TINYINT,
        firmwareversion TINYINT,
        humidity TINYINT,
        ozone TINYINT,
        altitude TINYINT,
        lat FLOAT,
        long FLOAT,
        PRIMARY KEY(rowid)
)
PARTITION BY HASH PARTITIONS 16
STORED AS KUDU;

ALTER TABLE sensordata ADD COLUMNS (firmwareversion TINYINT);
```

Drag a Stream Selector processor and two more Kudu destinations onto the canvas. Connect the XML Parser to the Stream Connector. You need to define three conditions: XML data with firmwareversion of 1 (or any value other than 2 or 3) goes to default stream (Kudu 1), XML data with firmwareversion 2 goes to stream 2 (Kudu 2), and finally XML data with firmwareversion 3 goes to stream 3 (Kudu 3).

Add two more conditions to the Stream Selector processor with these conditions.

```
${record:value('/text/firmwareversion[0]/value') == "3"}
${record:value('/text/firmwareversion[0]/value') == "2"}
```

CHAPTER 7 BATCH AND REAL-TIME DATA INGESTION AND PROCESSING

Your new pipeline should look like Figure 7-29.

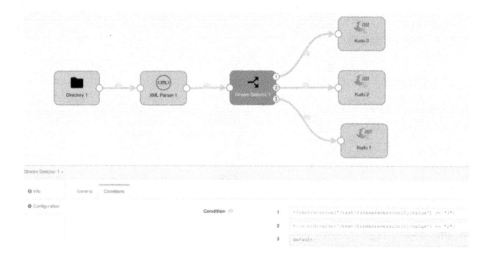

Figure 7-29. *Stream Selector*

For each Kudu destination, make sure you map the correct SDC field to the corresponding column names (Figure 7-30).

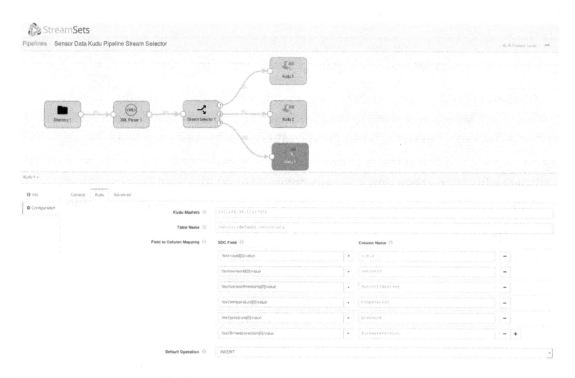

Figure 7-30. *First Kudu destination*

CHAPTER 7　BATCH AND REAL-TIME DATA INGESTION AND PROCESSING

Note that the second and third Kudu destinations will have additional columns (Figure 7-31). Don't forget to add the firmwareversion to the first Kudu destination (Figure 7-32).

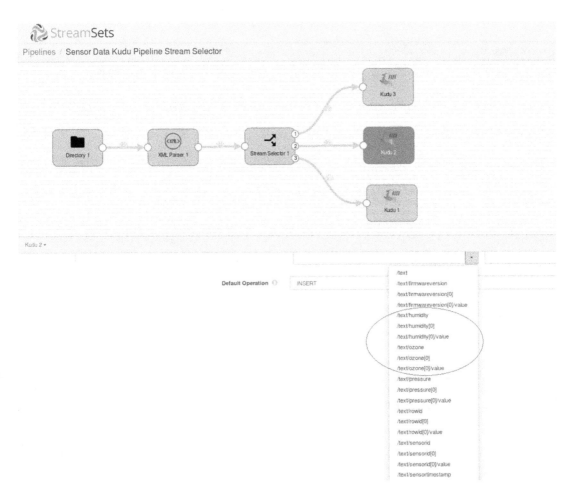

Figure 7-31. Second Kudu destination

CHAPTER 7 BATCH AND REAL-TIME DATA INGESTION AND PROCESSING

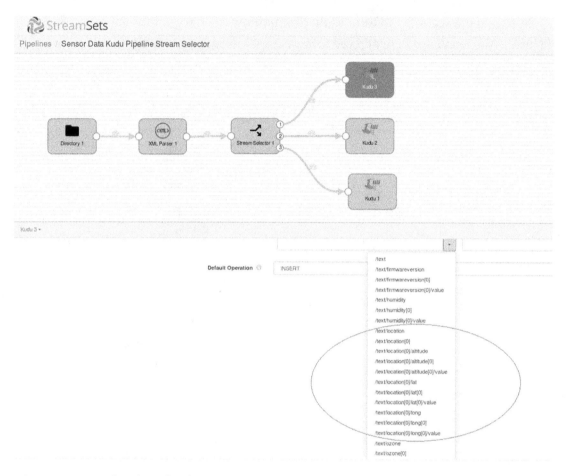

Figure 7-32. *Third Kudu destination*

CHAPTER 7 BATCH AND REAL-TIME DATA INGESTION AND PROCESSING

Validate and preview the pipeline you just created (Figure 7-33). Confirm that each Kudu destination is receiving the correct XML data (Figures 7-34 and 7-35).

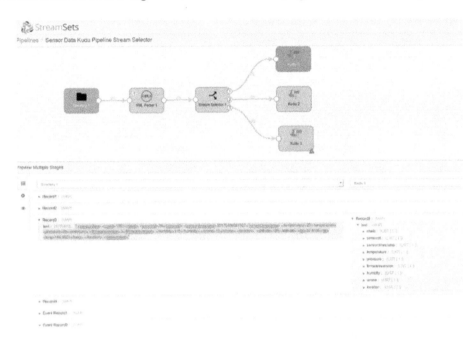

Figure 7-33. *Preview – First Kudu destination*

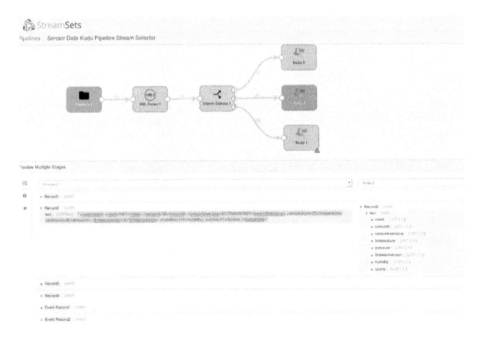

Figure 7-34. *Preview – Second Kudu destination*

CHAPTER 7 BATCH AND REAL-TIME DATA INGESTION AND PROCESSING

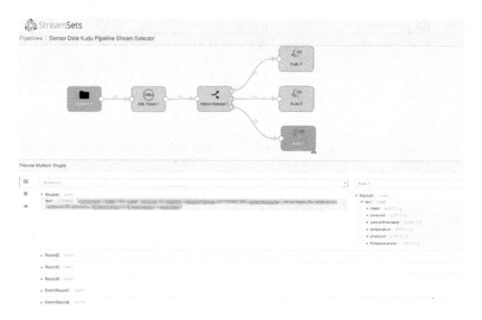

Figure 7-35. Preview – Third Kudu destination

Before you start the pipeline, confirm that the destination Kudu tables are empty from the impala-shell.

select count(*) from sensordata;

```
+----------+
| count(*) |
+----------+
| 0        |
+----------+
```

select count(*) from sensordata2;

```
+----------+
| count(*) |
+----------+
| 0        |
+----------+
```

select count(*) from sensordata3;

CHAPTER 7 BATCH AND REAL-TIME DATA INGESTION AND PROCESSING

```
+----------+
| count(*) |
+----------+
| 0        |
+----------+
```

Start the pipeline, then copy file002.xml to /sparkdata. After a few seconds, you'll see some updates on the charts in the monitoring panel. Click the Stream Selector processor. As you can see in the Record Count chart, Input has three records, while the three Output has one record each.

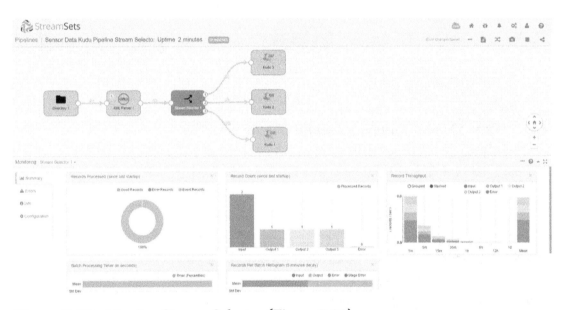

Figure 7-36. *Monitor Stream Selector (Figure 7-36)*

CHAPTER 7 BATCH AND REAL-TIME DATA INGESTION AND PROCESSING

Check each Kudu destination to see if the records were successfully ingested.

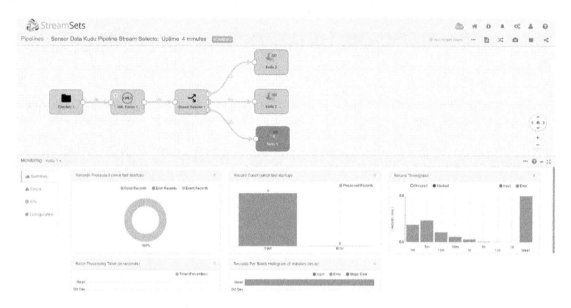

Figure 7-37. *Monitor Kudu destination (Figure 7-37)*

Check the Kudu tables to confirm that the records were successfully ingested.

```
select * from sensordata;
```

```
+--------+----------+------------------+-------------+
| rowid  | sensorid | sensortimestamp  | temperature |
+--------+----------+------------------+-------------+
| 1000   | 12       | 20170409001904   | 23          |
+--------+----------+------------------+-------------+

+-----------+------------------+
| pressure  | firmwareversion  |
+-----------+------------------+
| 30        | 1                |
+-----------+------------------+
```

```
select * from sensordata2;
```

```
+-------+----------+-----------------+-------------+
| rowid | sensorid | sensortimestamp | temperature |
+-------+----------+-----------------+-------------+
| 1001  | 39       | 20170409001927  | 25          |
+-------+----------+-----------------+-------------+

+----------+-----------------+----------+-------+
| pressure | firmwareversion | humidity | ozone |
+----------+-----------------+----------+-------+
| 28       | 2               | 110      | 31    |
+----------+-----------------+----------+-------+

select * from sensordata3;

+-------+----------+-----------------+-------------+
| rowid | sensorid | sensortimestamp | temperature |
+-------+----------+-----------------+-------------+
| 1001  | 39       | 20170409001927  | 25          |
+-------+----------+-----------------+-------------+

+----------+-----------------+----------+
| pressure | firmwareversion | humidity |
+----------+-----------------+----------+
| 28       | 3               | 115      |
+----------+-----------------+----------+

+-------+----------+-------------------+------------------+
| ozone | altitude | lat               | long             |
+-------+----------+-------------------+------------------+
| 12    | 121      | 37.8135986328125  | 144.9631042480469|
+-------+----------+-------------------+------------------+
```

Expression Evaluator

The Expression Evaluator allows users to perform data transformations and calculations and writes the results to new or existing fields. Users can also add or change field attributes and record header attributes.[xv] I often use the Expression Evaluator as a rules engine.

CHAPTER 7 BATCH AND REAL-TIME DATA INGESTION AND PROCESSING

Let's work on a new example. We'll use an Expression Evaluator to perform two data transformations. We'll convert the temperature from Celsius to Fahrenheit and save it to the existing temperature field. For the second transformation, we'll update a new status field with the values "NORMAL," "WARNING," or "CRITICAL" depending on the values of the pressure field. If pressure is over 40, the status field gets updated with "WARNING," and if pressure is over 50 it gets updated with "CRITICAL." Let's get started.

Add another "status" column to the Kudu sensordata table.

```
ALTER TABLE sensordata ADD COLUMNS (status STRING);

DESCRIBE sensordata;
```

```
+-----------------+----------+-------------+
| name            | type     | primary_key |
+-----------------+----------+-------------+
| rowid           | bigint   | true        |
| sensorid        | smallint | false       |
| sensortimestamp | string   | false       |
| temperature     | tinyint  | false       |
| pressure        | tinyint  | false       |
| firmwareversion | tinyint  | false       |
| status          | string   | false       |
+-----------------+----------+-------------+
```

We'll use the following XML data for this example.

```
cat file003.xml
```

```xml
<sensordata>
        <rowid>1000</rowid>
        <sensorid>12</sensorid>
        <sensortimestamp>20170409001904</sensortimestamp>
        <temperature>23</temperature>
        <pressure>30</pressure>
        <firmwareversion>1</firmwareversion>
</sensordata>
```

```xml
<sensordata>
        <rowid>1001</rowid>
        <sensorid>39</sensorid>
        <sensortimestamp>20170409001927</sensortimestamp>
        <temperature>25</temperature>
        <pressure>28</pressure>
        <firmwareversion>2</firmwareversion>
        <humidity>110</humidity>
        <ozone>31</ozone>
</sensordata>
<sensordata>
        <rowid>1002</rowid>
        <sensorid>39</sensorid>
        <sensortimestamp>20170409001927</sensortimestamp>
        <temperature>25</temperature>
        <pressure>28</pressure>
        <firmwareversion>3</firmwareversion>
        <humidity>115</humidity>
        <ozone>12</ozone>
        <location>
                <altitude>121</altitude>
                <lat>37.8136</lat>
                <long>144.9631</long>
        </location>
</sensordata>
<sensordata>
        <rowid>1003</rowid>
        <sensorid>58</sensorid>
        <sensortimestamp>20170409001930</sensortimestamp>
        <temperature>22</temperature>
        <pressure>44</pressure>
        <firmwareversion>2</firmwareversion>
        <humidity>112</humidity>
        <ozone>17</ozone>
</sensordata>
```

CHAPTER 7   BATCH AND REAL-TIME DATA INGESTION AND PROCESSING

```xml
<sensordata>
        <rowid>1004</rowid>
        <sensorid>72</sensorid>
        <sensortimestamp>20170409001934</sensortimestamp>
        <temperature>26</temperature>
        <pressure>59</pressure>
        <firmwareversion>2</firmwareversion>
        <humidity>115</humidity>
        <ozone>12</ozone>
</sensordata>
```

Make a copy of the original pipeline and add an Expression Evaluator processor in between the XML Parser processor and Kudu destination. Connect the stages. Your canvas should look like Figure 7-40.

Figure 7-38. *Expression Evaluator*

Click the Expression Evaluator processor (Figure 7-38) and navigate to the "Expressions" tab. We'll add two entries in the "Field Expressions" section. For the first entry, in the "Output Field" enter the following value:

/text/temperature[0]/value

268

CHAPTER 7   BATCH AND REAL-TIME DATA INGESTION AND PROCESSING

In the "Field Expression", enter the expression:

${record:value('/text/temperature[0]/value') * 1.8 + 32}

This is the formula to convert temperature from Celsius to Fahrenheit. Specifying an existing field in the "Output Field" will overwrite the value in the field.

For the second entry, we'll use an if-then-else expression. Enter a new field in the "Output Field":

/text/status

Enter the expression in "Field Expression":

${record:value('/text/pressure[0]/value') > 50?'CRITICAL': (record:value('/text/pressure[0]/value') > 40?'WARNING':'NORMAL')}

Your screen should look like Figure 7-41.

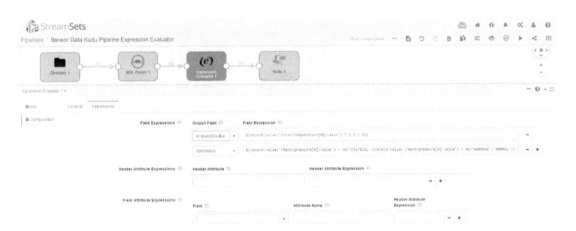

Figure 7-39. *Expression Evaluator - Expression*

CHAPTER 7 BATCH AND REAL-TIME DATA INGESTION AND PROCESSING

Map "/text/status" to the new status field in the Kudu destination.

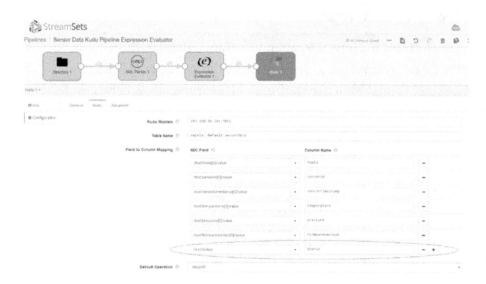

Figure 7-40. *Map SDC field to Kudu field*

Validate and preview the pipeline. Click the Expression Evaluator stage and confirm the values of the input and output records. Note that the value of temperature in the output Record 4 and Record 5 has been converted to Fahrenheit. The value of status on both output Record 4 and Record 5 were updated with "WARNING" and "CRITICAL" respectively, based on the value of the pressure field.

CHAPTER 7 BATCH AND REAL-TIME DATA INGESTION AND PROCESSING

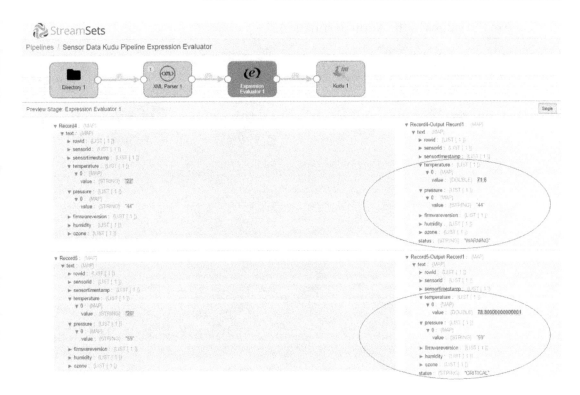

Figure 7-41. *Preview – Expression Evaluator*

CHAPTER 7 BATCH AND REAL-TIME DATA INGESTION AND PROCESSING

Confirm the values in the Kudu destination.

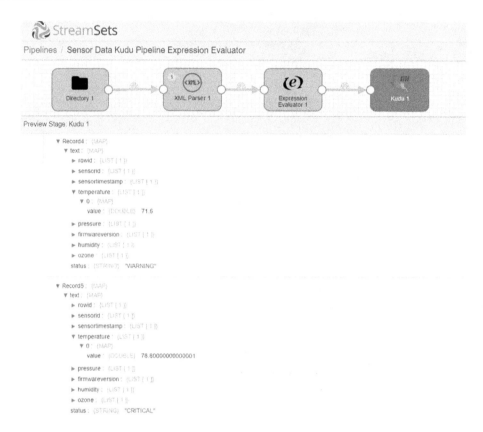

Figure 7-42. *Preview – Kudu destination*

CHAPTER 7 BATCH AND REAL-TIME DATA INGESTION AND PROCESSING

Run the pipeline and copy the XML file to /sparkdata.

It looks like all five records were successfully inserted to the Kudu sensordata table.

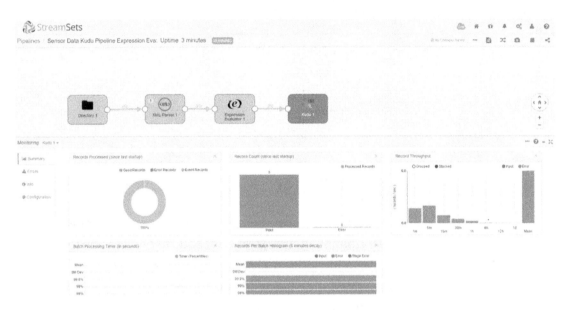

Figure 7-43. *Five records inserted into Kudu destination*

Confirm that the data were successfully inserted to Kudu. As you can see the status field was successfully updated based on pressure. Also, the temperatures are now in Fahrenheit.

```
SELECT * FROM sensordata;
```

rowid	sensorid	sensortimestamp	temperature
1002	39	20170409001927	77
1000	12	20170409001904	73
1001	39	20170409001927	77
1003	58	20170409001930	71
1004	72	20170409001934	78

```
+----------+-----------------+----------+
|pressure  | firmwareversion | status   |
+----------+-----------------+----------+
|28        | 3               | NORMAL   |
|30        | 1               | NORMAL   |
|28        | 2               | NORMAL   |
|44        | 2               | WARNING  |
|59        | 2               | CRITICAL |
+----------+-----------------+----------+
```

Using the JavaScript Evaluator

StreamSets Data Collector includes a JavaScript evaluator that you can use for more complex processing and transformation. It uses JavaScript code to process one record or one batch at a time.[xvi] As you can imagine, processing one record at a time will most likely be slower compared to processing by batch of data at a time. Processing by batch is recommended in production environments.

Although we're using JavaScript in this example, other languages are supported by Data Collector. In addition to JavaScript, you can use Java, Jython (Python), Groovy, and Spark Evaluators. These evaluators enable users to perform more advanced complex stream processing, taking advantage of the full capabilities of each programming languages.

Let's start with a simple example. We'll create two additional fields, action and uniqueid. We'll generate the value of action based on the value of the status field. Next, we'll generate a universally unique identifier (UUID) using a custom JavaScript function. We'll save this UUID in our uniqueid field.

We don't have to start from scratch. We'll make a copy of the previous pipeline and then drag a JavaScript Evaluator processor into the canvas and place in between the Expression Evaluator processor and the Kudu destination. Reconnect the stages. Your canvas should look like Figure 7-46.

CHAPTER 7 BATCH AND REAL-TIME DATA INGESTION AND PROCESSING

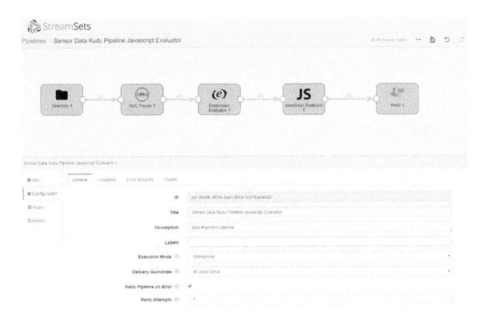

Figure 7-44. *JavaScript Evaluator*

Now let's configure the JavaScript Evaluator to use our custom JavaScript code in our pipeline. In the Properties panel, on the JavaScript tab, make sure that "Record Processing Mode" is set to "Batch by Batch." "Script" will contain your customer JavaScript code. As mentioned earlier, we'll create a function to generate a UUID.[xvii] The JavaScript Evaluator passes the batch to the script as an array. To access individual fields, use the format: records[arrayindex].value.text.columnname. To refer to the value of the status field use "records[i].value.text.status." We also created a new field, "record[i].value.text.action" using this format. See Listing 7-1.

Listing 7-1. Javascript Evaluator code

```
function generateUUID() {
    var d = new Date().getTime();
    var uuid = 'xxxxxxxx-xxxx-4xxx-yxxx-xxxxxxxxxxxx'.replace(/[xy]/g,
function(c) {
        var r = (d + Math.random()*16)%16 | 0;
        d = Math.floor(d/16);
        return (c=='x' ? r : (r&0x3|0x8)).toString(16);
    });
```

```
    return uuid;
};

for(var i = 0; i < records.length; i++) {
  try {

    var myUUID = generateUUID()

    if (records[i].value.text.status == "NORMAL")
            records[i].value.text.action = "None. Pressure is normal.";
      else if (records[i].value.text.status == "WARNING")
            records[i].value.text.action = "File a support ticket.";
      else if (records[i].value.text.status == "CRITICAL")
            records[i].value.text.action = "Inform your supervisor
            immediately!";

    records[i].value.text.uniqueid = myUUID;

    output.write(records[i]);
  } catch (e) {
    error.write(records[i], e);
  }
}
```

Your screen should look like Figure 7-47.

CHAPTER 7 BATCH AND REAL-TIME DATA INGESTION AND PROCESSING

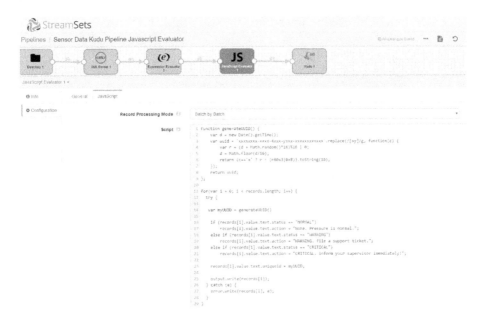

Figure 7-45. *JavaScript Evaluator code*

Add the action and uniqueid columns to the Kudu sensordata columns.

ALTER TABLE sensordata ADD COLUMNS (action STRING);

ALTER TABLE sensordata ADD COLUMNS (uniqueid STRING);

DESCRIBE sensordata;

```
+------------------+----------+--------------+
| name             | type     | primary_key  |
+------------------+----------+--------------+
| rowid            | bigint   | true         |
| sensorid         | smallint | false        |
| sensortimestamp  | string   | false        |
| temperature      | tinyint  | false        |
| pressure         | tinyint  | false        |
| firmwareversion  | tinyint  | false        |
| status           | string   | false        |
| action           | string   | false        |
| uniqueid         | string   | false        |
+------------------+----------+--------------+
```

CHAPTER 7   BATCH AND REAL-TIME DATA INGESTION AND PROCESSING

Map the SDC field with their corresponding Kudu table columns.

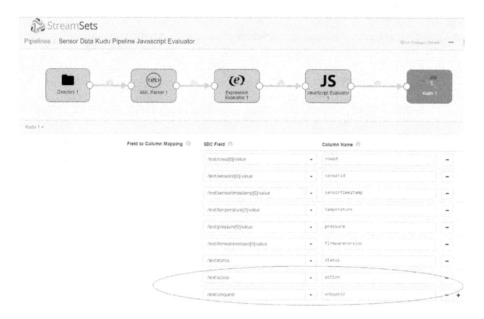

Figure 7-46. *Map SDC field*

Validate and preview your pipeline. Click the JavaScript Evaluator and confirm the values of the input and output records. Note that the action and unique fields have been populated.

Figure 7-47. *Preview values*

CHAPTER 7 BATCH AND REAL-TIME DATA INGESTION AND PROCESSING

Confirm the values in the Kudu destination.

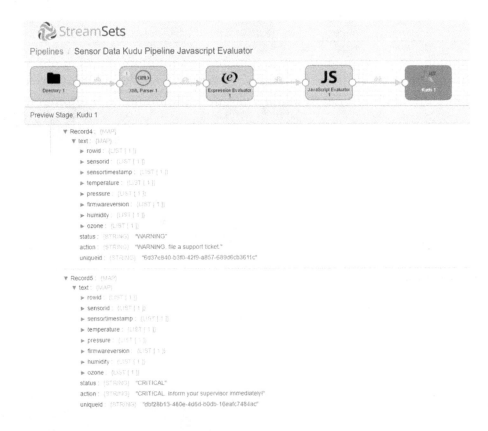

Figure 7-48. *Confirm values in Kudu destination*

Run the pipeline and copy the XML file to /sparkdata.

It looks like all five records were successfully processed by the JavaScript Evaluator.

279

CHAPTER 7 BATCH AND REAL-TIME DATA INGESTION AND PROCESSING

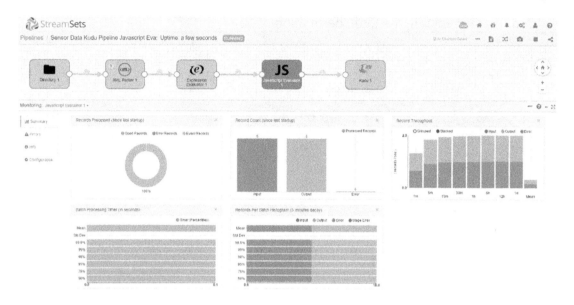

Figure 7-49. *Monitor JavaScript Evaluator*

All five records were also successfully inserted to the Kudu sensordata table.

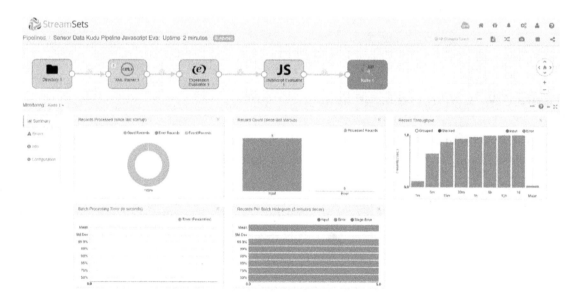

Figure 7-50. *Monitor Kudu destination*

CHAPTER 7   BATCH AND REAL-TIME DATA INGESTION AND PROCESSING

Confirm that the data was successfully inserted to Kudu. As you can see, the action and uniqueid fields were successfully populated.

```
SELECT rowid, status, action, substr(1,10,uniqueid) as uniqueid FROM sensordata;
```

```
+-------+----------+-------------------------------------------+------------+
| rowid | status   | action                                    |uniqueid    |
+-------+----------+-------------------------------------------+------------+
| 1002  | NORMAL   | None. Pressure is normal.                 |35c47cb3-e9 |
| 1000  | NORMAL   | None. Pressure is normal.                 |d3017138-09 |
| 1001  | NORMAL   | None. Pressure is normal.                 |d91416e0-1b |
| 1003  | WARNING  | WARNING. file a support ticket.           |30aa86be-71 |
| 1004  | CRITICAL | CRITICAL. inform your supervisor immediately!|f7161ea3-eb |
+-------+----------+-------------------------------------------+------------+
```

Congratulations! You've successfully used a JavaScript evaluator in your pipeline.

Ingesting into Multiple Kudu Clusters

Sometimes you'll be required to ingest data into two or more active-active clusters for high availability and scalability reasons. With Data Collector this is easily accomplished by simply adding Kudu destinations to the canvas and connecting them to the processor. In this example, we will simultaneously ingest XML data into two Kudu clusters, kuducluster01 and kuducluster02.

Drag a second Kudu destination to the canvas. You should now have two Kudu destinations. Make sure they're both connected to the XML processor.

CHAPTER 7 BATCH AND REAL-TIME DATA INGESTION AND PROCESSING

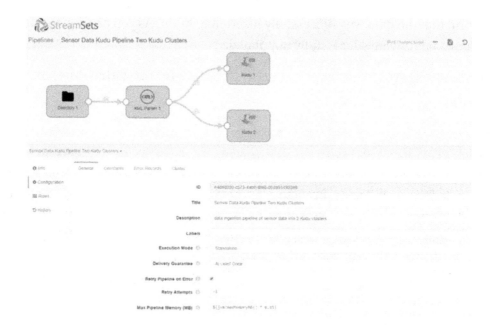

Figure 7-51. Multiple Kudu destinations

Click the first Kudu destination. Make sure the correct host name and ports are specified in the Kudu Masters field. Make sure the SDC fields also map to the Kudu table fields.

Figure 7-52. Configure first Kudu destination

CHAPTER 7 BATCH AND REAL-TIME DATA INGESTION AND PROCESSING

Next, click the second Kudu destination. Perform the same task as you did on the first Kudu destination.

Figure 7-53. *Configure second Kudu destination*

Validate and preview the pipeline. When you're done, start the pipeline and start adding data. Monitor the XML Parser. Note the number of input and output data.

CHAPTER 7 BATCH AND REAL-TIME DATA INGESTION AND PROCESSING

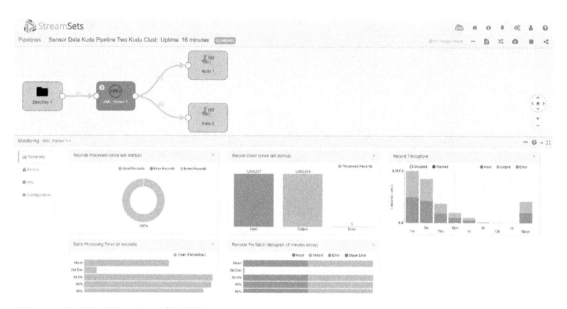

Figure 7-54. *Monitor XML Parser*

Monitor the first Kudu destination. Note that it processed 1,582, 554 records.

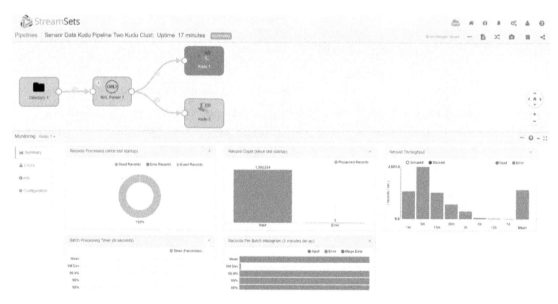

Figure 7-55. *Monitor first Kudu destination*

CHAPTER 7   BATCH AND REAL-TIME DATA INGESTION AND PROCESSING

Check the second Kudu destination and note that it also ingested the same amount of records.

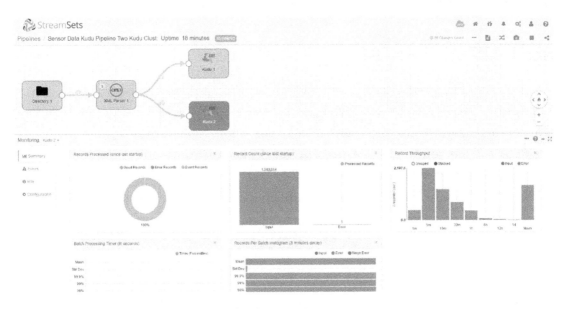

Figure 7-56. *Monitor second Kudu destination*

Verify in Impala. Perform a SELECT COUNT on the sensordata table on the first Kudu cluster.

```
select count(*) from sensordata;
```

```
+----------+
| count(*) |
+----------+
| 1582554  |
+----------+
```

Do the same SELECT COUNT on the second Kudu cluster.

```
select count(*) from sensordata;
```

```
+----------+
| count(*) |
+----------+
| 1582554  |
+----------+
```

CHAPTER 7 BATCH AND REAL-TIME DATA INGESTION AND PROCESSING

You ingested data into two Kudu clusters simultaneously. You can ingest to different combinations of platforms such as HBase, Cassandra, Solr, Kafka, S3, MongoDB, and so on. Consult the Data Collector User Guide for more details.

REST API

Data Collector has an easy-to-use web-based GUI for designing, running, and administering pipelines. However, you can also use the built-in REST API if you'd like to programmatically interact with Data Collector, for automation purposes, for example.[xviii] The REST API lets you access all aspects of Data Collector, from starting and stopping pipelines, returning configuration information and monitoring pipeline metrics.

You can access the REST API by clicking the Help icon and then selecting "RESTful API."

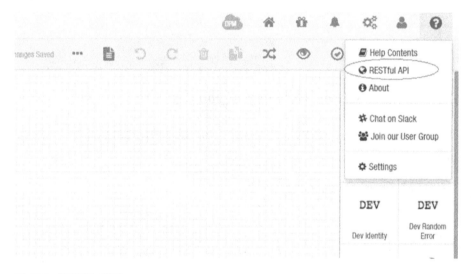

Figure 7-57. *REST API*

You'll be presented with a group of available operations. Expand the manager list by clicking on "manager." You can explore different operations under the "manager" group such as returning pipeline status, starting and stopping pipelines, and so on.

CHAPTER 7 BATCH AND REAL-TIME DATA INGESTION AND PROCESSING

Figure 7-58. *List of available operations*

Expand "Return All Pipeline Status." Choose the response content type "application/json" and click the "try it out" button.

The "response body" will contain details about your pipelines in JSON format similar to Listing 7-2. Depending on the number of pipelines and type of error messages returned, the response body could be quite large.

Listing 7-2. Return all pipeline status response body

```
{
  "47234db3-7a94-40ab-9465-2dc799e132e6": {
    "user": "admin",
    "name": "47234db3-7a94-40ab-9465-2dc799e132e6",
    "rev": "0",
    "status": "EDITED",
    "message": "Pipeline edited",
    "timeStamp": 1491742372750,
    "attributes": {
      "IS_REMOTE_PIPELINE": false
    },
```

```json
    "executionMode": "STANDALONE",
    "metrics": null,
    "retryAttempt": 0,
    "nextRetryTimeStamp": 0
  },
  "6c92cc6d-bdef-4b2b-ad62-69537e057128": {
    "user": "admin",
    "name": "6c92cc6d-bdef-4b2b-ad62-69537e057128",
    "rev": "0",
    "status": "EDITED",
    "message": "Pipeline edited",
    "timeStamp": 1491739709715,
    "attributes": {
      "IS_REMOTE_PIPELINE": false
    },
    "executionMode": "STANDALONE",
    "metrics": null,
    "retryAttempt": 0,
    "nextRetryTimeStamp": 0
  },
  "de3b27ed-0a92-47cb-8400-da5aa4cdf43e": {
    "user": "admin",
    "name": "de3b27ed-0a92-47cb-8400-da5aa4cdf43e",
    "rev": "0",
    "status": "STOPPED",
    "message": "The pipeline was stopped. The last committed source offset is 'file006.xml::-1'.",
    "timeStamp": 1492067839465,
    "attributes": {
      "IS_REMOTE_PIPELINE": false
    },
    "executionMode": "STANDALONE",
    "metrics": null,
    "retryAttempt": 0,
    "nextRetryTimeStamp": 0
  },
```

```
  "e4ded330-c573-4ab0-8fa8-004991493398": {
    "user": "admin",
    "name": "e4ded330-c573-4ab0-8fa8-004991493398",
    "rev": "0",
    "status": "STOPPED",
    "message": "The pipeline was stopped. The last committed source offset
is 'file006.xml::-1'.",
    "timeStamp": 1492176332877,
    "attributes": {
      "IS_REMOTE_PIPELINE": false
    },
    "executionMode": "STANDALONE",
    "metrics": null,
    "retryAttempt": 0,
    "nextRetryTimeStamp": 0
  }
}
```

You can make REST API requests by using curl, a utility that can download data using standard protocols. The username and password and Custom HTTP header attribute (X-Requested-By) are required.

```
curl -u admin:mypassword http://localhost:18630/rest/v1/pipelines/status -H
"X-Requested-By:myscript"
```

Event Framework

StreamSets has an Event Framework that allows users to kick off tasks in response to triggers or events that happen in the pipeline. StreamSets uses dataflow triggers to execute tasks such as send e-mails, execute a shell script, starting a JDBC query, or starting a Spark job after an event, for example, after the pipeline successfully completes a JDBC query.

Dataflow Performance Manager

One of the most powerful feature of StreamSets is StreamSets Dataflow Performance Manager. StreamSets Dataflow Performance Manager (DPM) is a management console that lets you manage complex data flows, providing a unified view of all running pipelines in your environment. DPM is extremely helpful in monitoring and troubleshooting pipelines. You'll appreciate its value more when you're tasked to monitor hundreds or even thousands of pipelines. Covering DPM is beyond the scope of this book. For more information, consult the Data Collector User Guide.

I've only covered a small subset of StreamSets features and capabilities. For more information about StreamSets, consult the StreamSets Data Collector User Guide online. In Chapter 9, I use StreamSets, Zoomdata, and Kudu to create a complete Internet of Things (IoT) application that ingests and visualizes sensor data in real time.

Other Next-Generation Big Data Integration Tools

There are other next-gen data integration tools available in the market. I discuss some of the most popular options in the Cask Data Application Platform.

The Cask Data Application Platform (CDAP) is an open source platform that you can use to develop big data applications and ETL jobs using Spark and the Hadoop stack. CDAP provides a GUI-based data ingestion studio for developing, deploying, and administering data pipelines. Its data preparation features provide an interactive method for data cleansing and transformation, a set of tasks commonly known as data wrangling. CDAP also has an application development framework, high-level Java APIs for rapid application development, and deployment. On top of that, it has metadata, data lineage, and security features that are important to enterprise environments. Like StreamSets, it has native support for Kudu. Let's develop a CDAP pipeline to ingest data to Kudu.

Data Ingestion with Kudu

The first thing we need to do is prepare test data for our example.

```
cat test01.csv
```

```
1,Jeff Wells,San Diego,71
2,Nancy Maher,Van Nuys,34
3,Thomas Chen,Rolling Hills,62
4,Earl Brown,Artesia,29
```

```
hadoop fs -put test01.csv /mydata
```

CHAPTER 7 BATCH AND REAL-TIME DATA INGESTION AND PROCESSING

To access CDAP, direct your browser to the hostname of the server where the application is installed, using port 11011. The first time you access CDAP, you're greeted with a welcome page.

Figure 7-59. *CDAP welcome page*

Close the window and click the button "Start by Adding Entities to CDAP." You're presented with five choices. Let's create a Pipeline.

Figure 7-60. *Add Entity*

291

CHAPTER 7   BATCH AND REAL-TIME DATA INGESTION AND PROCESSING

A canvas will appear. This is where you'll design, deploy, and administer your data pipelines. On the right side of the canvas, you can choose different sources, transformations, and sinks that you can use to design your pipeline.

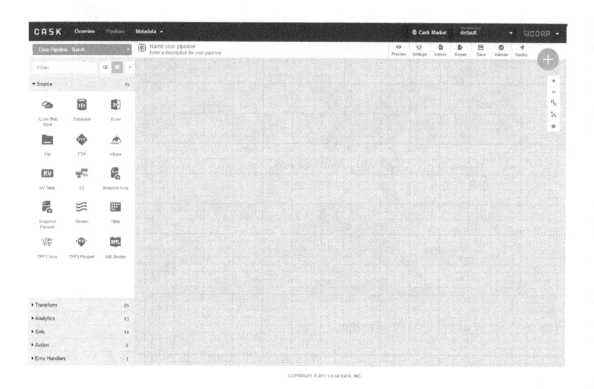

Figure 7-61. *CDAP canvas*

CHAPTER 7 BATCH AND REAL-TIME DATA INGESTION AND PROCESSING

You may notice that the Kudu source and sink are missing. We need to add the Kudu source and sink from the Cask Market. Click the "Cask Market" link near the upper-right corner of the application window.

Scroll down the list of items in the Cask Market until you find the Kudu source and sink.

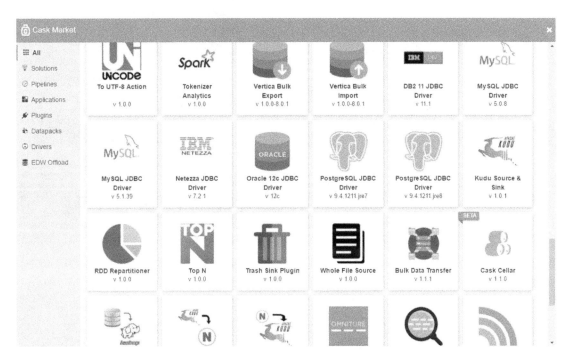

Figure 7-62. Cask Market

CHAPTER 7 BATCH AND REAL-TIME DATA INGESTION AND PROCESSING

Click the Kudu source and sink icon and then click "Deploy."

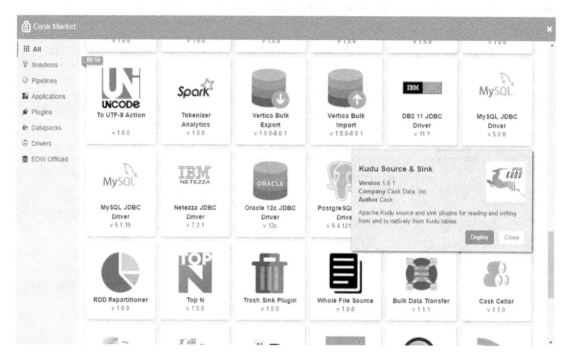

Figure 7-63. Cask Market – Kudu Source and Sink

Click "Finish."

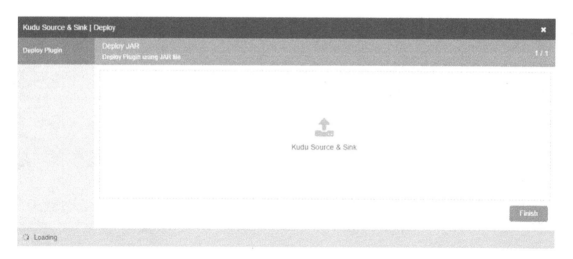

Figure 7-64. Finish installation

Click "Create a Pipeline." Notice the Kudu icon is now available as a data source and sink.

Figure 7-65. *Kudu data source and sink*

Our example CDAP pipeline will ingest a CSV file into a Kudu table. But before inserting the data, we will hash the name with a popular hashing algorithm. Let's drag a "File" source to the canvas.

CHAPTER 7 ■ BATCH AND REAL-TIME DATA INGESTION AND PROCESSING

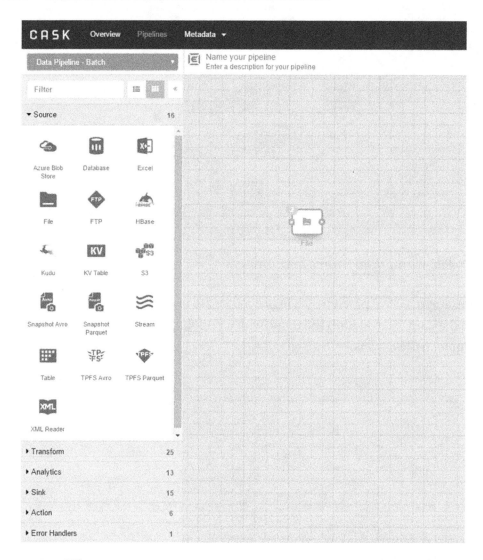

Figure 7-66. *File source*

CHAPTER 7 BATCH AND REAL-TIME DATA INGESTION AND PROCESSING

Double-click the file source icon. Enter the file properties such as the filename and the path of the file in HDFS.

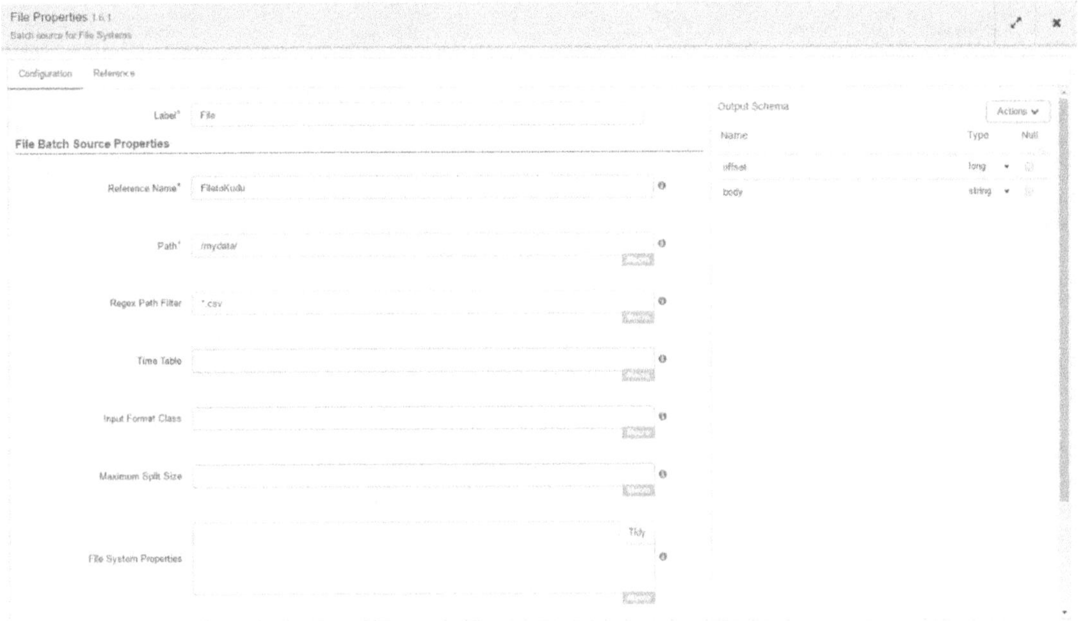

Figure 7-67. File properties

CHAPTER 7 BATCH AND REAL-TIME DATA INGESTION AND PROCESSING

CDAP makes it easy to access documentation about a particular sink, source, or transformation.

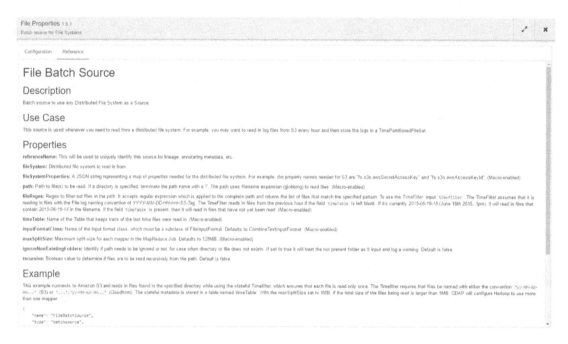

Figure 7-68. File Batch Source documentation

Next, drag a CSVParser transformation to the canvas and connect it to the File source.

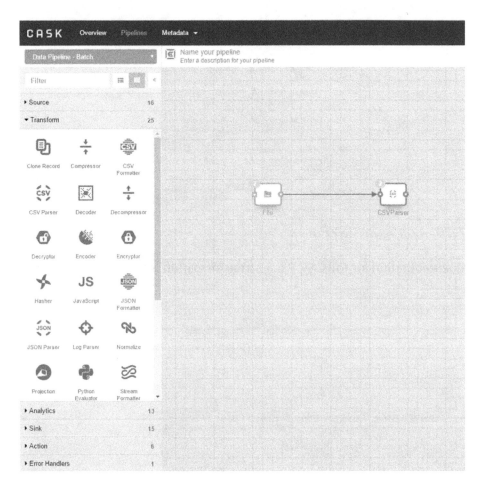

Figure 7-69. CSVParser transformation

CHAPTER 7 BATCH AND REAL-TIME DATA INGESTION AND PROCESSING

Enter the CSVParser properties. Make sure the Output Schema has the correct columns.

Figure 7-70. Configure CSVParser

The documentation for the CSV Parser transformation is readily available if you need assistance.

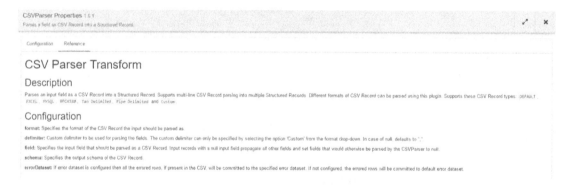

Figure 7-71. CSVParser documentation

CHAPTER 7 BATCH AND REAL-TIME DATA INGESTION AND PROCESSING

Drag a Hasher transformation to the canvas and connect it with the CSVParser.

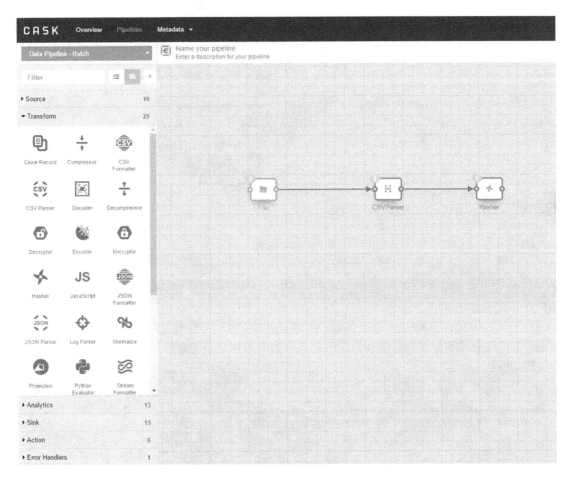

Figure 7-72. *Hasher transformation*

CHAPTER 7   BATCH AND REAL-TIME DATA INGESTION AND PROCESSING

Configure the Hasher by choosing the hashing algorithm and specifying the field you want to hash. For this example, we'll choose MD5 as our hashing algorithm.

Figure 7-73. Hasher configuration

Drag a Kudu sink and connect it to the Hasher.

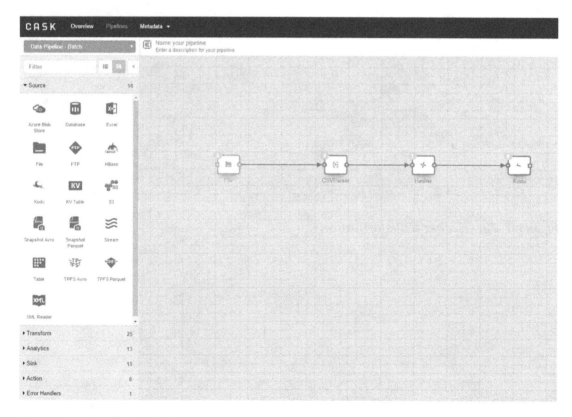

Figure 7-74. Kudu sink

CHAPTER 7 BATCH AND REAL-TIME DATA INGESTION AND PROCESSING

Configure the Kudu sink. Note that CDAP uses the Kudu native APIs to insert data to the tables instead of going through Impala. CDAP will also create the destination table so you just need to specify the table name, instead of using the "impala::database_name.table_name" format. In this example, we'll use the table name "users_table." Later, we'll create an external Impala table on top of this Kudu table.

Figure 7-75. Kudu sink configuration

303

The Kudu sink documentation is available just in case you need help with some of the options.

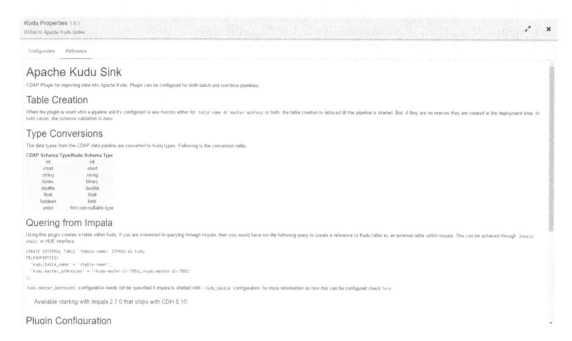

Figure 7-76. Kudu sink documentation

Your canvas should look like Figure 7-77. You can now deploy the pipeline. You can preview and validate the pipeline first to make sure there are no errors.

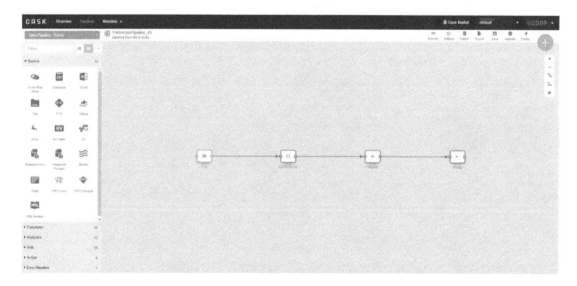

Figure 7-77. Complete pipeline

CHAPTER 7   BATCH AND REAL-TIME DATA INGESTION AND PROCESSING

Finally, the Canvas will show you the number of records transferred and processed from the File source all the way to the Kudu sink. You'll see an indicator near the upper-left corner of the canvas whether the job as successful or not. See Figure 7-78.

Figure 7-78. *Number of records transferred and processed*

Check the logs and confirm that the job succeeded as shown in Figure 7-79.

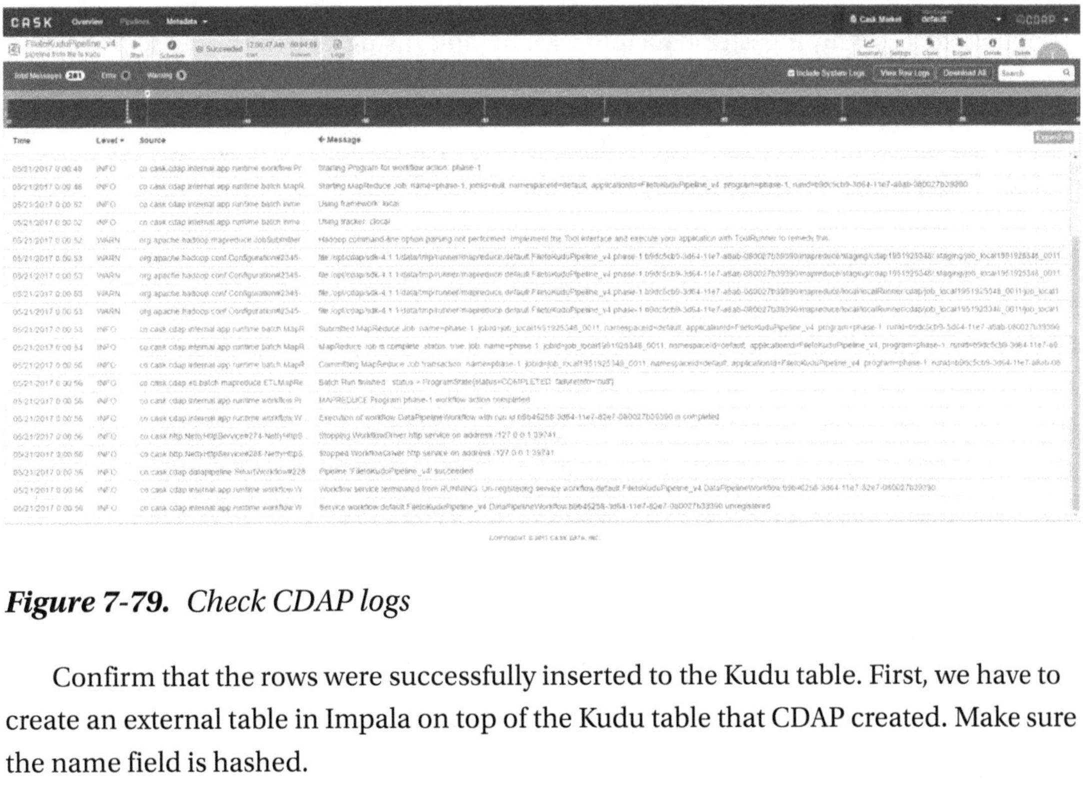

Figure 7-79. *Check CDAP logs*

Confirm that the rows were successfully inserted to the Kudu table. First, we have to create an external table in Impala on top of the Kudu table that CDAP created. Make sure the name field is hashed.

```
impala-shell

CREATE EXTERNAL TABLE cdap_users
STORED AS KUDU
TBLPROPERTIES (
'kudu.table_name' = 'users_table'
);
```

305

SELECT * FROM cdap_users;

```
+----+----------------------------------+---------------+-----+
| id | name                             | city          | age |
+----+----------------------------------+---------------+-----+
| 3  | dd500fc6d39cde55b6b4858e9854a21d | Rolling Hills | 62  |
| 1  | 228b855279d81c5251cff62e2b503079 | San Diego     | 71  |
| 4  | 332035b1942026174865ede2021dad2a | Artesia       | 29  |
| 2  | 8283a7fa1a09657dcc62125f8d734a7e | Van Nuys      | 34  |
+----+----------------------------------+---------------+-----+
```

You've successfully ingested data into a Kudu table, using a Field Hasher to perform in-stream transformation to your data.

Pentaho Data Integration

Pentaho offers a complete line of products for data integration, big data processing, and business analytics. We'll focus on Pentaho Data Integration (PDI) in this chapter. There is a community version called Pentaho Community Edition (CE), which includes Kettle, the open source version of Pentaho Data Integration. There is an Enterprise version, which includes Native YARN integration, analyzer and dashboard enhancements, advanced security, and high availability features.[xix]

Pentaho PDI has an intuitive user interface with ready-built and easy-to-use components to help you develop data ingestion pipelines. Similar to most ETL tools in the market, Pentaho PDI allows you to connect to different types of data sources ranging from popular RDBMS such as Oracle, SQL Server, MySQL, and Teradata to Big Data and NoSQL platforms such as HDFS, HBase, Cassandra, and MongoDB. PDI includes an integrated enterprise orchestration and scheduling capabilities for coordinating and managing workflows. You can effortlessly switch between Pentaho's native execution engine and Apache Spark to scale your pipelines to handle large data volumes.

Ingest CSV into HDFS and Kudu

Unlike other data integration tools described in this chapter, Pentaho does not include native support for Kudu yet. In order for us to insert data into Kudu, we will need to use Pentaho's generic Table Output component using Impala's JDBC driver. Using Table Output directly may not be fast enough depending on the size of your data and data ingestion requirements. One way to improve performance is to stage data to HDFS first, then using Table Output to ingest data from HDFS to Kudu. In some cases, this may be faster than directly ingesting into Kudu.

The first thing we need to do is prepare test data for our example.

```
cd /mydata

cat test01.csv

id,name,city,age
1,Jeff Wells,San Diego,71
2,Nancy Maher,Van Nuys,34
3,Thomas Chen,Rolling Hills,62
4,Earl Brown,Artesia,29
```

Start Pentaho PDI. We'll use the Community Edition (Kettle) for this example. Navigate to where you installed the binaries and execute ./spoon.sh. See Figure 7-80.

Figure 7-80. *Start Pentaho Data Integration*

CHAPTER 7 BATCH AND REAL-TIME DATA INGESTION AND PROCESSING

The Spoon graphical user interface is shown in Figure 7-81. This is where you design and build your jobs and transformation.

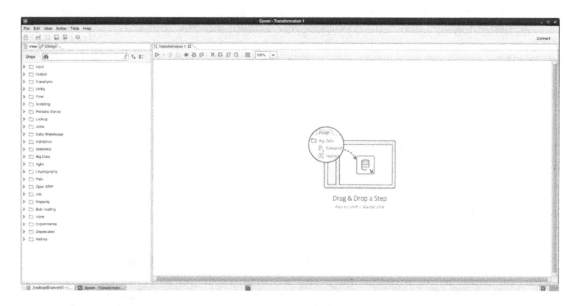

Figure 7-81. *Graphical view*

CHAPTER 7 BATCH AND REAL-TIME DATA INGESTION AND PROCESSING

On the left side of the window, you can find a list of all the supported input, output and transformation steps available for ETL development. Expand "Input" and drag "CSV file input" step to the canvas as shown in Figure 7-82.

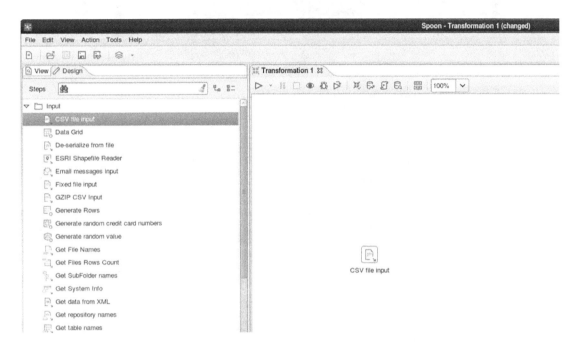

Figure 7-82. *CSV file input*

CHAPTER 7 BATCH AND REAL-TIME DATA INGESTION AND PROCESSING

Double-click the icon. Enter the configuration details such as the file name, delimiters, and so on. See Figure 7-83.

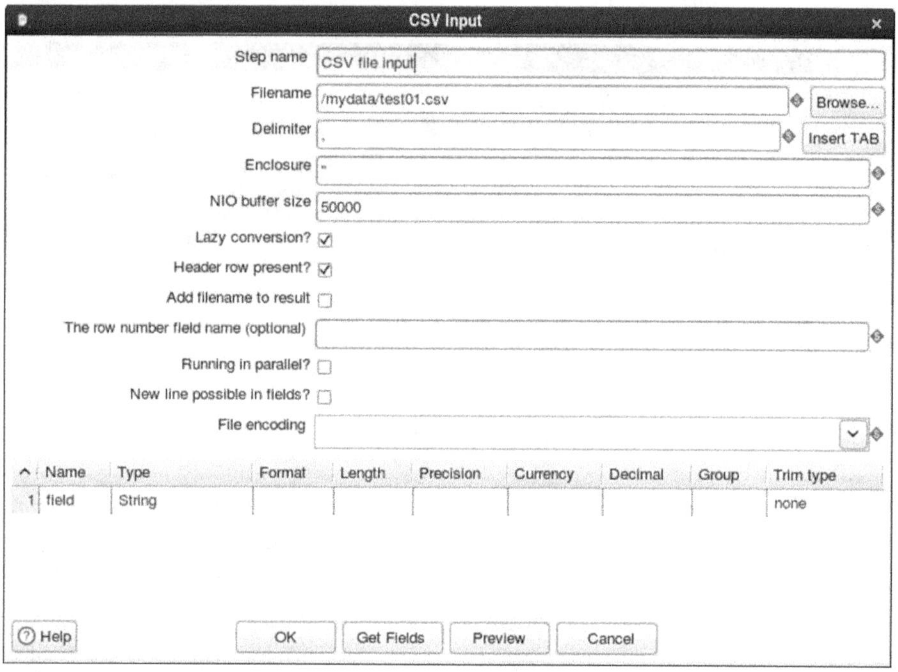

Figure 7-83. Configure CSV file input

CHAPTER 7 BATCH AND REAL-TIME DATA INGESTION AND PROCESSING

Click "Get Fields" to let Pentaho infer the field type and size of the fields. See Figure 7-84.

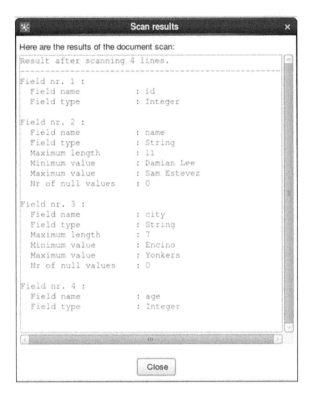

Figure 7-84. *Get fields*

CHAPTER 7 BATCH AND REAL-TIME DATA INGESTION AND PROCESSING

Expand "Big Data" and drag "Hadoop File Output" to the canvas as shown in Figure 7-85.

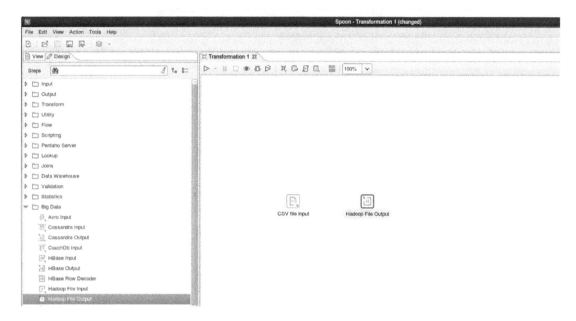

Figure 7-85. *Hadoop file output*

Connect "CSV file input" to "Hadoop File Output," as shown in Figure 7-86.

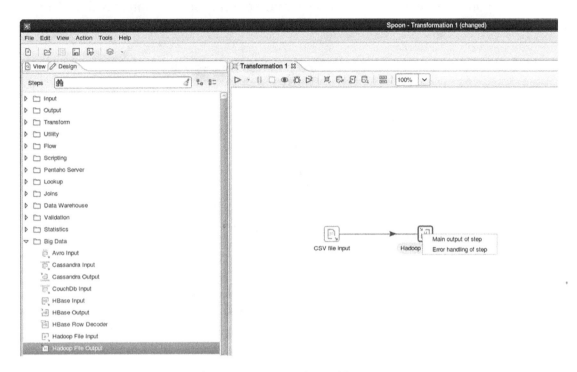

Figure 7-86. *Connect CSV file input to Hadoop file output*

CHAPTER 7 BATCH AND REAL-TIME DATA INGESTION AND PROCESSING

Double-click the "Hadoop File Output" component to configure the destination. Enter all required information such as the name of the cluster, the address of the HDFS namenode, and so on. See Figure 7-87.

Figure 7-87. *Configure Hadoop file output*

CHAPTER 7 BATCH AND REAL-TIME DATA INGESTION AND PROCESSING

Enter additional parameters to the Hadoop File Output stage. In our example, we'll rename the file based on our own Date time format. See Figure 7-88.

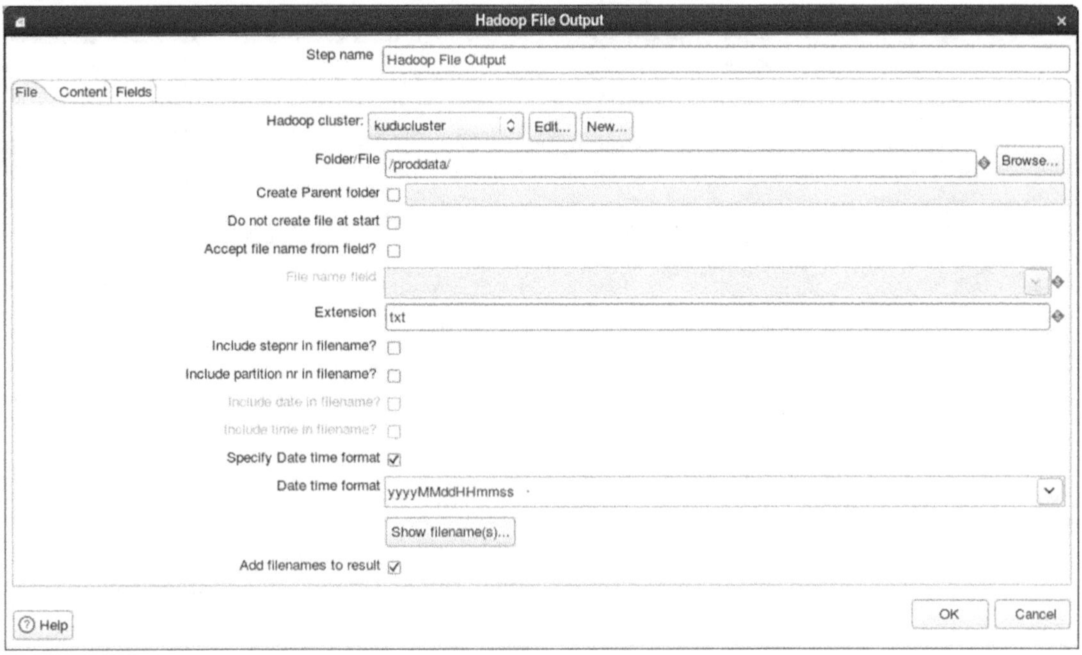

***Figure 7-88.** Rename file based on date time format*

CHAPTER 7 BATCH AND REAL-TIME DATA INGESTION AND PROCESSING

Specify details about the fields such as data type, format, length, precision and so on. See Figure 7-89.

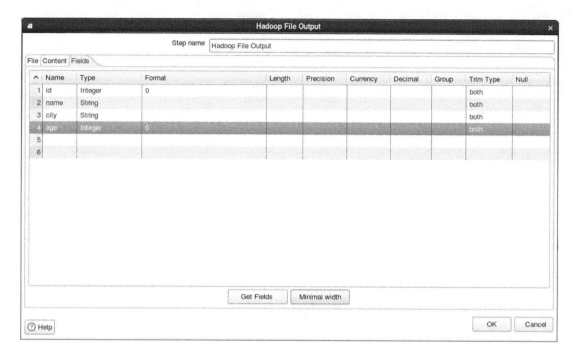

Figure 7-89. *Configure fields*

CHAPTER 7 BATCH AND REAL-TIME DATA INGESTION AND PROCESSING

You can preview and sanity check your data. See Figure 7-90.

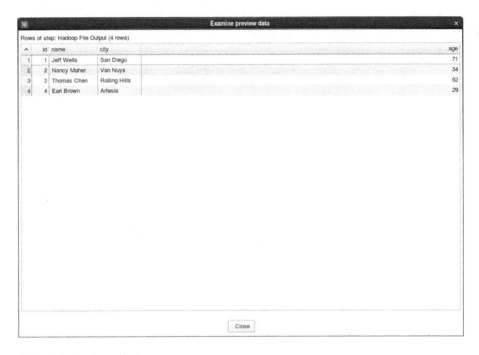

Figure 7-90. *Preview data*

CHAPTER 7 BATCH AND REAL-TIME DATA INGESTION AND PROCESSING

One of the most helpful features of PDI is the ability to monitor execution metrics of your job. Try doing a test run to make sure that the data is getting transferred from your source to destination. See Figure 7-91.

Figure 7-91. *Run the job*

CHAPTER 7 BATCH AND REAL-TIME DATA INGESTION AND PROCESSING

Navigate to "metrics" in the Execution Results panel as shown in Figure 7-92.

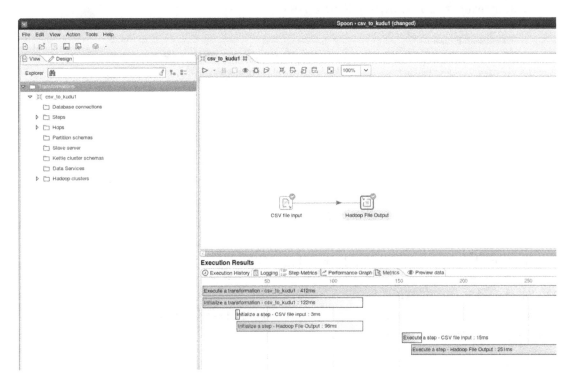

Figure 7-92. *Execution Results*

Looking closer, it shows you how long each step in the job takes. In this example, the entire job took 412 ms to execute. Initializing the transformation took 122 ms, part of the 122 ms, initializing the CSV file input step took 3 ms while initializing the Hadoop File Output step took 96 ms. The actual execution of the CSV file input step took 15 ms, while the execution of the Hadoop File Output took 251 ms. See Figure 7-93.

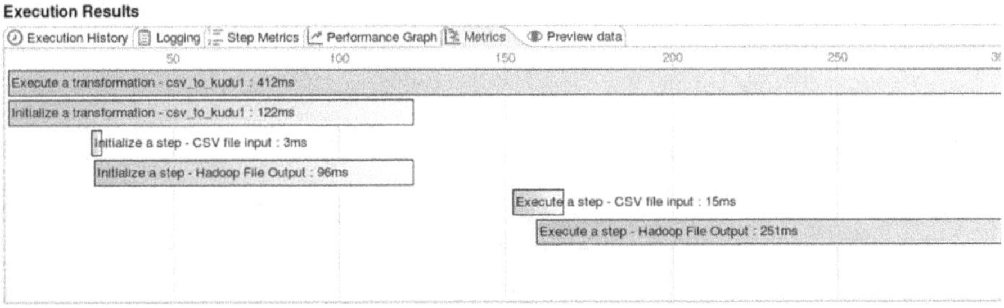

Figure 7-93. *Metrics – Execution Results*

CHAPTER 7 BATCH AND REAL-TIME DATA INGESTION AND PROCESSING

Navigate to the "Logging" tab to inspect the application log. As you can see the job successfully executed as shown in Figure 7-94.

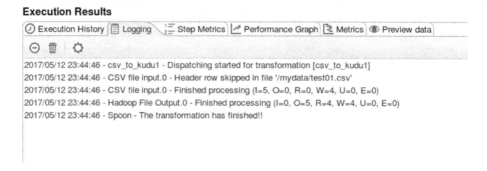

Figure 7-94. Logging – Execution Results

Confirm that the file was indeed copied to HDFS.

```
hadoop fs -ls /proddata
-rw-r--r--   3 hadoop supergroup        129 2017-05-13 00:05/
proddata/20170513000514.txt

hadoop fs -cat /proddata/20170513000514.txt
id,name,city,age
1,Jeff Wells,San Diego,71
2,Nancy Maher,Van Nuys,34
3,Thomas Chen,Rolling Hills,62
4,Earl Brown,Artesia,29
```

CHAPTER 7 BATCH AND REAL-TIME DATA INGESTION AND PROCESSING

Let's configure the final destination. Drag a Table Output step to the canvas. Double-click the icon and start configuring the step. Let's configure the Impala driver. Download the Impala JDBC driver from Cloudera.com and copy it to <install directory>/data-integration/lib. See Figure 7-95.

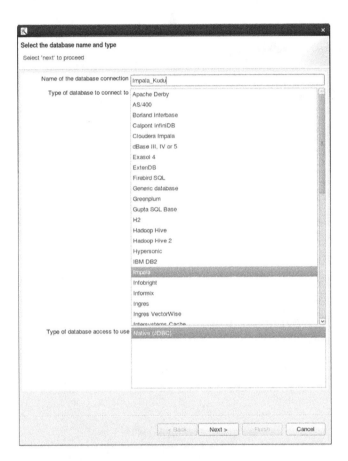

Figure 7-95. Configure database connection

CHAPTER 7 BATCH AND REAL-TIME DATA INGESTION AND PROCESSING

Configure the JDBC driver. Enter the hostname or IP address of the Impala server, the TCP/UP port and the database name as shown in Figure 7-96.

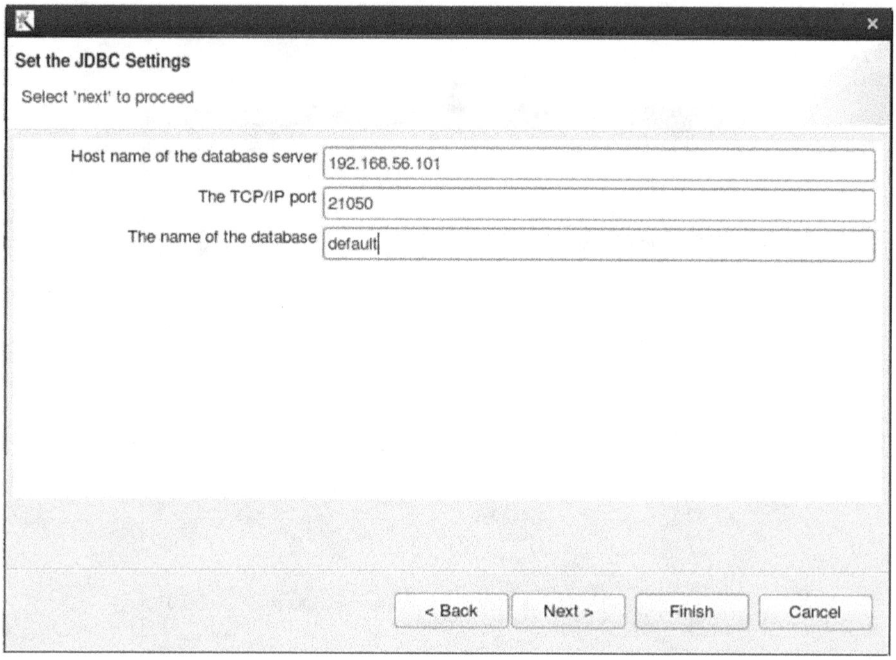

Figure 7-96. Configure JDBC settings

CHAPTER 7 BATCH AND REAL-TIME DATA INGESTION AND PROCESSING

Test the database connection as shown in Figure 7-97.

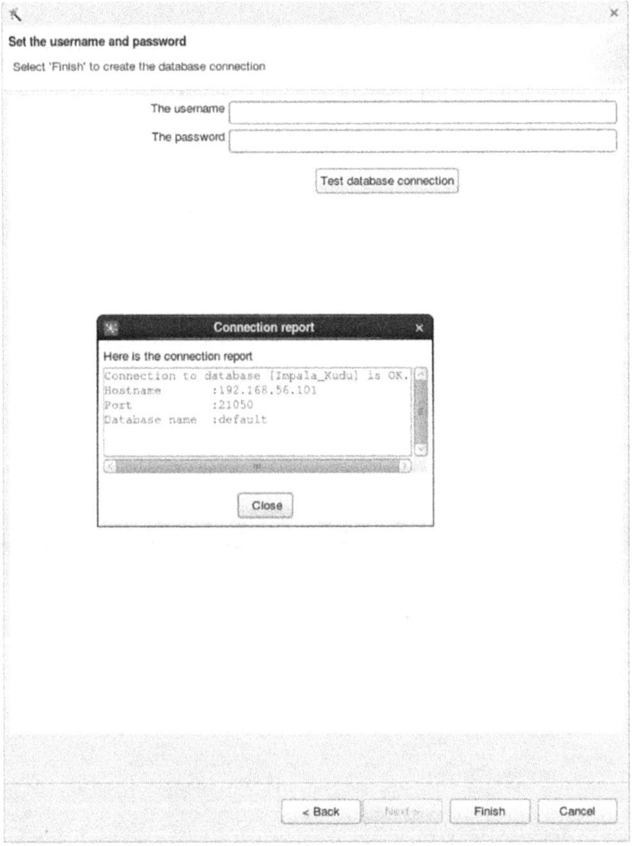

Figure 7-97. Test the database connection

CHAPTER 7 BATCH AND REAL-TIME DATA INGESTION AND PROCESSING

Create the destination table.

impala-shell

```
CREATE TABLE pentaho_users
(
 id BIGINT,
 name STRING,
 city STRING,
 age TINYINT,
 PRIMARY KEY(id)
)
PARTITION BY HASH PARTITIONS 4
STORED AS KUDU;
```

Enter the destination table and other relevant configuration options. See Figure 7-98.

Figure 7-98. *Configure table output*

CHAPTER 7 BATCH AND REAL-TIME DATA INGESTION AND PROCESSING

Click "Get fields" and make sure the source and destination fields match as shown in Figure 7-99.

Figure 7-99. *Get fields*

Preview and sanity check the data as shown in Figure 7-100.

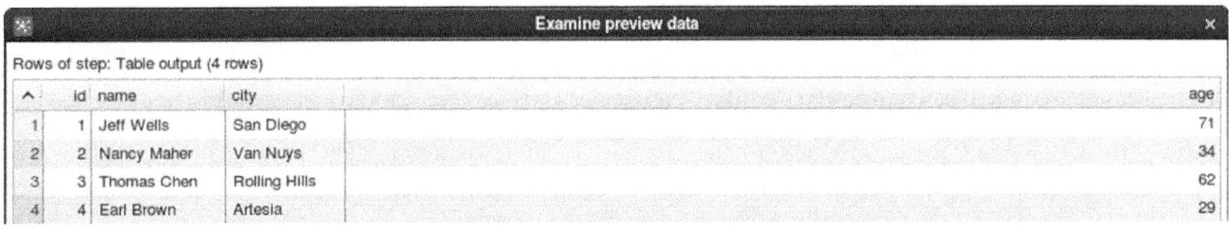

^	id	name	city	age
1	1	Jeff Wells	San Diego	71
2	2	Nancy Maher	Van Nuys	34
3	3	Thomas Chen	Rolling Hills	62
4	4	Earl Brown	Artesia	29

Rows of step: Table output (4 rows)

Figure 7-100. *Preview data*

CHAPTER 7 BATCH AND REAL-TIME DATA INGESTION AND PROCESSING

Execute the job. Monitor the logs and make sure the job successfully executes. See Figure 7-101.

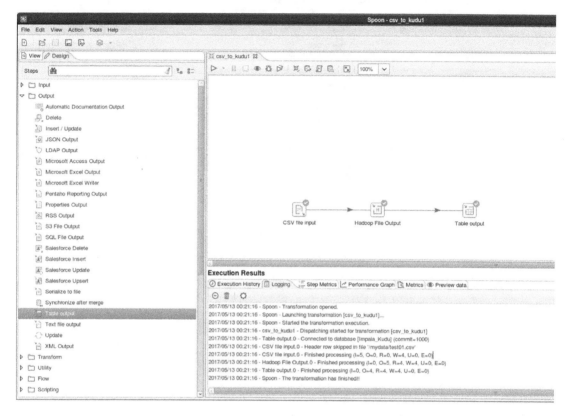

Figure 7-101. Run the job

CHAPTER 7 BATCH AND REAL-TIME DATA INGESTION AND PROCESSING

The job ran successfully.

```
2017/05/13 00:21:16 - Spoon - Transformation opened.
2017/05/13 00:21:16 - Spoon - Launching transformation [csv_to_kudu1]...
2017/05/13 00:21:16 - Spoon - Started the transformation execution.
2017/05/13 00:21:16 - csv_to_kudu1 - Dispatching started for transformation
[csv_to_kudu1]
2017/05/13 00:21:16 - Table output.0 - Connected to database [Impala_Kudu]
(commit=1000)
2017/05/13 00:21:16 - CSV file input.0 - Header row skipped in
file '/mydata/test01.csv'
2017/05/13 00:21:16 - CSV file input.0 - Finished processing
(I=5, O=0, R=0, W=4, U=0, E=0)
2017/05/13 00:21:16 - Hadoop File Output.0 - Finished processing
(I=0, O=5, R=4, W=4, U=0, E=0)
2017/05/13 00:21:16 - Table output.0 - Finished processing
(I=0, O=4, R=4, W=4, U=0, E=0)
2017/05/13 00:21:16 - Spoon - The transformation has finished!!
```

Confirm that the data was successfully inserted into the Kudu table.

impala-shell

select * from pentaho_users;

```
+----+--------------+----------------+-----+
| id | name         | city           | age |
+----+--------------+----------------+-----+
| 2  | Nancy Maher  | Van Nuys       | 34  |
| 3  | Thomas Chen  | Rolling Hills  | 62  |
| 1  | Jeff Wells   | San Diego      | 71  |
| 4  | Earl Brown   | Artesia        | 29  |
+----+--------------+----------------+-----+
```

CHAPTER 7 BATCH AND REAL-TIME DATA INGESTION AND PROCESSING

Data Ingestion to Kudu with Transformation

Let's start with another example. This time we'll use a string replace transformation to replace the string "Yonkers" with "Berkeley."

Prepare the data.

```
ls /mydata
test01.csv  test02.csv

cat test01.csv

id,name,city,age
1,Jeff Wells,San Diego,71
2,Nancy Maher,Van Nuys,34
3,Thomas Chen,Rolling Hills,62
4,Earl Brown,Artesia,29

cat test02.csv

id,name,city,age
5,Damian Lee,Yonkers,27
6,John Lohan,Encino,55
7,Lily Tran,Reseda,50
8,Sam Estevez,Tucson,81
```

CHAPTER 7 BATCH AND REAL-TIME DATA INGESTION AND PROCESSING

Enter the source directory and regular expression (Wildcard) to search for files. See Figure 7-102.

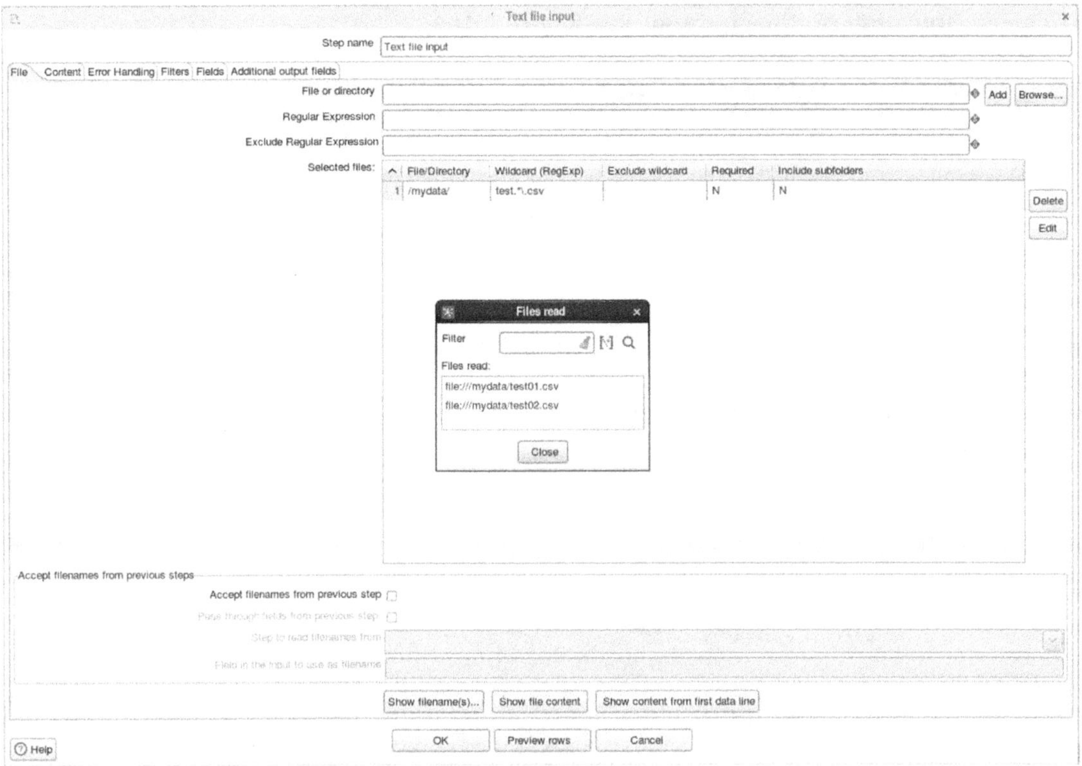

Figure 7-102. Specify source directory and file

Preview and sanity check the data. See Figure 7-103.

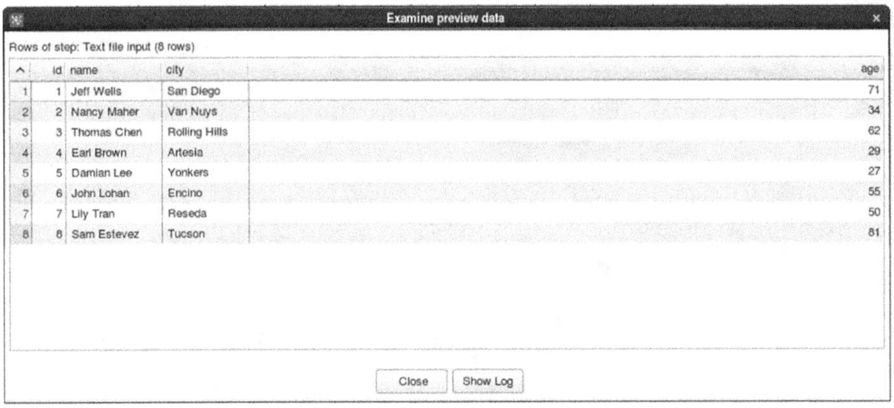

Figure 7-103. Preview data

CHAPTER 7   BATCH AND REAL-TIME DATA INGESTION AND PROCESSING

Drag a "Replace string" transformation step to the canvas. Double-click the icon and configure it by specifying the string to search and the string to replace it with. As we mentioned earlier, we'll replace the string "Yonkers" with "Berkeley" in the city field. You can ignore the other options for now. See Figure 7-104.

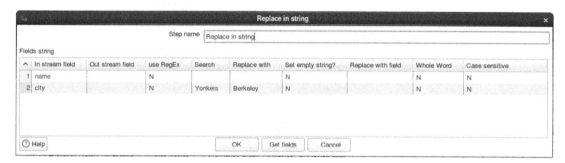

Figure 7-104.   Replace a string transformation

Run the job as shown in Figure 7-105.

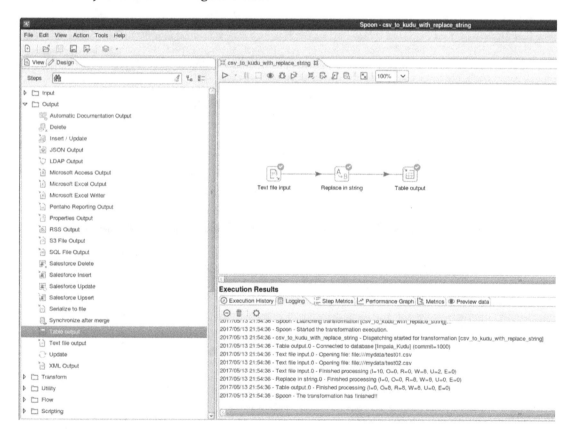

Figure 7-105.   Run the job

330

Check the table and make sure "Yonkers" is replaced with "Berkeley." Compare the value of the city in row id 5 with the original source text file.

impala-shell

select * from pentaho_users order by id;

```
+----+--------------+----------------+-----+
| id | name         | city           | age |
+----+--------------+----------------+-----+
| 1  | Jeff Wells   | San Diego      | 71  |
| 2  | Nancy Maher  | Van Nuys       | 34  |
| 3  | Thomas Chen  | Rolling Hills  | 62  |
| 4  | Earl Brown   | Artesia        | 29  |
| 5  | Damian Lee   | Berkeley       | 27  |
| 6  | John Lohan   | Encino         | 55  |
| 7  | Lily Tran    | Reseda         | 50  |
| 8  | Sam Estevez  | Tucson         | 81  |
+----+--------------+----------------+-----+
```

SQL Server to Kudu

In this example, we'll show how to ingest data from an RDBMS (SQL Server 2016) to Kudu.

Make sure you have data in your source SQL Server 2016 database. We'll use the same table we used in other examples. Inside a salesdb database, there's a table called users with the following rows. Use SQL Server Management Studio to confirm. See Figure 7-106.

CHAPTER 7 BATCH AND REAL-TIME DATA INGESTION AND PROCESSING

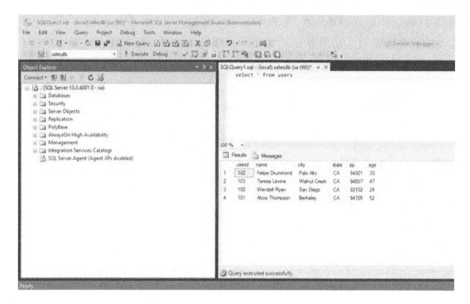

Figure 7-106. *Check the table in SQL Server Management Studio*

In Spoon, drag a Table Input step to the canvas as shown in Figure 7-107.

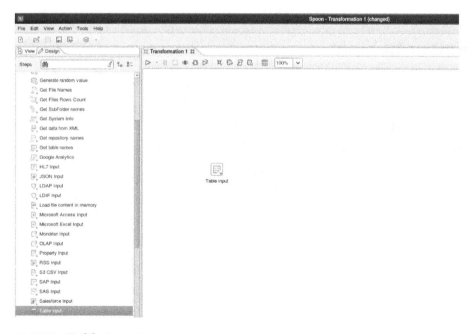

Figure 7-107. *Table input*

CHAPTER 7 BATCH AND REAL-TIME DATA INGESTION AND PROCESSING

Configure the Table Input step. See Figure 7-108.

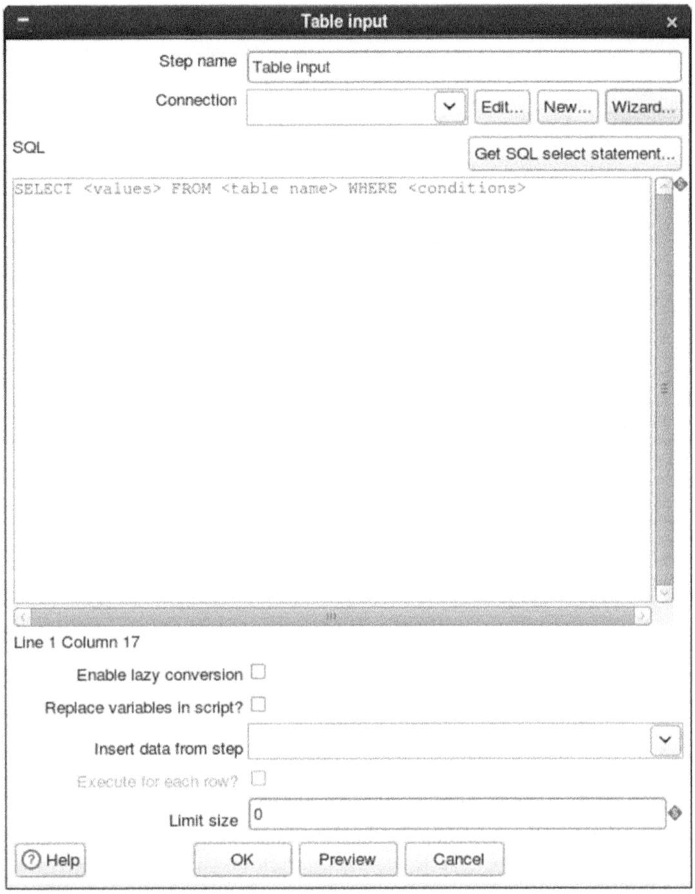

Figure 7-108. Configure Table input

Make sure to install the SQL Server JDBC driver by copying sqljdbc41.jar (JDK 1.7) or sqljdbc42.jar (JDK 1.8) to <install directory>/data-integration/lib. Download the JDBC driver from Microsoft.com if you haven't already. Choose the MS SQL Server (native) connection type as shown in Figure 7-109.

CHAPTER 7 ■ BATCH AND REAL-TIME DATA INGESTION AND PROCESSING

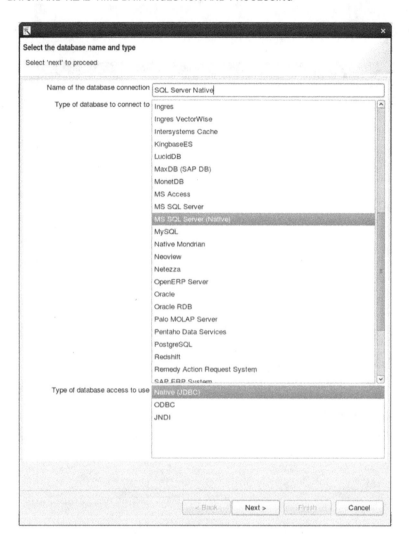

Figure 7-109. *Configure database connection*

Configure the JDBC setting (Figure 7-110).

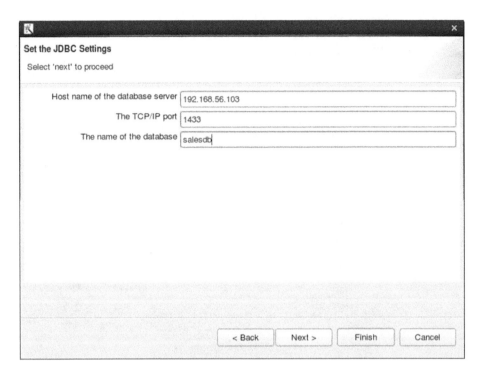

Figure 7-110. *Configure JDBC settings*

Test the connection to the source SQL Server 2016 database (Figure 7-111).

CHAPTER 7 BATCH AND REAL-TIME DATA INGESTION AND PROCESSING

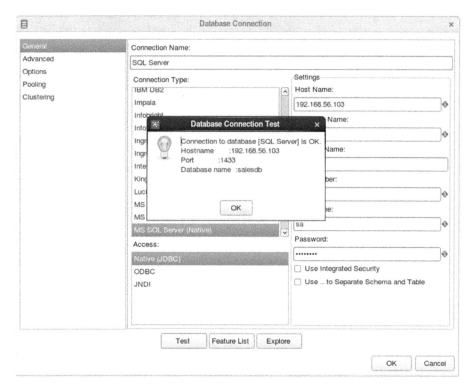

Figure 7-111. *Test database connection*

Now that you've configured the connection to your source database, specify a SQL query as your data source. See Figure 7-112.

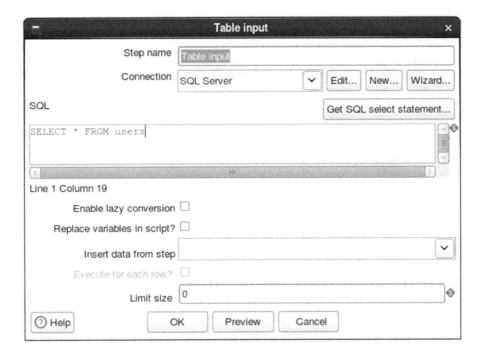

***Figure 7-112.** Specify SQL query as data source*

Preview your data to ensure your SQL query is valid (Figure 7-113).

^	userid	name	city	state	zip	age
1	102	Felipe Drummond	Palo Alto	CA	94301	33
2	103	Teresa Levine	Walnut Creek	CA	94507	47
3	100	Wendell Ryan	San Diego	CA	92102	24
4	101	Alicia Thompson	Berkeley	CA	94705	52

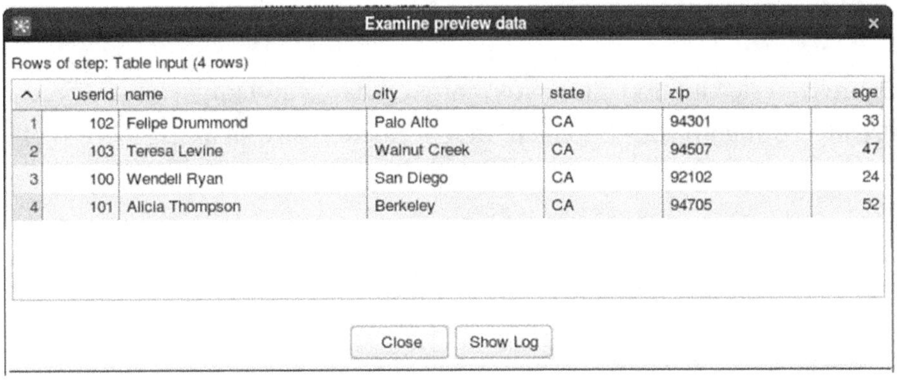

***Figure 7-113.** Preview data*

CHAPTER 7 BATCH AND REAL-TIME DATA INGESTION AND PROCESSING

Drag a Table Output step to your canvas and connect it to the Table Input step (Figure 7-114).

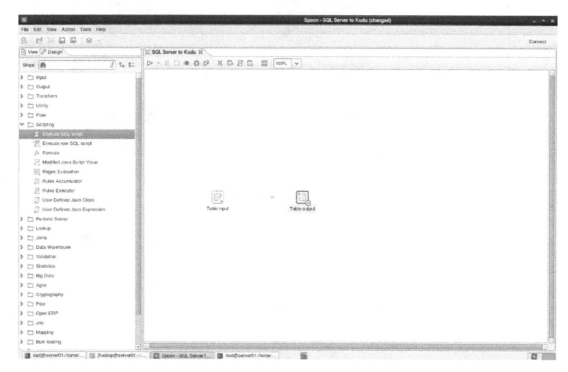

Figure 7-114. Table output

Configure your destination. Enter your target schema and target table. If you haven't done it already, configure your Impala JDBC connection as well as in Figure 7-115.

CHAPTER 7 BATCH AND REAL-TIME DATA INGESTION AND PROCESSING

Figure 7-115. *Configure Table output*

Run the job (Figure 7-116).

CHAPTER 7   BATCH AND REAL-TIME DATA INGESTION AND PROCESSING

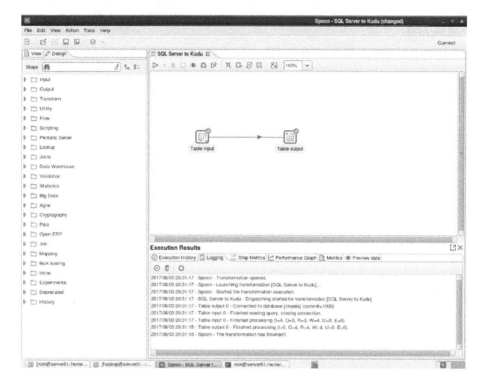

Figure 7-116. *Run the job*

Inspect the logs for errors.

2017/06/03 20:31:17 - Spoon - Transformation opened.
2017/06/03 20:31:17 - Spoon - Launching transformation [SQL Server to Kudu]...
2017/06/03 20:31:17 - Spoon - Started the transformation execution.
2017/06/03 20:31:17 - SQL Server to Kudu - Dispatching started for transformation [SQL Server to Kudu]
2017/06/03 20:31:17 - Table output.0 - Connected to database [Impala] (commit=1000)
2017/06/03 20:31:17 - Table input.0 - Finished reading query, closing connection.
2017/06/03 20:31:17 - Table input.0 - Finished processing (I=4, O=0, R=0, W=4, U=0, E=0)
2017/06/03 20:31:18 - Table output.0 - Finished processing (I=0, O=4, R=4, W=4, U=0, E=0)
2017/06/03 20:31:18 - Spoon - The transformation has finished!!

Confirm that the data was successfully inserted to the Kudu table.

impala-shell

select * from users;

```
+--------+------------------+--------------+-------+-------+-----+
| userid | name             | city         | state | zip   | age |
+--------+------------------+--------------+-------+-------+-----+
| 102    | Felipe Drummond  | Palo Alto    | CA    | 94301 | 33  |
| 100    | Wendell Ryan     | San Diego    | CA    | 92102 | 24  |
| 101    | Alicia Thompson  | Berkeley     | CA    | 94705 | 52  |
| 103    | Teresa Levine    | Walnut Creek | CA    | 94507 | 47  |
+--------+------------------+--------------+-------+-------+-----+
```

Congratulations! You've successfully ingested data from SQL Server 2016 to Kudu.

Talend Open Studio

Talend is one of the leading software companies that specializes in big data integration. Talend offers free open source data ingestion tools called Open Studio for Big Data and Open Studio for Data Integration. Both tools provide modern graphical user interface, YARN support, HDFS, HBase, Hive and Kudu support, connectors to Oracle, SQL Server and Teradata, and fully open source under Apache License v2.[xx] The main difference between Open Studio for Data Integration and Open Studio for Big Data is that Data Integration can only generate native Java code, while Big Data can generate both native Java, Spark and MapReduce code.

The commercial version gives you access to live Talend support, guaranteed response times, upgrades, and product patches. The open source version only provides community support.[xxi] If you are processing terabytes or petabytes of data, I suggest you use the Big Data edition. For traditional ETL-type workloads such as moving data from an RDBMS to Kudu, with some light data transformation in between, the Data Integration edition is sufficient. We will use Talend Open Studio for Data Integration in this chapter.

CHAPTER 7 BATCH AND REAL-TIME DATA INGESTION AND PROCESSING

> **Note** Talend Kudu components are provided by a third-party company, One point Ltd. These components are free and downloadable from the Talend Exchange – `https://exchange.talend.com/`. The Kudu Output and Input components need to be installed before you can use Talend with Kudu.

Ingesting CSV Files to Kudu

Let's start with a familiar example of ingesting CSV files to Kudu.

Prepare the test data.

```
ls /mydata

test01.csv

cat test01.csv

id,name,city,age
1,Jeff Wells,San Diego,71
2,Nancy Maher,Van Nuys,34
3,Thomas Chen,Rolling Hills,62
4,Earl Brown,Artesia,29
```

Create the destination Kudu table in Impala.

```
impala-shell

CREATE TABLE talend_users
(
 id INTEGER,
 name STRING,
 city STRING,
 age INTEGER,
 PRIMARY KEY(id)
)
PARTITION BY HASH PARTITIONS 4
STORED AS KUDU;
```

CHAPTER 7   BATCH AND REAL-TIME DATA INGESTION AND PROCESSING

Start Talend Open Studio for Data Integration as shown in Figure 7-117.

Figure 7-117. *Start Talend Open Studio for Data Integration*

You can select to create a new project or open an existing one. Let's start a new project (Figure 7-118).

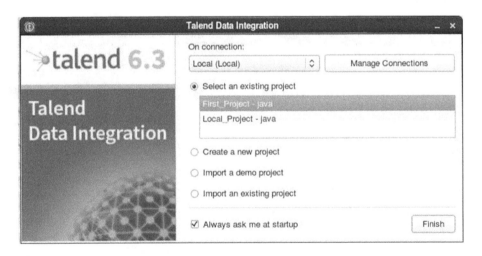

Figure 7-118. *Select an existing project or create a new project*

You're presented with a GUI similar to Figure 7-119.

343

CHAPTER 7   BATCH AND REAL-TIME DATA INGESTION AND PROCESSING

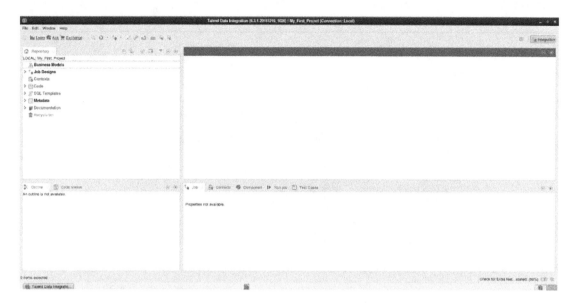

Figure 7-119. *Talend Open Studio Graphical User Interface*

Let's create a new job (Figure 7-120).

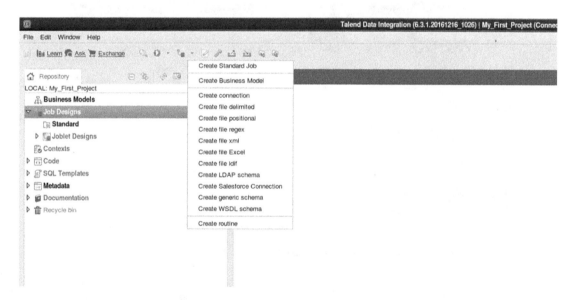

Figure 7-120. *Create job*

CHAPTER 7 BATCH AND REAL-TIME DATA INGESTION AND PROCESSING

Specify a name for the job. You can specify other properties such as purpose, description, author, and so on. Click "Finish" when you're done (Figure 7-121).

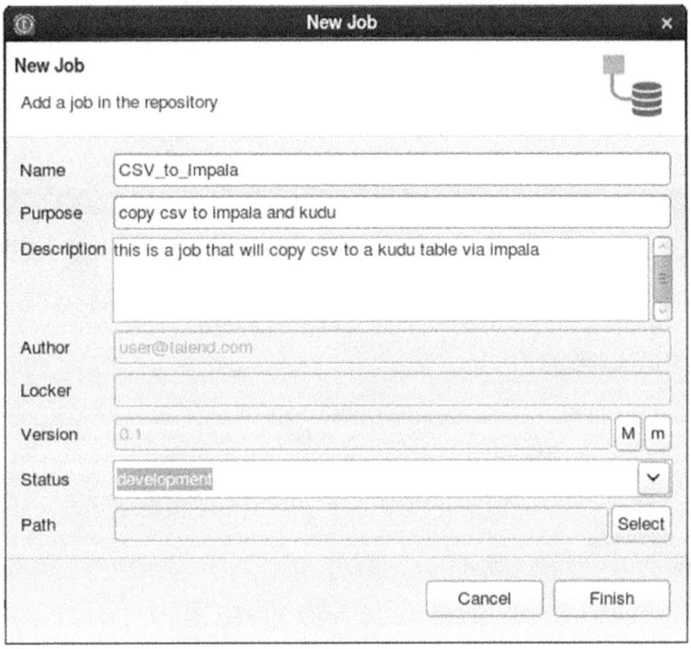

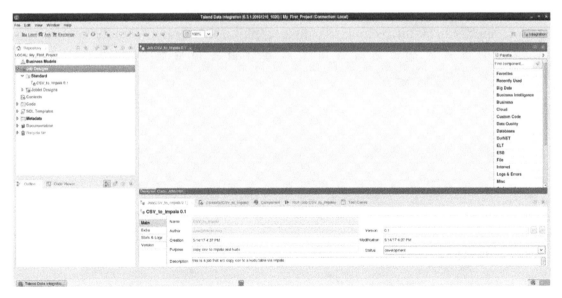

Figure 7-121. Specify job name and other job properties

You'll see a canvas similar to Figure 7-122. This is where you'll design and run jobs.

Figure 7-122. Job canvas

345

CHAPTER 7 BATCH AND REAL-TIME DATA INGESTION AND PROCESSING

Similar to previous data ingestion tools discussed in earlier in this chapter, you have to specify a source, transformation steps, and destination when designing data ingestion pipelines. On the right-hand side of the canvas you will see a list of inputs that you can use as your data source. Because we're ingesting a CSV file, drag and drop a tFileInputDelimited source into the canvas. Configure the source by specifying the file name of the CSV file. See Figure 7-123.

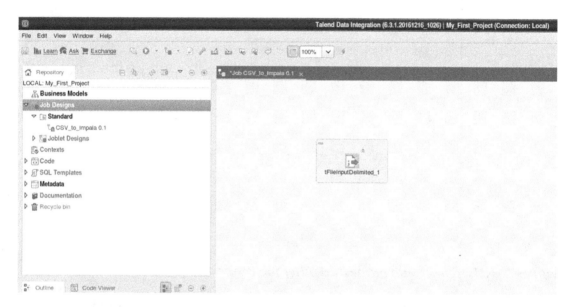

Figure 7-123. *tFileInputDelimited source*

Next, specify an output by dragging and dropping a Kudu output into the canvas. Configure the Kudu output by specifying the table name, connection information, and so on (Figure 7-124).

CHAPTER 7   BATCH AND REAL-TIME DATA INGESTION AND PROCESSING

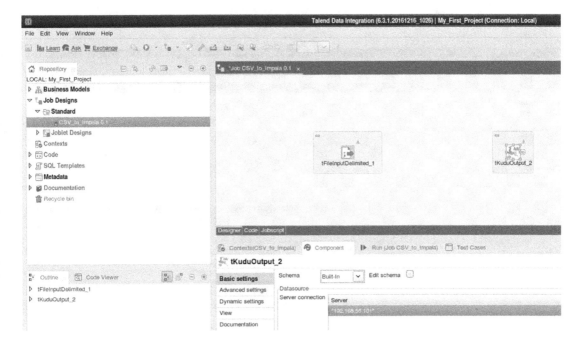

Figure 7-124. *Kudu output*

Don't forget to connect the input and output icons (Figure 7-125).

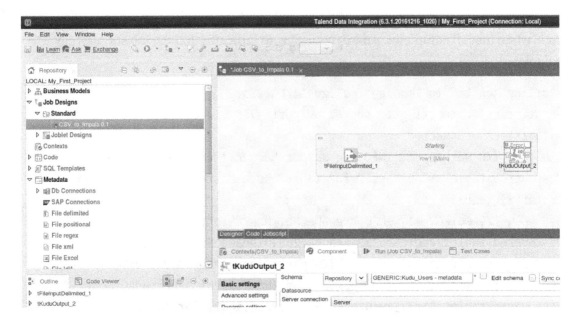

Figure 7-125. *Connect File Input delimited and Kudu output*

CHAPTER 7 BATCH AND REAL-TIME DATA INGESTION AND PROCESSING

Run the job as shown in Figure 7-126.

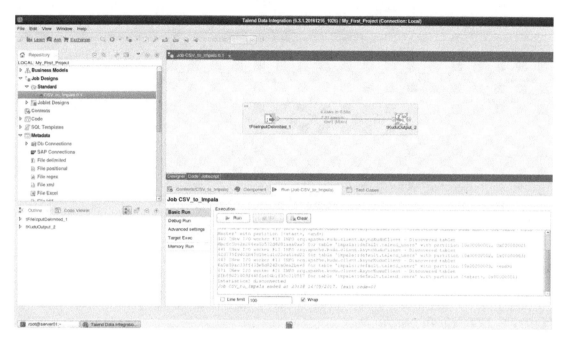

Figure 7-126. *Run the job*

The logs will show information about the job while it's running. You'll see an exit code when the job completes. An exit code of zero indicates that the job successfully completed.

```
Starting job CSV_to_Impala at 23:20 14/05/2017.
[statistics] connecting to socket on port 3725
[statistics] connected
330 [New I/O worker #1] INFO org.apache.kudu.client.AsyncKuduClient -
Discovered tablet Kudu Master for table 'Kudu Master' with partition
[<start>, <end>)
399 [New I/O worker #1] INFO org.apache.kudu.client.AsyncKuduClient -
Discovered tablet 4bcfc5b62a284ea0b572d8201aea0aa5 for table
'impala::default.talend_users' with partition [0x00000001, 0x00000002)
400 [New I/O worker #1] INFO org.apache.kudu.client.AsyncKuduClient -
Discovered tablet 42d775f9402b45d18e1d1c22ca61ed22 for table
'impala::default.talend_users' with partition [0x00000002, 0x00000003)
```

```
400 [New I/O worker #1] INFO org.apache.kudu.client.AsyncKuduClient -
Discovered tablet 6a0e39ac33ff433e8d8242ca0ea2bee8 for table
'impala::default.talend_users' with partition [0x00000003, <end>)
453 [New I/O worker #1] INFO org.apache.kudu.client.AsyncKuduClient -
Discovered tablet ffbf9021409f445fae04b1f35c318567 for table
'impala::default.talend_users' with partition [<start>, 0x00000001)
[statistics] disconnected
Job CSV_to_Impala ended at 23:20 14/05/2017. [exit code=0]
```

Confirm that the rows were successfully inserted into the Kudu table.

impala-shell

select * from talend_users;

```
+----+-------------+---------------+-----+
| id | name        | city          | age |
+----+-------------+---------------+-----+
| 3  | Thomas Chen | Rolling Hills | 62  |
| 4  | Earl Brown  | Artesia       | 29  |
| 1  | Jeff Wells  | San Diego     | 71  |
| 2  | Nancy Maher | Van Nuys      | 34  |
+----+-------------+---------------+-----+
```

SQL Server to Kudu

For our second example, let's ingest data from SQL Server to Kudu using Talend Open Studio. Create the Kudu table DimGeography in Impala if you haven't done it already.

Create a new job (Figure 7-127).

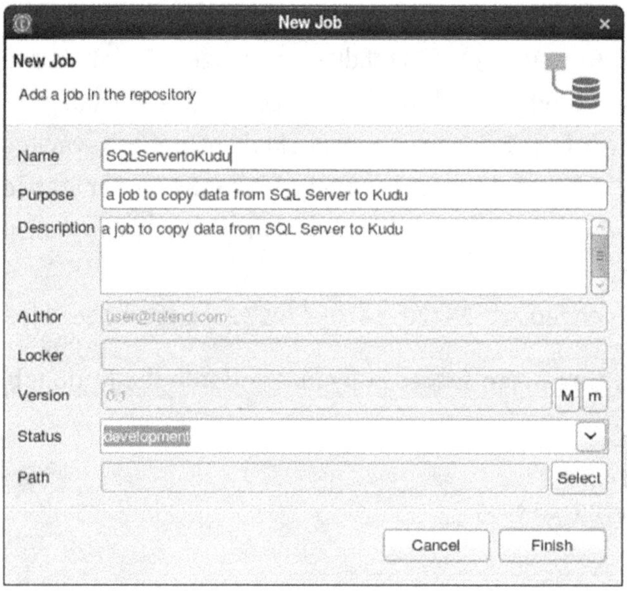

CHAPTER 7   BATCH AND REAL-TIME DATA INGESTION AND PROCESSING

Figure 7-127. New job

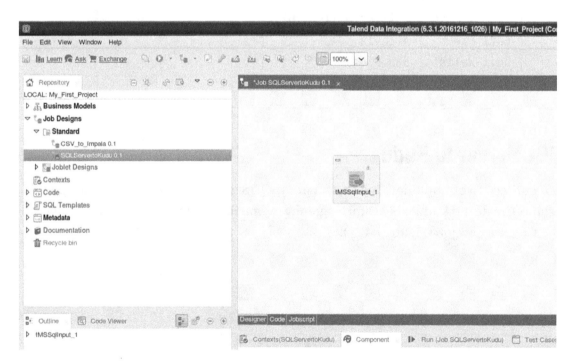

Figure 7-128. tMSSQLinput. See Figure 7-128.

Configure the component by making sure the correct data types are specified and so on (Figure 7-129).

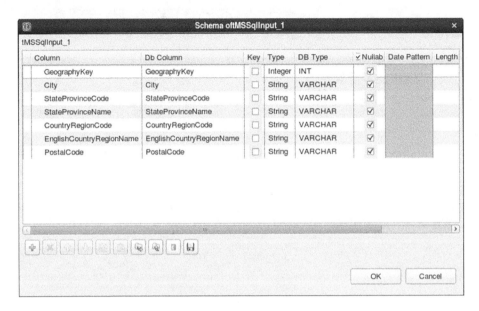

***Figure 7-129.** Configure MS SQL input*

Drag and drop a Kudu output component into the canvas. Connect it to the tMSSQLinput component as shown in Figure 7-130.

CHAPTER 7 BATCH AND REAL-TIME DATA INGESTION AND PROCESSING

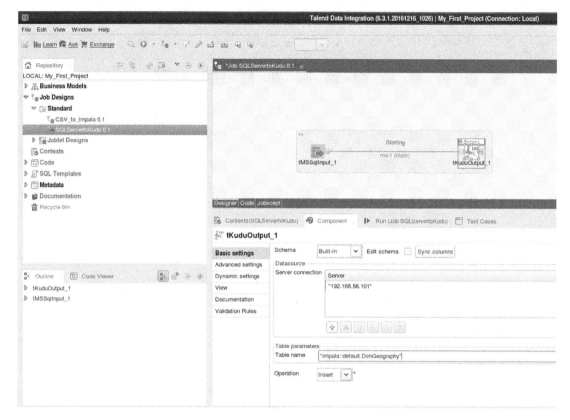

Figure 7-130. Kudu output

Configure the Kudu output and make sure the data types matches with the SQL Server table as shown in Figure 7-131.

CHAPTER 7 BATCH AND REAL-TIME DATA INGESTION AND PROCESSING

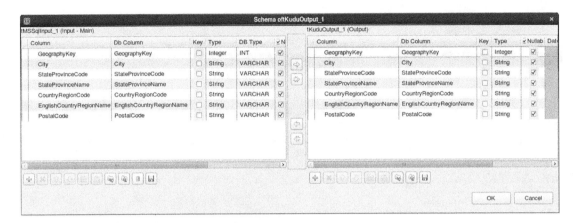

Figure 7-131. *Configure Kudu output*

Sync the schemas if they do not match (Figure 7-132).

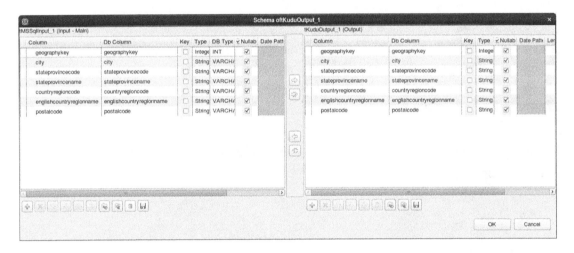

Figure 7-132. *Sync the schema*

CHAPTER 7 BATCH AND REAL-TIME DATA INGESTION AND PROCESSING

Run the job as shown in Figure 7-133.

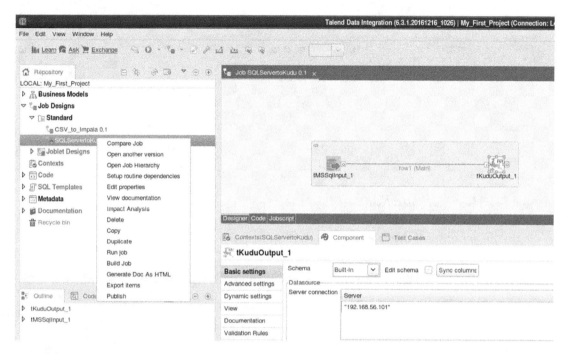

Figure 7-133. *Run the job*

Verify if the job ran successfully by comparing source and destination row counts (Figure 7-134).

CHAPTER 7 BATCH AND REAL-TIME DATA INGESTION AND PROCESSING

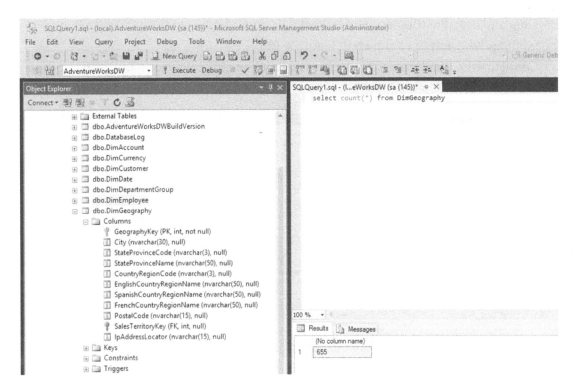

Figure 7-134.* Verify if the job ran successfully*

```
impala-shell

SELECT count(*) FROM DimGeography;

+----------+
| count(*) |
+----------+
| 655      |
+----------+
```

Data Transformation

Let's now use some of Talend's built-in features for data transformation. I'll use the tReplace component to replace a value in the specified input column. We'll replace the value "United Kingdom" with "UK." I'll also use the tFilterRow component to filter results to only include records where city is equal to "London" or "Berkshire."

CHAPTER 7   BATCH AND REAL-TIME DATA INGESTION AND PROCESSING

Drag and drop the tReplace and tFilterRow components into the canvas, in between the input and the output as shown in Figure 7-136.

Configure the tReplace component by selecting the field that contains the value that you want to replace (Figure 7-135).

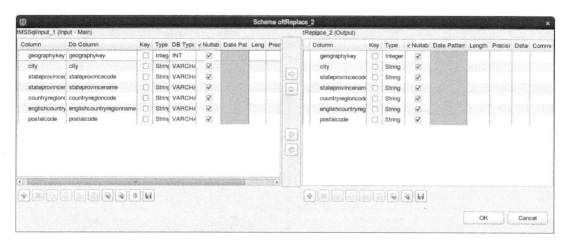

Figure 7-135. Select field

Configure the tReplace component by specifying the value you want to replace and the replacement value. In this case we'll replace "United Kingdom" with "UK" (Figure 7-136).

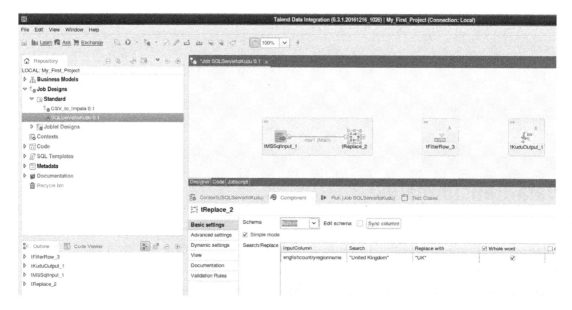

Figure 7-136. Configure tReplace

CHAPTER 7 BATCH AND REAL-TIME DATA INGESTION AND PROCESSING

Configure the tFilterRow component. We'll return only the rows that equals "London" and "Berkshire" in the city field (Figure 7-137).

Figure 7-137. *Configure tFilterRow*

Don't forget to connect all the components as shown in Figure 7-138.

Figure 7-138. *Connect tFilterRow to Kudu output*

357

CHAPTER 7 BATCH AND REAL-TIME DATA INGESTION AND PROCESSING

Run the job as shown in Figure 7-139.

Figure 7-139. Run the job

Inspect the data in the Kudu table to ensure that the job successfully executed. As you can see from the results, only records where city is equal to London and Berkshire were returned. United Kingdom was also replaced with UK.

```
impala-shell

SELECT geographykey as gkey, city, stateprovincecode as
spc, stateprovincename as spn, countryregioncode as crc,
englishcountryregionname as ecr, postalcode as pc FROM DimGeography;

+------+-----------+-----+---------+-----+-----+----------+
| gkey | city      | spc | spn     | crc | ecr | pc       |
+------+-----------+-----+---------+-----+-----+----------+
| 246  | London    | ENG | England | GB  | UK  | EC1R 0DU |
| 250  | London    | ENG | England | GB  | UK  | SW6 SBY  |
| 254  | London    | ENG | England | GB  | UK  | W1N 9FA  |
| 257  | London    | ENG | England | GB  | UK  | W1Y 3RA  |
| 230  | Berkshire | ENG | England | GB  | UK  | RG11 5TP |
| 244  | London    | ENG | England | GB  | UK  | C2H 7AU  |
| 248  | London    | ENG | England | GB  | UK  | SW19 3RU |
| 249  | London    | ENG | England | GB  | UK  | SW1P 2NU |
| 251  | London    | ENG | England | GB  | UK  | SW8 1XD  |
| 252  | London    | ENG | England | GB  | UK  | SW8 4BG  |
| 253  | London    | ENG | England | GB  | UK  | W10 6BL  |
| 256  | London    | ENG | England | GB  | UK  | W1X3SE   |
| 245  | London    | ENG | England | GB  | UK  | E17 6JF  |
| 247  | London    | ENG | England | GB  | UK  | SE1 8HL  |
| 255  | London    | ENG | England | GB  | UK  | W1V 5RN  |
+------+-----------+-----+---------+-----+-----+----------+
Fetched 15 row(s) in 4.50s
```

Other Big Data Integration Players

This chapter would not be complete if I didn't mention the traditional ETL players. They're tools that have been enhanced to include big data integration. Even so, they still lag in terms of native features compared to the newer big data integration tools I just discussed. Most lack native Spark support and connectors to popular big data sources, for example.

Informatica

Informatica is the largest software development company in the world that specializes in data integration. Founded in 1993, the company is based in Redwood City, California. Informatica also develops software for master data management, data quality, b2b data exchange, data virtualization, and more.[xxii] Informatica PowerCenter Big Data Edition is the company's flagship product for big data integration. Like the other ETL tools described in this chapter, Informatica PowerCenter Big Data Edition features an easy-to-use visual development environment. Informatica PowerCenter Big Data Edition has tight integration with the leading traditional and big data platforms making it easy for you schedule, manage and monitor processes, and workflows across your enterprise.[xxiii] As of this writing, Informatica PowerCenter doesn't have native Kudu support; however you may be able to ingest data into Kudu using Informatica PowerCenter via Impala and JDBC/ODBC.

Microsoft SQL Server Integration Services

This list would not be complete if I didn't mention the top three largest enterprise software development companies, Microsoft, IBM, and Oracle. SQL Server Integration Services (SSIS) includes features that supports Hadoop and HDFS data integration. SSIS provides the Hadoop Connection Manager and the following Control Flow Tasks: Hadoop File System Task, Hadoop Hive Task, and Hadoop Pig Task. SSIS supports the following data source and destination: HDFS File Source and HDFS File Destination. As with all the ETL tools described in this chapter, SSIS also sports a GUI development environment.[xxiv] As of this writing, SSIS doesn't have native Kudu support; however you may be able to ingest data into Kudu via Impala and JDBC/ODBC.

Oracle Data Integrator for Big Data

Oracle Data Integrator for Big Data provides advanced data integration capabilities for big data platforms. It supports a diverse set of workloads, including Spark, Spark Streaming, and Pig transformations and connects to various big data sources such as Kafka and Cassandra. For Orchestration of ODI jobs, users have a choice between using ODI agents or Oozie as orchestration engines.[xxv] As of this writing, Oracle Data Integrator doesn't have native Kudu support; however you may be able to ingest data into Kudu using Oracle Data Integrator via Impala and JDBC/ODBC.

CHAPTER 7   BATCH AND REAL-TIME DATA INGESTION AND PROCESSING

IBM InfoSphere DataStage

IBM InfoSphere DataStage is IBM's data integration tool that comes with IBM InfoSphere Information Server. It supports extract, transform, and load (ETL) and extract, load, and transform (ELT) patterns.[xxvi] It provides some big data support such as accessing and processing files on HDFS and moving tables from Hive to an RDBMS.[xxvii] The IBM InfoSphere Information Server operations console can be used to monitor and administer DataStage jobs, including monitoring their job logs and resource usage. The operations console also aids in troubleshooting issues when runtime issues occur.[xxviii] As of this writing, DataStage doesn't have native Kudu support; however you may be able to ingest data into Kudu using DataStage via Impala and JDBC/ODBC.

Syncsort

Syncsort is another leading software development company specializing in big data integration. They are well known for their extensive support for mainframes. By 2013, Syncsort had positioned itself as a "Big Iron to Big Data" data integration company.[xxix]

Syncsort DMX-h, Syncsort's big data integration platform, has a batch and streaming interfaces that can collect different types of data from multiple sources such as Kafka, HBase, RDBMS, S3, and Mainframes.[xxx] Like the other ETL solutions described in this chapter, Syncsort features an easy-to-use GUI for ETL development.[xxxi] As of this writing, DMX-h doesn't have native Kudu support; however you may be able to ingest data into Kudu using DMX-h via Impala and JDBC/ODBC.

Apache NIFI

Apache NIFI is an open source real-time data ingestion tool mostly used in the Hortonworks environment, although it can be used to ingest data to and from other platforms such as Cloudera and MapR. It's similar to StreamSets in many aspects. One of the main limitations of NIFI is its lack of YARN support. NIFI runs as an independent JVM process or multiple JVM processes if configured in cluster mode.[xxxii] NIFI does not have native Kudu support at the time of this writing, although the open source community is working on this. Check out NIFI-3973 for more details.[xxxiii]

361

CHAPTER 7 BATCH AND REAL-TIME DATA INGESTION AND PROCESSING

Data Ingestion with Native Tools

Ten years ago, you had to be Java developer in order to ingest and process data with Hadoop. Dozens of lines of code were often required to perform simple data ingestion or processing. Today, thanks to the thousands of committers and contributors to various Apache projects, the Hadoop ecosystem has a whole heap of (and some say too many) native tools for data ingestion and processing. Some of these tools can be used to ingest data into Kudu. I'll show examples using Flume, Kafka, and Spark. If you want to develop your own data ingestion routines, Kudu provides APIs for Spark, Java, C++, and Python.

Kudu and Spark

Kudu provides a Spark API that you can use to ingest data into Kudu tables. In the following example, we'll join a table stored in a SQL Server database with another table stored in an Oracle database and insert the joined data into a Kudu table. Chapter 6 provides a more thorough discussion of Kudu and Spark.

Start the spark-shell. Don't forget to include the necessary drivers and dependencies.

```
spark-shell --packages org.apache.kudu:kudu-spark_2.10:1.1.0 --driver-class-path ojdbc6.jar:sqljdbc41.jar --jars ojdbc6.jar,sqljdbc41.jar
```

Set up the Oracle connection

```
val jdbcURL = "jdbc:oracle:thin:sales/cloudera@//192.168.56.30:1521/EDWPDB"
val connectionProperties = new java.util.Properties()
```

Create a dataframe from the Oracle table.

```
val oraDF = sqlContext.read.jdbc(jdbcURL, "users", connectionProperties)

oraDF.show
```

```
+------+---------------+------------+-----+-----+---+
|USERID|           NAME|        CITY|STATE|  ZIP|AGE|
+------+---------------+------------+-----+-----+---+
|   102|Felipe Drummond|   Palo Alto|   CA|94301| 33|
|   103|  Teresa Levine|Walnut Creek|   CA|94507| 47|
|   100|   Wendell Ryan|   San Diego|   CA|92102| 24|
|   101|Alicia Thompson|    Berkeley|   CA|94705| 52|
+------+---------------+------------+-----+-----+---+
```

Register the table so we can run SQL against it.

```
oraDF.registerTempTable("ora_users")
```

Set up the SQL Server connection.

```
val jdbcURL = "jdbc:sqlserver://192.168.56.103;databaseName=salesdb;user=sa;password=cloudera"
val connectionProperties = new java.util.Properties()
```

Create a dataframe from the SQL Server table.

```
val sqlDF = sqlContext.read.jdbc(jdbcURL, "userattributes", connectionProperties)

sqlDF.show
```

```
+------+------+------+------------------+
|userid|height|weight|        occupation|
+------+------+------+------------------+
|   100|   175|   170|        Electrician|
|   101|   180|   120|          Librarian|
|   102|   180|   215|     Data Scientist|
|   103|   178|   132| Software Developer|
+------+------+------+------------------+
```

Register the table so that we can join it to the Oracle table.

```
sqlDF.registerTempTable("sql_userattributes")
```

Join both tables. We'll insert the results to a Kudu table.

```
val joinDF = sqlContext.sql("select ora_users.userid,ora_users.name,ora_users.city,ora_users.state,ora_users.zip,ora_users.age,sql_userattributes.height,sql_userattributes.weight,sql_userattributes.occupation from ora_users  INNER JOIN sql_userattributes ON ora_users.userid=sql_userattributes.userid")
```

joinDF.show

```
+------+---------------+------------+-----+-----+---+------+------+----------+
|userid|           name|        city|state|  zip|age|height|weight|occupation|
+------+---------------+------------+-----+-----+---+------+------+----------+
|   100|   Wendell Ryan|   San Diego|   CA|92102| 24|   175|   170|Electrician|
|   101|Alicia Thompson|    Berkeley|   CA|94705| 52|   180|   120| Librarian|
|   102| Felipe Drummond|  Palo Alto|   CA|94301| 33|   180|   215|Data
                                                                  Scientist|
|   103|   Teresa Levine|Walnut Creek|  CA|94507| 47|   178|   132|Software
                                                                  Developer|
+------+---------------+------------+-----+-----+---+------+------+----------+
```

You can also join the dataframes using the following method.

`val joinDF2 = oraDF.join(sqlDF,"userid")`

joinDF2.show

```
+------+---------------+------------+-----+-----+---+------+------+----------+
|userid|           NAME|        CITY|STATE|  ZIP|AGE|height|weight|occupation |
+------+---------------+------------+-----+-----+---+------+------+----------+
|   100|   Wendell Ryan|   San Diego|   CA|92102| 24|   175|   170|Electrician|
|   101|Alicia Thompson|    Berkeley|   CA|94705| 52|   180|   120|Librarian |
|   102| Felipe Drummond|  Palo Alto|   CA|94301| 33|   180|   215|Data
                                                                  Scientist|
|   103|   Teresa Levine|Walnut Creek|  CA|94507| 47|   178|   132|Software
                                                                  Developer|
+------+---------------+------------+-----+-----+---+------+------+----------+
```

Create the destination Kudu table in Impala.

impala-shell

```
create table users2 (
userid BIGINT PRIMARY KEY,
name STRING,
city STRING,
state STRING,
```

```
zip STRING,
age STRING,
height STRING,
weight STRING,
occupation STRING
)
PARTITION BY HASH PARTITIONS 16
STORED AS KUDU;
```

Go back to the spark-shell and set up the Kudu connection.

```
import org.apache.kudu.spark.kudu._

val kuduContext = new KuduContext("kuducluster:7051")
```

Insert the data to Kudu.

```
kuduContext.insertRows(JoinDF, "impala::default.users2")
```

Confirm that the data was successfully inserted into the Kudu table.

```
impala-shell

select * from users2;
```

```
+------+----------------+------------+-----+------+---+------+------+-----------+
|userid|name            |city        |state|zip   |age|height|weight|occupation |
+------+----------------+------------+-----+------+---+------+------+-----------+
|102   |Felipe Drummond |Palo Alto   |CA   |94301 |33 |180   |215   |Data
                                                                    Scientist|
|103   |Teresa Levine   |Walnut Creek|CA   |94507 |47 |178   |132   |Software
                                                                    Developer|
|100   |Wendell Ryan    |San Diego   |CA   |92102 |24 |175   |170   |Electrician|
|101   |Alicia Thompson |Berkeley    |CA   |94705 |52 |180   |120   |Librarian  |
+------+----------------+------------+-----+------+---+------+------+-----------+
```

CHAPTER 7 BATCH AND REAL-TIME DATA INGESTION AND PROCESSING

Flume, Kafka, and Spark Streaming

Using Flume, Kafka, and Spark Streaming for real-time data ingestion and event processing is a common architectural pattern.Apache Flume

Allows users to collect, transform, and move large amounts of streaming and event data into HDFS, HBase, and Kafka to name a few.[xxxiv] It's an integrated component of most Hadoop distributions. Flume's architectures can be divided into sources, channels, and sinks. Sources are your data sources, channels provide an intermediate buffer between sources and sinks and provides reliability, and sinks represent your destination. Flume has a simple architecture as you can see in Figure 7-140.

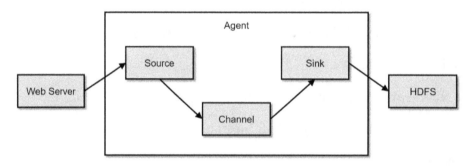

Figure 7-140. Flume Architecture

Flume is configured using a simple configuration file. The example configuration file lets you generate events from a script and then logs them to the console.

```
# Name the components
agent1.sources = source1
agent1.sinks = sinks1
agent1.channels = channel1

# Configure the source
agent1.sources.source1.type = exec
agent1.sources.source1.command = /tmp/myscript.sh
agent1.sources.source1.channels = channel1

# Configure the channel
agent1.channels.channel1.type = memory
agent1.channels.channel1.capacity = 10000
agent1.channels.channel1.transactionCapacity = 1000
```

```
# Configure the sink
agent1.sinks.sinks1.type = logger
agent1.sinks.sinks1.channel = channel1
```

To learn more about Flume, *Using Flume* (O'Reilly, 2014) by Hari Shreedharan is the definitive guide.

Apache Kafka

Kafka is a fast, scalable, and reliable distributed publish-subscribe messaging system. Kafka is now a standard component of architectures that requires real-time data ingestion and streaming. Although not required, Kafka is frequently used with Apache Flume, Storm, Spark Streaming, and StreamSets.

Kafka runs as a cluster on one or more brokers and has built-in replication and partitioning as shown in Figure 7-141.

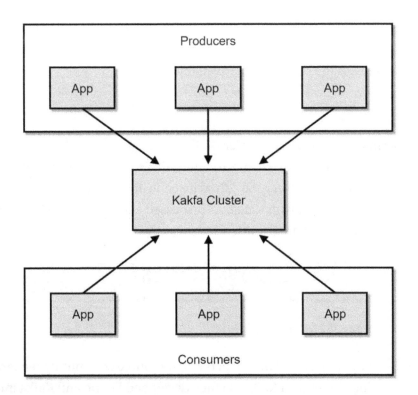

Figure 7-141. Kafka Producers and Consumers

Kafka records are published as topics. Each topic is partitioned and continuously appended to. Each record in the partition is assigned a unique offset that identifies each record.[xxxv] Think of the topic as a table and the offset as a primary key (Figure 7-142).

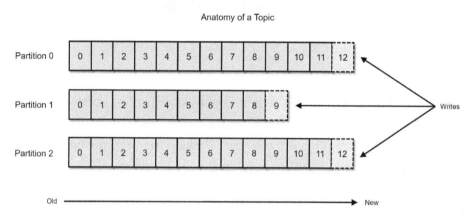

Figure 7-142. Kafka Topic

To learn more about Kafka, *Kafka: The Definitive Guide* (O'Reilly, 2017) by Shapiro, Narkhede, and Paling, is the best resource available.

Flafka

When Kafka was first released in 2011, users had to write Java code in order to ingest data into and read data out of Kafka. Unfortunately, this requirement slowed the adoption of Kafka because not everyone can write code in Java. There has to be an easier way for applications to interact with Kafka. As discussed earlier, these tasks are easily accomplished with Flume (Figure 7-143).

Figure 7-143. A typical Flafka pipeline

The engineers at Cloudera and others in the open source community recognized the benefits of integrating Flume with Kafka, so they developed Flume and Kafka integration frequently referred to as Flafka. Starting in CDH 5.2, Flume can act as consumer and producer for Kafka. With further development included in Flume 1.6 and CDH 5.3, the ability for Kafka to act as a Flume channel has been added.[xxxvi]

CHAPTER 7 BATCH AND REAL-TIME DATA INGESTION AND PROCESSING

Spark Streaming

Historically, Kafka did not provide any mechanism for stream processing. Kafka Streams was recently released to provide basic stream processing capabilities. Flume also has Interceptors that can be used for light-weight stream processing. However, Flume Interceptors have become out of vogue lately due to the popularity of Apache Spark. Spark Streaming is more powerful than Kafka Streams and Flume Interceptors due to Spark Streaming's integration with Spark SQL, Spark MLlib, and other features of Spark (Figure 7-144). Unlike Interceptors, Spark Streaming doesn't require Flume and can integrate with other data sources such as MQTT, Kafka, and Kinesis to name a few. If you have an unsupported data source that you want to integrate with Spark Streaming, you can implement a custom receiver that can handle that particular data source. Consult Chapter 2 for an example on how read and write to Kudu with Spark Streaming.

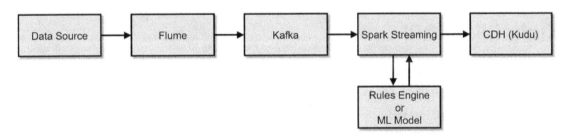

Figure 7-144. A typical Flafka pipeline with Spark Streaming and Kudu

Sqoop

Sqoop is not technically compatible with Kudu. You cannot use Sqoop to transfer data from an RDBMS to Kudu and vice versa. However, Sqoop may be the only tool some users may have for ingesting data from an RDBMS.

What you can do is to use Sqoop to ingest data from an RDBMS into HDFS and then use Spark or Impala to read the data from HDFS and insert it into Kudu. Here's a few examples of what you can do with Sqoop. Make sure you install and configure the correct drivers.

Get a list of available databases in SQL Server.

```
sqoop list-databases --connect jdbc:sqlserver://10.0.1.124:1433 --username myusername --password mypassword
```

Copy a table from SQL Server to Hadoop.

```
sqoop import --connect "jdbc:sqlserver://10.0.1.124:1433;database=Adventu
reWorksDW2012;username=myusername;password=mypassword" --table DimProduct
--hive-import --hive-overwrite
```

Copy a table from Hive to SQL Server.

```
sqoop export --connect "jdbc:sqlserver://10.0.1.124:1433;database=Adventu
reWorksDW2012;username=myusername;password=mypassword" --table salesfact_
hadoop --hcatalog-database default --hcatalog-table sales_data
```

Then you can just simply execute an Impala insert into…select statement:

```
INSERT INTO my_kudu_table SELECT * FROM sales_data;
```

Kudu Client API

Kudu provides NoSQL-style Java, C++, and Python client APIs. Applications that requires the best performance from Kudu should use the client APIs. In fact, some of the data ingestion tools discussed earlier, such as StreamSets, CDAP, and Talend utilizes the client APIs to ingest data into Kudu. DML changes via the API is available for querying in Impala immediately without the need to execute INVALIDATE METADATA. Refer Chapter 2 for more details on the Kudu client API.

MapReduce and Kudu

If your organization still uses MapReduce, you may be delighted to know that Kudu integrates with MapReduce. Example MapReduce code can be found on Kudu's official website.[xxxvii]

Summary

There are several ways to ingest data into Kudu. You can use third-party commercials tools such as StreamSets and Talend. You can create your own applications using native tools such as Apache Spark and Kudu's client APIs offers. Kudu enables users to ingest data by batch or real time while running analytic queries at the same time, making it the ideal platform for IoT and advanced analytics. Now that you've ingested data into Kudu, you need to extract value out of them. In Chapters 8 and 9, I will discuss common ways to analyze data stored in your Kudu tables.

References

i. Business Wire; "StreamSets Raises $12.5 Million in Series A Funding Led by Battery Ventures and NEA," Business Wire, 2015, https://www.businesswire.com/news/home/20150924005143/en/StreamSets-Raises-12.5-Million-Series-Funding-Led

ii. StreamSets; "Conquer Dataflow Chaos," StreamSets, 2018, https://streamsets.com/

iii. Apache Kudu; "Example Use Cases," Apache Kudu, 2018, https://kudu.apache.org/docs/#kudu_use_cases

iv. Tom White; "The Small Files Problem," Cloudera, 2009, https://blog.cloudera.com/blog/2009/02/the-small-files-problem/

v. Aaron Fabri; "Introducing S3Guard: S3 Consistency for Apache Hadoop," Cloudera, 2017, http://blog.cloudera.com/blog/2017/08/introducing-s3guard-s3-consistency-for-apache-hadoop/

vi. Eelco Dolstra; "S3BinaryCacheStore is eventually consistent," Github, 2017, https://github.com/NixOS/nix/issues/1420

vii. Sumologic; "10 Things You Might Not Know About Using S3," Sumologic, 2018, https://www.sumologic.com/aws/s3/10-things-might-not-know-using-s3/

viii. SteamSets; "MQTT Subscriber," StreamSets, 2018, https://streamsets.com/documentation/datacollector/latest/help/index.html#Origins/MQTTSubscriber.html#concept_ukz_3vt_lz

ix. StreamSets; "CoAP Client," StreamSets, 2018, https://streamsets.com/documentation/datacollector/latest/help/index.html#Destinations/CoAPClient.html#concept_hw5_s3n_sz

x. StreamSets; "Read Order," StreamSets, 2018, https://streamsets.com/documentation/datacollector/latest/help/#datacollector/UserGuide/Origins/Directory.html#concept_qcq_54n_jq

xi. StreamSets; "Processing XML Data with Custom Delimiters," StreamSets, 2018, https://streamsets.com/documentation/datacollector/latest/help/#Pipeline_Design/TextCDelim.html#concept_okt_kmg_jx

xii. StreamSets; "Processing XML Data with Custom Delimiters," StreamSets, 2018, https://streamsets.com/documentation/datacollector/latest/help/#Pipeline_Design/TextCDelim.html#concept_okt_kmg_jx

xiii. Arvind Prabhakar; "Tame unruly big data flows with StreamSets," InfoWorld, 2016, http://www.infoworld.com/article/3138005/analytics/tame-unruly-big-data-flows-with-streamsets.html

xiv. StreamSets; "Stream Selector," StreamSets, 2018, https://streamsets.com/documentation/datacollector/latest/help/#Processors/StreamSelector.html#concept_tqv_t5r_wq

xv. StreamSets; "Expression," StreamSets, 2018, https://streamsets.com/documentation/datacollector/latest/help/#Processors/Expression.html#concept_zm2_pp3_wq

xvi. StreamSets; "JavaScript Evaluator," StreamSets, 2018, https://streamsets.com/documentation/datacollector/latest/help/index.html#Processors/JavaScript.html#concept_n2p_jgf_lr

xvii. JSFiddle; "generateUUID," JSFiddle, 2018, https://jsfiddle.net/briguy37/2MVFd/

xviii. Pat Patterson; "Retrieving Metrics via the StreamSets Data Collector REST API," StreamSets, 2016, https://streamsets.com/blog/retrieving-metrics-via-streamsets-data-collector-rest-api/

xix. Pentaho; "Product Comparison," Pentaho, 2017, https://support.pentaho.com/hc/en-us/articles/205788659-PENTAHO-COMMUNITY-COMMERCIAL-PRODUCT-COMPARISON

xx. Talend; "Open Source Integration Software," Talend, 2018, https://www.talend.com/download/talend-open-studio/

xxi. Talend; "Why Upgrade?," Talend, 2018, https://www.talend.com/products/why-upgrade/

xxii. Crunchbase; "Informatica," Crunchbase, 2018, https://www.crunchbase.com/organization/informatica#/entity

xxiii. Informatica; "Informatica PowerCenter Big Data Edition," Informatica, 2018, https://www.informatica.com/content/dam/informatica-com/global/amer/us/collateral/data-sheet/powercenter-big-data-edition_data-sheet_2194.pdf

xxiv. Viral Kothari; "Microsoft SSIS WITH Cloudera BIGDATA," YouTube, 2016, https://www.youtube.com/watch?v=gPLfcL2zDX8

xxv. Oracle; "Oracle Data Integrator For Big Data," Oracle, 2018, http://www.oracle.com/us/products/middleware/data-integration/odieebd-ds-2464372.pdf;

xxvi. IBM; "Overview of InfoSphere DataStage," IBM, 2018, https://www.ibm.com/support/knowledgecenter/SSZJPZ_11.5.0/com.ibm.swg.im.iis.ds.intro.doc/topics/what_is_ds.html

xxvii. IBM; "Big data processing," IBM, 2018, https://www.ibm.com/support/knowledgecenter/SSZJPZ_11.5.0/com.ibm.swg.im.iis.ds.intro.doc/topics/ds_samples_bigdata.html

xxviii. IBM Analytic Skills; "Monitoring DataStage jobs using the IBM InfoSphere Information Server Operations Console," IBM, 2013, https://www.youtube.com/watch?v=qOl_6HqyVes

CHAPTER 7 BATCH AND REAL-TIME DATA INGESTION AND PROCESSING

xxix. SyncSort; "Innovative Software from Big Iron to Big Data," SyncSort, 2018, http://www.syncsort.com/en/About/Syncsort-History

xxx. SyncSort; "Syncsort DMX-h," SyncSort, 2018, http://www.syncsort.com/en/Products/BigData/DMXh

xxxi. SyncSort; "Introduction to Syncsort DMX-h 8," SyncSort, 2015, https://www.youtube.com/watch?v=7e_8YadLa9E

xxxii. Hortonworks; "APACHE NIFI," Hortonworks, 2018, https://hortonworks.com/apache/nifi/

xxxiii. Jira; "Create a new Kudu Processor to ingest data," Jira, 2017, https://issues.apache.org/jira/browse/NIFI-3973

xxxiv. Apache Flume; "Flume 1.8.0 User Guide," Apache Flume, 2018, https://flume.apache.org/FlumeUserGuide.html

xxxv. Apache Kafka; "Introduction," Apache Kafka, 2018, https://kafka.apache.org/intro

xxxvi. Gwen Shapira, Jeff Holoman; "Flafka: Apache Flume Meets Apache Kafka for Event Processing," Cloudera, 2014, http://blog.cloudera.com/blog/2014/11/flafka-apache-flume-meets-apache-kafka-for-event-processing/

xxxvii. Dan Burkert; "ImportCsv.java," Apache Kudu, 2017, https://github.com/apache/kudu/blob/master/java/kudu-client-tools/src/main/java/org/apache/kudu/mapreduce/tools/ImportCsv.java

CHAPTER 8

Big Data Warehousing

For decades the Enterprise Data Warehouse has been the central repository for corporate data. It is an integral part of every organization's business intelligence infrastructure.

Data warehouses come in various shapes and forms. For years, Microsoft and Oracle sold relational database management software for data warehouses on SMP (Symmetric Multiprocessing) systems. Others, such as Teradata, sell large integrated MPP (Massively Parallel Processing) systems. A cloud provider such as Microsoft offers a fully managed cloud data warehouse such as Azure SQL Data Warehouse. The Apache Hive project was developed to provide a data warehouse infrastructure built on top of Hadoop.[i] Whatever form or shape it takes, the data warehouse's purpose remains the same – to provide a platform where structured data can be efficiently stored, processed, and analyzed.

In this chapter, I'll discuss how big data (and more specifically Cloudera Enterprise) is disrupting data warehousing. I assume the readers are familiar with data warehousing so I won't cover the basics and theory of data warehousing. For a definitive guide to data warehousing, consult Ralph Kimball's *The Data Warehouse Toolkit*.[ii]

Note I will be using terms "big data" and "Hadoop" interchangeably in this chapter (and throughout the book). Technically big data refers to a whole ecosystem of technologies and frameworks of which Hadoop is a part of. I will also use the terms "data lake," "data hub," and "big data platform" interchangeably throughout the book.

CHAPTER 8   BIG DATA WAREHOUSING

Enterprise Data Warehousing in the Era of Big Data

The era of big data has disrupted data warehousing as we know it. The pioneers of big data invented new methods in data processing and analysis that make traditional data warehousing seem outdated. The ability of big data platforms to process and analyze both structured and unstructured data as well as new techniques like ELT (extract, load, and transform), schema on read, data wrangling, and self-service data analysis exposed the limits of data warehousing, making traditional data modeling and ETL development life cycles look slow and rigid.

Big data platforms are generally more scalable than traditional data warehousing platforms from a cost and technical perspective. Generally, the largest data warehouses can only handle up to hundreds of terabytes of data, requiring significant hardware investment to scale beyond that. Big data platforms can easily process and analyze petabytes of data using low-cost commodity servers. Perhaps the most practical concern for most organizations is the financial burden of implementing and maintaining a commercial data warehouse. Companies spend millions of dollars in licensing, infrastructure, and development costs in order to have a fully functional enterprise data warehouse. Open source MPP Apache projects such as Impala, Drill, and Presto offer comparable performance and scalability at a fraction of the cost.

Structured Data Still Reigns Supreme

It's not all doom and gloom for data warehousing, far from it. Despite all the hype around big data, a survey conducted by Dell showed that more than two-thirds of companies reported that structured data represents 75% of data being processed, with one-third of the companies reported not analyzing unstructured data at all. The survey also shows that structured data is growing at a faster rate than unstructured data.[iii]

EDW Modernization

Although Hadoop platforms such as Cloudera Enterprise come close to replicating some of the capabilities of the data warehouse, commercial data warehouse platforms are still the best platform for analyzing structured data. Impala and Kudu (and Hive) still lack many of the (proprietary) capabilities and features of a commercial data warehouse platform. You will find that some of the popular commercial business intelligence and

OLAP (Online Analytical Processing) tools are dependent on these proprietary features. Some of the capabilities not found in Impala and Kudu are multi-row and multi-table transactions, OLAP support, advanced backup and recovery features, secondary indexes, materialized views, support for decimal data type, and auto-increment columns to mention a few. Table 8-1 contains a side-by-side comparison of Impala and Kudu versus traditional data warehouse platforms.

Table 8-1. Impala and Kudu vs. a Traditional Data Warehouse

Feature	Impala and Kudu	Data Warehouse Platform
Multi-row and multi-table transactions	No	Yes
Auto-increment column	No	Yes
Foreign key constraint	No	Yes
Secondary indexes	No	Yes
Materialized views	No	Yes
Triggers	No	Yes
Stored procedures	No	Yes
Database caching	No	Yes
OLAP support	No	Yes
Procedural extension to SQL	No	Yes
Advanced backup and recovery	No	Yes
Decimal and Complex Data Types	No	Yes

Hadoop vendors usually market their platform as a "data hub" or "data lake" that can be used to augment or modernize data warehouses, instead of replacing it. In fact, most companies these days have both a data warehouse and a data lake. There are various ways a data lake can be used to modernize an enterprise data warehouse: ETL Offloading, Active Archive and Analytics Offloading, and Data Consolidation. These use cases are relatively easy to implement. The return on investment is also easy to measure and justify to management.

CHAPTER 8 BIG DATA WAREHOUSING

ETL Offloading

In a typical data warehouse environment, IO- and CPU-intensive data transformation and loading is executed on the data warehouse (Figure 8-1). Over time, as more data is ingested and stored in the data warehouse, its ETL window will start eating into the business hours, causing critical reports and dashboards to be unavailable to business users. An obvious workaround is to upgrade the data warehouse. This is a short-term workaround. Over time, you'll have the same performance and scalability issue as you ingest more data into your data warehouse.

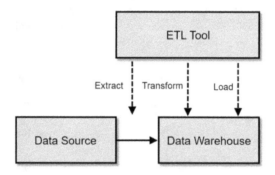

Figure 8-1. *Traditional ETL*

The correct solution is to augment your data warehouse with a data lake via ETL offloading. As you can see in Figure 8-2, most of the heavy data transformation is executed on the big data platform that, thanks to highly distributed data processing frameworks such as Spark and MapReduce, can churn through massive amounts of data in relatively short amounts of time. The data lake can act as a staging area or an operational data store. Processed and aggregated data is then loaded into the data warehouse, which is now free to be used for data analytics.

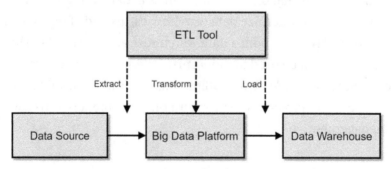

Figure 8-2. *ETL Offloading with Big Data*

CHAPTER 8 BIG DATA WAREHOUSING

Analytics Offloading and Active Archiving

You can also use your data lake to offload some of your expensive reports and ad hoc queries. By offloading some of your analytics, you free up your data warehouse for more important tasks. Analytics offloading also allows your data warehouse to handle more data and concurrent users while keeping a small data warehouse footprint – saving your organization millions in data warehouse upgrades.

If you already use your big data platform for ETL offloading, then you already have a copy of your data stored in the big data platform. You can also help make your data warehouse more efficient by moving or archiving older data to the big data platform. Once you've got the data you need in the big data platform, you can just direct your BI or data visualization tool such as Tableau or MS Power BI to Impala via JDBC (Figure 8-3). In some cases, BI tools has a semantic layer that provides a way to connect, read, and cache data from different data sources. It hides the details of the data source from users, providing a unified view of your data.

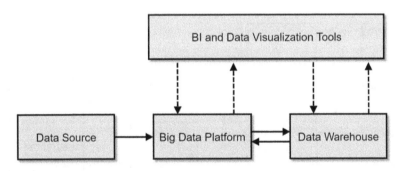

Figure 8-3. Some of the analytics are offloaded to the big data platform

Data Consolidation

One of Hadoop's best features is its ability to store massive amounts of data. By combining storage from commodity hardware and providing distributed storage engines such as HDFS, HBase, and Kudu, Hadoop made storage and processing of large data sets not just possible but also practical. Unlike the data warehouse, which can only store structured data, data lakes can store and process structured and unstructured data sets, making it a true repository of corporate data.

379

Consolidating your data in a centralized data lake delivers several benefits. First, it enhances data analysis by making it easier to join and correlate data. Second, because your data is in one central location, it makes it easier for users to run SQL joins across various data sets providing, for example, a 360-degree view of your customer. Third, feature engineering – choosing and processing attributes and creating feature sets for your predictive models is also easier. Fourth, data governance and master data management becomes straightforward once data is consolidated in one central location, providing you with one golden copy of the data instead of having multiple copies of the same piece of data spread across multiple data silos.

Perhaps the most tangible benefit from consolidating your data is the potentially massive reduction in licensing, operational, and infrastructure costs. It is fairly common for large organizations to accumulate hundreds or even thousands of databases scattered across its enterprise. By consolidating these databases into a centralized data lake (Figure 8-4), you can decommission the old servers hosting these databases and stop paying for the database software license.

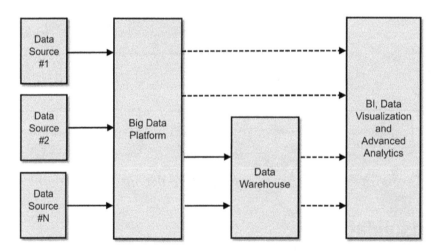

Figure 8-4. Consolidating data in a data lake or data hub

Replatforming the Enterprise Data Warehouse

This is probably the reason why you're reading this book or chapter. It is technically possible to replace or replatform data warehouses with Cloudera Enterprise. Plenty of companies have successfully done it, and in the rest of the chapter we'll discuss how to do it.

> **Note** A word of caution. Unless you have a really compelling reason, I suggest you think twice before you replatform your data warehouse with Cloudera Enterprise. There are a multitude of things to consider. You need to carefully evaluate your company's data management and analytic needs. You also need to assess Cloudera Enterprise's capabilities and weaknesses. Will your BI and OLAP tools work with Impala? Will you need to migrate thousands of ETL jobs? How hard is it to port PL/SQL or T-SQL to Impala SQL? These are just some of the questions you need to ask yourself before you go through the ordeal of migrating your data warehouse to Cloudera Enterprise.

Big Data Warehousing 101

We'll use the AdventureWorksDW free sample database from Microsoft[iv] for our examples. You can download the database from Microsoft's website if you want to follow our examples.

Dimensional Modeling

Dimensional modeling is a way of logically modeling database tables for ease of use and fast SQL query performance. Dimensional models serve as the logical basis for most business intelligence and OLAP systems in the market today. In fact, most data warehouse platforms are designed and optimized with dimensional modeling in mind. Dimensional modeling has several concepts:

Facts

A dimensional model has fact tables, which contains measures, or numerical values that represent a business metric such as sales or cost. A data warehouse will usually have several fact tables for different business processes. Before Kudu, Fact tables were usually stored in Parquet format for performance reasons.

CHAPTER 8 BIG DATA WAREHOUSING

Dimension Tables

Dimension tables contains attributes that describe or give context to the measures stored in the fact tables. Typical attributes are dates, locations, ages, religions, nationality, and gender to mention a few.

Star Schema

There are two ways to physically implement a dimensional model: by using a star or snowflake schema. A star schema contains a fact table in the middle, surrounded by dimension tables. The fact and dimension tables are connected by the tables' foreign and primary keys. The ER diagram in Figure 8-5 is an example of a star schema.

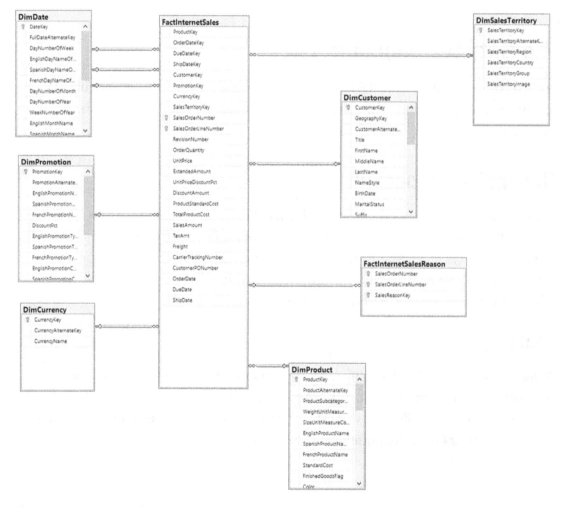

Figure 8-5. *Star Schema*

Snowflake Schema

A snowflake schema is very similar to a star schema, with a central fact table but with dimension tables that are further normalized into subdimension tables. Ralph Kimball, the data warehousing expert, generally recommends using a star schema. He recommends using a snowflake schema in cases where some measures refer to a host of attributes not shared by other measures.[v] The ER diagram shown in Figure 8-6 is an example of a snowflake schema.

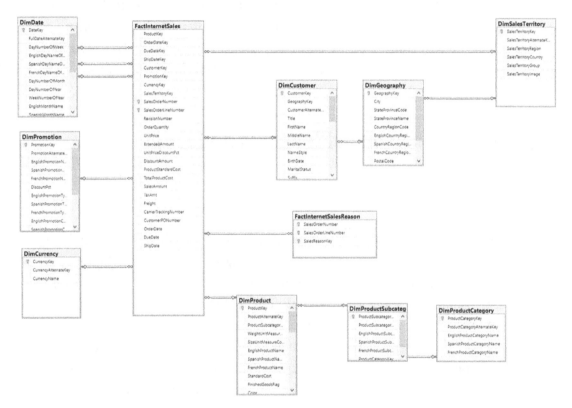

Figure 8-6. Snowflake Schema

Slowly Changing Dimensions

One of the biggest stumbling blocks for effectively using Hive or Impala for data warehousing was its inability to perform one simple thing – update a table. Columnar file formats such as Parquet stored in HDFS cannot be updated. This might come as a shock to people who have an RDBMS background. Users need to overwrite the whole table or partition in order to change values stored in the Parquet table. This limitation becomes a major headache when dealing with slowly changing dimensions. Before Kudu, the best way to handle slowly changing dimensions was to store dimension tables in HBase and then creating an external Hive table on top of the table to make it accessible from Impala. Impala doesn't support the SQL update statement but because HBase supports versioning, updates can be imitated by executing insert statements using the same rowkey.[vi]

Big Data Warehousing with Impala and Kudu

For years, Hive was used to implement rudimentary data warehouses in Hadoop. It was painfully slow and unwieldy, but it was way easier to use than MapReduce and it got the job done so users tolerated it.

Years later Impala and other open source MPP engines appeared. Used together with columnar file formats such as Parquet, it paved the way for Hadoop in the Enterprise, enabling BI and OLAP tools to connect to Hadoop clusters and get fast performance. ETL and dimensional modeling with Impala and Parquet were still cumbersome, but the benefits outweighed the costs, so users endured using it.

Impala and Kudu just made it a whole lot easier to store, process, and analyze relational data on Cloudera Enterprise. While it still doesn't have all of the advanced features of a commercial data warehouse platform, its capabilities are now close enough to a relational database.

Let's work on an example by implementing a simple dimensional star schema model in Kudu. I'll replicate a subset of the AdventureWorksDW database from SQL Server to Kudu and run some SQL queries against it using Impala. We'll copy three dimension tables and one fact table. We'll also copy a subset of the columns. The schema should look like Figure 8-7.

CHAPTER 8   BIG DATA WAREHOUSING

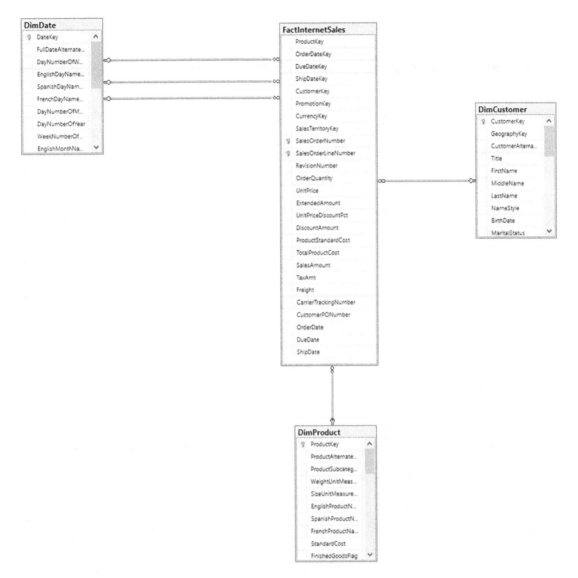

Figure 8-7. Example schema

I'll transfer the following columns: ProductKey, OrderDateKey, CustomerKey, OrderQuantity, UnitPrice, SalesAmount, TaxAmt, TotalProductCost, Freight, OrderDate, SalesOrderNumber, and SalesOrderLineNumber from the FactInternetSales table as shown in Figure 8-8.

CHAPTER 8 BIG DATA WAREHOUSING

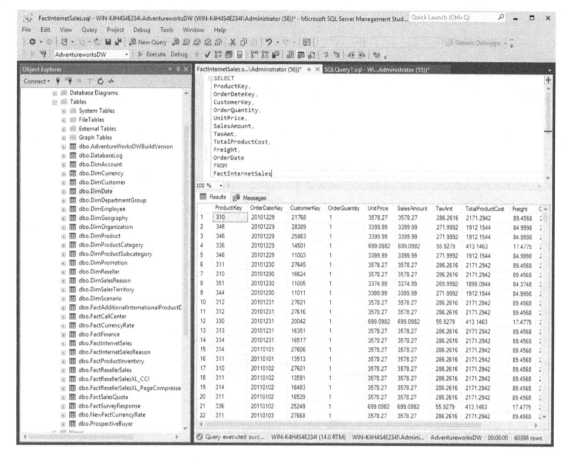

Figure 8-8. *Fact Internet Sales*

I'll copy the following columns: CustomerKey, FirstName, LastName, BirthDate, YearlyIncome, TotalChildren, EnglishEducation, EnglishOccupation, HouseOwnerFlag, and NumberofCarsOwned from the DimCustomer table as shown in Figure 8-9.

CHAPTER 8 BIG DATA WAREHOUSING

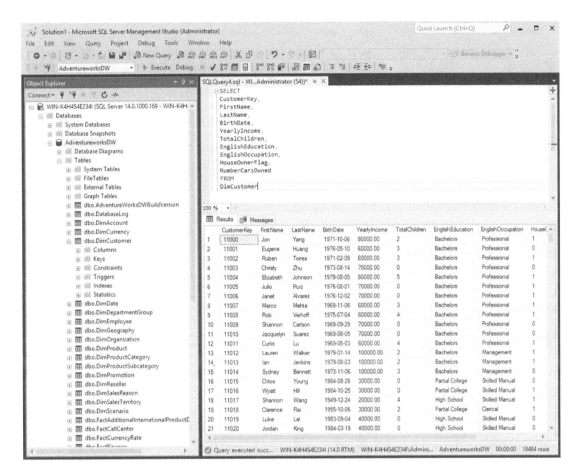

Figure 8-9. *DimCustomer*

We'll pick four columns from DimProduct: ProductKey, EnglishProductName, EnglishDescription, and Color from the DimProduct table as shown in Figure 8-10.

CHAPTER 8 ■ BIG DATA WAREHOUSING

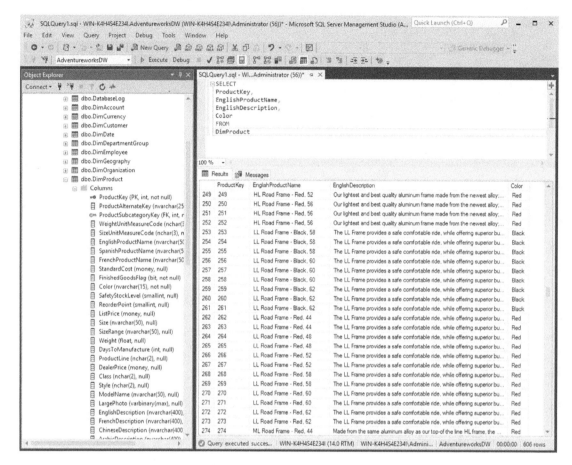

***Figure 8-10.** DimProduct*

We'll pick a few columns from the DimDate table: DateKey, FullDateAlternateKey, DayNumberOfWeek, EnlighsDayNameOfWeek, DayNumberOfMonth, DayNumberOfYear, WeekNumberOfYear, EnglishMonthName, MonthNumberOfYear, CalendarQuarter, CalendarYear, FiscalQuarter, and FiscalYear as shown in Figure 8-11.

CHAPTER 8 BIG DATA WAREHOUSING

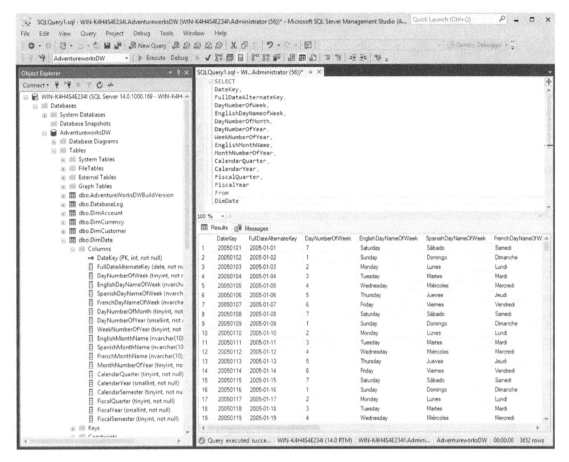

Figure 8-11. *DimDate*

I need to copy the tables from SQL Server to Impala. To do that, I manually generate a CSV file for each table. You can use any of the ETL tools we covered in Chapter 7. In this case, I used SQL Server's data export and import tool to generate the CSV files. Don't forget to exclude the column headers. I then move the files to HDFS and create an external table on top of each file. Listing 8-1 shows the steps.

389

Listing 8-1. Copy the tables from SQL Server to Impala

```
hadoop fs -mkdir /tmp/factinternetsales
hadoop fs -mkdir /tmp/dimcustomer
hadoop fs -mkdir /tmp/dimproduct
hadoop fs -mkdir /tmp/dimdate

hadoop fs -put FactInternetSales.csv /tmp/factinternetsales/
hadoop fs -put DimCustomer.csv /tmp/dimcustomer/
hadoop fs -put DimProduct.csv /tmp/dimproduct/
hadoop fs -put DimDate.csv /tmp/dimdate/

impala-shell

CREATE EXTERNAL TABLE FactInternetSales_CSV (
OrderDate STRING,
ProductKey BIGINT,
OrderDateKey BIGINT,
CustomerKey BIGINT,
OrderQuantity FLOAT,
UnitPrice FLOAT,
SalesAmount FLOAT,
TaxAmt FLOAT,
TotalProductCost FLOAT,
Freight FLOAT
)
ROW FORMAT
DELIMITED FIELDS TERMINATED BY ','
LINES TERMINATED BY '\n' STORED AS TEXTFILE
LOCATION '/tmp/factinternetsales';

CREATE EXTERNAL TABLE DimCustomer_CSV (
CustomerKey BIGINT,
FirstName STRING,
LastName STRING,
BirthDate STRING,
YearlyIncome FLOAT,
TotalChildren INT,
```

```
EnglishEducation STRING,
EnglishOccupation STRING,
HouseOwnerFlag INT,
NumberCarsOwned INT
)
ROW FORMAT
DELIMITED FIELDS TERMINATED BY ','
LINES TERMINATED BY '\n' STORED AS TEXTFILE
LOCATION '/tmp/dimcustomer';

CREATE EXTERNAL TABLE DimDate_CSV (
DateKey BIGINT,
FullDateAlternateKey STRING,
DayNumberOfWeek INT,
EnglishDayNameOfWeek STRING,
DayNumberOfMonth INT,
DayNumberOfYear INT,
WeekNumberOfYear INT,
EnglishMonthName STRING,
MonthNumberOfYear INT,
CalendarQuarter INT,
CalendarYear INT,
FiscalQuarter INT,
FiscalYear INT
)
ROW FORMAT
DELIMITED FIELDS TERMINATED BY ','
LINES TERMINATED BY '\n' STORED AS TEXTFILE
LOCATION '/tmp/dimdate';

CREATE EXTERNAL TABLE DimProduct_CSV (
ProductKey BIGINT,
EnglishProductName STRING,
EnglishDescription STRING,
Color STRING
)
```

```
ROW FORMAT
DELIMITED FIELDS TERMINATED BY ','
LINES TERMINATED BY '\n' STORED AS TEXTFILE
LOCATION '/tmp/dimproduct';

invalidate metadata;

select count(*) from FactInternetSales_csv;

60398

select count(*) from DimCustomer_csv;

18484

select count(*) from DimProduct_csv;

606

select count(*) from DimDate_csv;

3652
```

Now that the data is accessible from Impala, We can create the Kudu tables and populate it with the data from the external tables. The steps are shown in Listing 8-2.

Note As discussed in Chapter 2, Kudu does not support multi-row and multi-table ACID-compliant transactions. An ETL job will continue to run successfully even if some rows are rejected due to primary key or constraint violations. Additional data validation must be performed after an ETL process to ensure data is in a consistent state. Also, consider denormalizing your tables to reduce the possibility of data inconsistencies in Kudu.[vii]

Listing 8-2. Build Kudu tables and populate it with data

```
CREATE TABLE FactInternetSales
(
ID STRING NOT NULL,
OrderDate BIGINT NOT NULL,
ProductKey BIGINT,
```

```
OrderDateKey BIGINT,
CustomerKey BIGINT,
OrderQuantity FLOAT,
UnitPrice FLOAT,
SalesAmount FLOAT,
TaxAmt FLOAT,
TotalProductCost FLOAT,
Freight FLOAT,
SalesOrderNumber STRING,
SalesOrderLineNumber TINYINT,
PRIMARY KEY(ID, OrderDate)
)
PARTITION BY HASH (ID) PARTITIONS 16,
RANGE (OrderDate)
(
PARTITION unix_timestamp('2017-01-01') <= VALUES < unix_timestamp
('2017-02-01'),
PARTITION unix_timestamp('2017-02-01') <= VALUES < unix_timestamp
('2017-03-01'),
PARTITION unix_timestamp('2017-03-01') <= VALUES < unix_timestamp
('2017-04-01'),
PARTITION unix_timestamp('2017-04-01') <= VALUES < unix_timestamp
('2017-05-01'),
PARTITION unix_timestamp('2017-05-01') <= VALUES < unix_timestamp
('2017-06-01'),
PARTITION unix_timestamp('2017-06-01') <= VALUES < unix_timestamp
('2017-07-01'),
PARTITION unix_timestamp('2017-07-01') <= VALUES < unix_timestamp
('2017-08-01'),
PARTITION unix_timestamp('2017-08-01') <= VALUES < unix_timestamp
('2017-09-01'),
PARTITION unix_timestamp('2017-09-01') <= VALUES < unix_timestamp
('2017-10-01'),
PARTITION unix_timestamp('2017-10-01') <= VALUES < unix_timestamp
('2017-11-01'),
```

CHAPTER 8 BIG DATA WAREHOUSING

```
PARTITION unix_timestamp('2017-11-01') <= VALUES < unix_timestamp
('2017-12-01'),
PARTITION unix_timestamp('2017-12-01') <= VALUES < unix_timestamp
('2018-01-01')
)
STORED AS KUDU;

INSERT INTO FactInternetSales
SELECT
uuid(),
unix_timestamp('2017-01-20'),
ProductKey,
OrderDateKey,
CustomerKey,
OrderQuantity,
UnitPrice,
SalesAmount,
TaxAmt,
TotalProductCost,
Freight,
SalesOrderNumber,
SalesOrderLineNumber
FROM
FactInternetSales_CSV;

SELECT COUNT(*) FROM FactInternetSales_csv

+----------+
| count(*) |
+----------+
| 60398    |
+----------+
```

```
SELECT COUNT(*) FROM FactInternetSales;

+----------+
| count(*) |
+----------+
| 60398    |
+----------+
CREATE TABLE DimCustomer (
ID STRING,
CustomerKey BIGINT,
FirstName STRING,
LastName STRING,
BirthDate STRING,
YearlyIncome FLOAT,
TotalChildren INT,
EnglishEducation STRING,
EnglishOccupation STRING,
HouseOwnerFlag INT,
NumberCarsOwned INT,
PRIMARY KEY(ID)
)
PARTITION BY HASH (ID) PARTITIONS 16
STORED AS KUDU);

INSERT INTO DimCustomer
SELECT
uuid(),
CustomerKey,
FirstName,
LastName,
BirthDate,
YearlyIncome,
TotalChildren,
EnglishEducation,
EnglishOccupation,
HouseOwnerFlag,
```

```
NumberCarsOwned
FROM
DimCustomer_CSV;

SELECT COUNT(*) FROM DimCustomer;

+----------+
| count(*) |
+----------+
| 18484    |
+----------+

SELECT COUNT(*) FROM DimCustomer_CSV;

+----------+
| count(*) |
+----------+
| 18484    |
+----------+

CREATE TABLE DimDate (
ID STRING,
DateKey BIGINT,
FullDateAlternateKey STRING,
DayNumberOfWeek INT,
EnglishDayNameOfWeek STRING,
DayNumberOfMonth INT,
DayNumberOfYear INT,
WeekNumberOfYear INT,
EnglishMonthName STRING,
MonthNumberOfYear INT,
CalendarQuarter INT,
CalendarYear INT,
FiscalQuarter INT,
FiscalYear INT,
PRIMARY KEY(ID)
)
```

```sql
PARTITION BY HASH (ID) PARTITIONS 16
STORED AS KUDU;

INSERT INTO DimDate
SELECT
uuid(),
DateKey,
FullDateAlternateKey,
DayNumberOfWeek,
EnglishDayNameOfWeek,
DayNumberOfMonth,
DayNumberOfYear,
WeekNumberOfYear,
EnglishMonthName,
MonthNumberOfYear,
CalendarQuarter,
CalendarYear,
FiscalQuarter,
FiscalYear
FROM DimDate_CSV;

SELECT count(*) from DimDate;
+----------+
| count(*) |
+----------+
| 3652     |
+----------+

SELECT count(*) from Dimdate_CSV;
+----------+
| count(*) |
+----------+
| 3652     |
+----------+
```

CHAPTER 8 BIG DATA WAREHOUSING

```
CREATE TABLE DimProduct (
ID STRING,
ProductKey BIGINT,
EnglishProductName STRING,
EnglishDescription STRING,
Color STRING
)
PARTITION BY HASH (ID) PARTITIONS 16
STORED AS KUDU;

INSERT INTO DimProduct
SELECT
uuid(),
productkey,
englishproductname,
englishdescription,
color
FROM
DimProduct_CSV

select COUNT(*) from DimProduct

+----------+
| count(*) |
+----------+
| 606      |
+----------+

select COUNT(*) from DimProduct_CSV

+----------+
| count(*) |
+----------+
| 606      |
+----------+
```

CHAPTER 8 BIG DATA WAREHOUSING

There are a few important things to discuss here. Notice that I used the Impala built-in function uuid() to generate a unique ID for the fact and dimension tables. Impala and Kudu do not include an auto-increment feature, which is actually considered an anti-pattern on distributed environments. With Kudu, Spanner, and similar distributed systems, range partitioning on a column whose values are monotonically increasing will direct all inserts to be written to a single tablet at a time, causing the tablet to grow much larger than other tablets and limiting the scalability of inserts. This is also known as hot spotting. Hot spotting also happens in HBase when inserting rows with monotonically increasing row keys.

While using UUIDs as a primary key on a database running on a single SMP server is a disaster, in distributed MPP environments, hash partitioning using a unique key that is not derived from the business or data contained in the table will spread the workload across the nodes in your cluster. Hash partitioning protects from data and workload skew and increases scalability of data ingestion by enabling inserts to go to multiple tablets in parallel.[viii]

Some database systems such as Microsoft SQL Server have a native UUID data type that provide a more efficient storage for UUIDs. Kudu doesn't have a UUID data type, therefore you have to use the STRING data type to store the UUID. The 36-character UUID does take up some space; however benchmarks indicate that having a 36-character primary key in Kudu doesn't affect performance for most queries, although your experience may vary. If you are concerned about using UUIDs, then another option is to hash the UUID to generate a non-sequential BIGINT value to serve as your primary key. You can use Impala's built-in hashing function fnv_hash() (which is not cryptographically secure, which means collision may occur when generating unique ID's) or a more cryptographically secure hashing algorithm. Just make sure you test the performance of your insert statements. Calling built-in functions every insert statement may potentially cause performance bottlenecks. Of course, there are other ways to generate a non-sequential unique key.

As you can see, the fact table is hash and range partitioned. By combining the two partitioning strategies, we gain the benefit of both, while reducing the limitations of each.[ix] By using range partitioning on the date column, your query can take advantage of partition pruning when scanning by date ranges. In addition, new partitions can be added as new data is added to the table (which will result in 16 additional tablets). By using hash partitioning on the ID column (which is a non-sequential UUID), inserts are spread evenly across multiple tablets up to the number of hash partitions (which in our case is 16), increasing throughput and performance of data ingestion. Comparing this strategy with the current AdventureWorksDW schema, you will find a couple of differences. First off, FactInternetSales' primary key is on SalesOrderNumber and

399

SalesOrderLineNumber. This is adequate for an SQL Server database running on one server. However, this breaks several of the rules we've just discussed about partitioning and choosing primary key values on distributed systems. SalesOrderNumber is monotonically increasing, and the majority of SalesOrderLineNumber's value is 1. If we use these two columns as our primary key, all inserts will be directed to one partition, causing data and workload skew and slowing down data ingestion.

Dimensions tables are usually small and mostly read-only. Slowly changing dimensions doesn't usually require the kind of performance consideration as fact tables so we'll just hash partition the dimension tables using a unique UUID as the primary key. Larger dimension tables may require a different partitioning strategy.

While Kudu requires all tables to have primary keys, it doesn't have foreign key constraints. Therefore, data inconsistencies are possible in a Kudu data warehouse. However, it is common practice for data warehouses to have foreign key constraints disabled due to performance and manageability reasons. Listing 8-3 shows the structure of the Kudu tables we just created.

Listing 8-3. Structure of Kudu tables

```
describe FactInternetSales;
```

name	type	comment	primary_key	nullable
id	string		true	false
orderdate	bigint		true	false
productkey	bigint		false	true
orderdatekey	bigint		false	true
customerkey	bigint		false	true
orderquantity	float		false	true
unitprice	float		false	true
salesamount	float		false	true
taxamt	float		false	true
totalproductcost	float		false	true
freight	float		false	true
salesordernumber	string		false	true
salesorderlinenumber	tinyint		false	true

describe DimDate;

```
+----------------------+--------+---------+-------------+----------+
| name                 | type   | comment | primary_key | nullable |
+----------------------+--------+---------+-------------+----------+
| id                   | string |         | true        | false    |
| datekey              | bigint |         | false       | true     |
| fulldatealternatekey | string |         | false       | true     |
| daynumberofweek      | int    |         | false       | true     |
| englishdaynameofweek | string |         | false       | true     |
| daynumberofmonth     | int    |         | false       | true     |
| daynumberofyear      | int    |         | false       | true     |
| weeknumberofyear     | int    |         | false       | true     |
| englishmonthname     | string |         | false       | true     |
| monthnumberofyear    | int    |         | false       | true     |
| calendarquarter      | int    |         | false       | true     |
| calendaryear         | int    |         | false       | true     |
| fiscalquarter        | int    |         | false       | true     |
| fiscalyear           | int    |         | false       | true     |
+----------------------+--------+---------+-------------+----------+
```

describe DimCustomer;

```
+-------------------+--------+---------+-------------+----------+
| name              | type   | comment | primary_key | nullable |
+-------------------+--------+---------+-------------+----------+
| id                | string |         | true        | false    |
| customerkey       | bigint |         | false       | true     |
| firstname         | string |         | false       | true     |
| lastname          | string |         | false       | true     |
| birthdate         | string |         | false       | true     |
| yearlyincome      | float  |         | false       | true     |
| totalchildren     | int    |         | false       | true     |
| englisheducation  | string |         | false       | true     |
| englishoccupation | string |         | false       | true     |
| houseownerflag    | int    |         | false       | true     |
| numbercarsowned   | int    |         | false       | true     |
+-------------------+--------+---------+-------------+----------+
```

CHAPTER 8 BIG DATA WAREHOUSING

```
describe DimProduct;
```

```
+---------------------+---------+---------+-------------+----------+
| name                | type    | comment | primary_key | nullable |
+---------------------+---------+---------+-------------+----------+
| id                  | string  |         | true        | false    |
| productkey          | bigint  |         | false       | true     |
| englishproductname  | string  |         | false       | true     |
| englishdescription  | string  |         | false       | true     |
| color               | string  |         | false       | true     |
+---------------------+---------+---------+-------------+----------+
```

Note As mentioned earlier in this chapter, Kudu does not support the decimal data type. The float and double data types only store a very close approximation of the value instead of the exact value as defined in the IEEE 754 specification.[x] Because of this, behavior float and double are not appropriate for storing financial data. At the time of writing, support for decimal data type is still under development (Apache Kudu 1.5 / CDH 5.13). Refer to KUDU-721 for more details. A workaround is to store financial data as string then use Impala to cast the value to decimal every time you need to read the data. Since Parquet supports decimals, another workaround would be to use Parquet for your fact tables and Kudu for dimension tables.

We can now run Impala queries against our data warehouse in Kudu.

For example, here's a query to get a top 20 list of everyone who earns more than $100,000 and are also home owners.

```
SELECT
DC.FirstName,
DC.LastName,
DC.BirthDate,
DC.HouseOwnerFlag,
DC.YearlyIncome
FROM FactInternetSales FIS
INNER JOIN DimCustomer DC ON FIS.customerkey = DC.customerkey
```

```
WHERE
DC.YearlyIncome >= 100000
AND
DC.HouseOwnerFlag = 1
ORDER BY
DC.YearlyIncome DESC
LIMIT 20;
```

firstname	lastname	birthdate	houseownerflag	yearlyincome
Shannon	Navarro	1971-05-03	1	170000
Devin	Anderson	1959-08-22	1	170000
Luis	Washington	1946-05-07	1	170000
Ian	Watson	1950-12-08	1	170000
Jenny	Chander	1955-02-04	1	170000
Nathan	Wright	1965-12-20	1	170000
Trisha	Wu	1959-01-03	1	170000
Dale	Holt	1952-02-01	1	170000
Craig	Gutierrez	1952-04-11	1	170000
Devin	Martin	1960-02-01	1	170000
Katrina	Luo	1950-04-22	1	170000
Alvin	Nara	1972-11-16	1	170000
Megan	Morgan	1966-02-13	1	170000
Armando	Munoz	1961-10-31	1	170000
Arturo	Zheng	1975-04-06	1	170000
Vincent	Sun	1959-11-03	1	170000
Colleen	Anand	1963-09-23	1	170000
Caleb	Jenkins	1970-04-22	1	170000
Devin	Johnson	1965-12-16	1	170000
Shannon	Liu	1966-03-08	1	170000

CHAPTER 8 BIG DATA WAREHOUSING

To get a list of everyone who bought a "Sport-100 Helmet" and also owns more than two cars.

```
SELECT
DC.FirstName,
DC.LastName,
DC.BirthDate,
DC.NumberCarsOwned,
DP.EnglishProductName
FROM FactInternetSales FIS
INNER JOIN DimCustomer DC ON FIS.customerkey = DC.customerkey
INNER JOIN DimProduct DP ON FIS.productkey = DP.productkey
WHERE
DP.EnglishProductName = 'Sport-100 Helmet'
AND
DC.NumberCarsOwned > 2
LIMIT 20;
```

firstname	lastname	birthdate	numbercarsowned	englishproductname
Colleen	Goel	1954-04-06	4	Sport-100 Helmet
Timothy	Stewart	1948-12-22	4	Sport-100 Helmet
Janelle	Arthur	1971-01-14	3	Sport-100 Helmet
Colin	Lal	1976-11-30	4	Sport-100 Helmet
Melody	Ruiz	1979-08-01	4	Sport-100 Helmet
Nicolas	Andersen	1972-11-10	3	Sport-100 Helmet
Hunter	Lopez	1958-02-06	3	Sport-100 Helmet
Jake	Zhu	1978-06-24	3	Sport-100 Helmet
Caleb	Yang	1985-08-13	3	Sport-100 Helmet
Jeffery	Ma	1969-12-03	3	Sport-100 Helmet
Miranda	Patterson	1952-01-07	4	Sport-100 Helmet
Isaac	Ward	1957-11-09	3	Sport-100 Helmet
Roy	Chandra	1953-02-09	4	Sport-100 Helmet
William	Harris	1979-09-30	3	Sport-100 Helmet
Xavier	Williams	1974-01-18	3	Sport-100 Helmet

```
| Mason   | Mitchell | 1952-11-03 | 4 | Sport-100 Helmet |
| Claudia | Zhou     | 1961-12-04 | 3 | Sport-100 Helmet |
| Tamara  | Lal      | 1961-06-17 | 3 | Sport-100 Helmet |
| Misty   | Tang     | 1947-08-30 | 3 | Sport-100 Helmet |
| Alisha  | Zeng     | 1977-10-31 | 3 | Sport-100 Helmet |
+---------+----------+------------+---+------------------+
```

Summary

Congratulations! You've just used Impala and Kudu for data warehousing! You can connect your favorite business intelligence, data visualization, and OLAP tools to Impala via JDBC/ODBC to create dashboards, reports, and OLAP cubes against the tables we just created. I refer you to Chapter 9 for more information on data visualization tools. You can use Spark to process and ingest massive amounts of data into Kudu. That is discussed in detail in Chapter 6. You can also use commercial third-party ETL applications such as StreamSets and Talend to load data into Kudu. That's covered in Chapter 7. It looks like you have everything you need to build a data warehouse!

This chapter discussed several ways to take advantage of Impala and Kudu. You can modernize your data warehouse or replace it altogether. Just make sure that you have carefully evaluated its strengths and weaknesses, and whether it is the correct solution for your organization.

References

i. Hortonworks; "Data Warehousing with Apache Hive," Hortonworks, 2018, https://docs.hortonworks.com/HDPDocuments/HDP2/HDP-2.6.2/bk_data-access/content/ch_using-hive.html

ii. The Kimball Group; "The Data Warehouse Toolkit, 3rd Edition," The Kimball Group, 2018, http://www.kimballgroup.com/data-warehouse-business-intelligence-resources/books/data-warehouse-dw-toolkit/

iii. Dell; "Dell Survey: Structured Data Remains Focal Point Despite Rapidly Changing Information Management Landscape," Dell, 2018, http://www.dell.com/learn/us/en/uscorp1/press-releases/2015-04-15-dell-survey

iv. Microsoft; "AdventureWorks Databases and Scripts for SQL Server 2016 CTP3," Microsoft, 2018, https://www.microsoft.com/en-us/download/details.aspx?id=49502

v. Oracle; "Star and Snowflake Schemas," Oracle, 2018, http://www.oracle.com/webfolder/technetwork/tutorials/obe/db/10g/r2/owb/owb10gr2_gs/owb/lesson3/starandsnowflake.htm

vi. Cloudera; "Using Impala to Query HBase Tables," Cloudera, 2018, https://www.cloudera.com/documentation/enterprise/latest/topics/impala_hbase.html

vii. Cloudera; "Using Impala to Query Kudu Tables," Cloudera, 2018, https://www.cloudera.com/documentation/enterprise/5-12-x/topics/impala_kudu.html

viii. Apache Kudu; "Kudu: Frequently Asked Questions," Apache Kudu, 2018, https://kudu.apache.org/faq.html

ix. Apache Kudu; "Hash and Range Partitioning Example," Apache Kudu, 2018, https://kudu.apache.org/docs/schema_design.html#hash-range-partitioning-example

x. Microsoft; "Using decimal, float, and real Data," Microsoft, 2018, https://technet.microsoft.com/en-us/library/ms187912(v=sql.105).aspx

CHAPTER 9

Big Data Visualization and Data Wrangling

It's easy to understand why self-service data analysis and visualization have become popular these past few years. It made users more productive by giving them the ability to perform their own analysis and allowing them to interactively explore and manipulate data based on their own needs without relying on traditional business intelligence developers to develop reports and dashboards, a task that can take days, weeks, or longer. Users can perform ad hoc analysis and run follow-up queries to answer their own questions. They're also not limited by static reports and dashboards. Output from self-service data analysis can take various forms depending on the type of analysis. The output can take the form of interactive charts and dashboards, pivot tables, OLAP cubes, predictions from machine learning models, or query results returned by a SQL query.

Big Data Visualization

Several books have already been published about popular data visualization tools such as Tableau, Qlik, and Power BI. These tools all integrate well with Impala and some of them have rudimentary support for big data. They're adequate for typical data visualization tasks, and most of the time that's all typical users will ever need. However, when an organization's analytic requirements go beyond what traditional tools can handle, it's time to look at tools that were specifically designed for big data. Note that when I talk about big data, I am referring to the three V's of big data. I am not only referring to the size (volume) of data, but also the variety (Can these tools analyze structured, unstructured, semi-structured data) and velocity of data (Can these tools perform real-time or near real-time data visualization?). I explore data visualization tools in this chapter that are specifically designed for big data.

SAS Visual Analytics

SAS has been one of the leaders in advanced analytics for more than 40 years. The SAS software suite has hundreds of components designed for various use cases ranging from data mining and econometrics to six sigma and clinical trial analysis. For this chapter, we're interested in the self-service analytic tool known as SAS Visual Analytics.

SAS Visual Analytics is a web-based, self-service interactive data analytics application from SAS. SAS Visual Analytics allows users to explore their data, write reports, and process and load data into the SAS environment. It has in-memory capabilities and was designed for big data. SAS Visual Analytics can be deployed in distributed and non-distributed mode. In distributed mode, the high-performance SAS LASR Analytic Server daemons are installed on the Hadoop worker nodes. It can take advantage of the full processing capabilities of your Hadoop cluster, enabling it to crank through large volumes of data stored in HDFS.

SAS Visual Analytics distributed deployment supports Cloudera Enterprise, Hortonworks HDP, and the Teradata Data Warehouse Appliance. The non-distributed SAS VA deployment runs on a single machine with a locally installed SAS LASR Analytic Server and connects to Cloudera Impala via JDBC.[i]

Zoomdata

Zoomdata is a data visualization and analytics company based in Reston, Virginia.[ii] It was designed to take advantage of Apache Spark, making it ideal for analyzing extremely large data sets. Spark's in-memory and streaming features enable Zoomdata to support real-time data visualization capabilities. Zoomdata supports various data sources such as Oracle, SQL Server, Hive, Solr, Elasticsearch, and PostgreSQL to name a few. One of the most exciting features of Zoomdata is its support for Kudu and Impala, making it ideal for big data and real-time data visualization. Let's take a closer look at Zoomdata.

Self-Service BI and Analytics for Big Data

Zoomdata is a modern BI and data analytics tool designed specifically for big data. Zoomdata supports a wide range of data visualization and analytic capabilities, from exploratory self-service data analysis, embedded visual analytics to dashboarding. Exploratory self-service data analysis has become more popular during the past few years. It allows users to interactively navigate and explore data with the goal of discovering hidden patterns and insights from data. Compared to traditional business intelligence tools, which rely heavily on displaying predefined metrics and KPIs using static dashboards.

Zoomdata also supports dashboards. Dashboards are still an effective way of communicating information to users who just wants to see results. Another nice feature is the ability to embed charts and dashboards created in Zoomdata into other web applications. Utilizing standard HTML5, CSS, WebSockets, and JavaScript, embedded analytics can run on most popular web browsers and mobile devices such as iPhones, iPads, and Android tablets. Zoomdata provides a JavaScript SDK if you need the extra flexibility and you need to control how your applications look and behave. A REST API is available to help automate the management and operation of the Zoomdata environment.

Real-Time Data Visualization

Perhaps the most popular feature of Zoomdata is its support for real-time data visualization. Charts and KPIs are updated in real time from live streaming sources such as MQTT and Kafka or an SQL-based data source such as Oracle, SQL Server, or Impala.

Architecture

At the heart of Zoomdata is a stream processing engine that divides and processes all incoming data as data streams. The data source doesn't have to be live, as long as the data has a time attribute, Zoomdata can stream it. Zoomdata has a technology called Data DVR, which not only lets users stream data, but also rewind, replay, and pause real-time streaming data, just like a DVR. Zoomdata pushes most of the data processing down to the data source where the data resides, taking advantage of data locality by minimizing data movement.[iii] Zoomdata uses Apache Spark behind the scenes as a complementary data processing layer. It caches streamed data as Spark DataFrames and stores it in its result set cache. Zoomdata always inspects and tries to retrieves the data from the result set cache first before going to the original source.[iv] Figure 9-1 shows the high-level architecture of Zoomdata. Apache Spark enables most of Zoomdata's capabilities such as Zoomdata Fusion and Result Set Caching.

CHAPTER 9 BIG DATA VISUALIZATION AND DATA WRANGLING

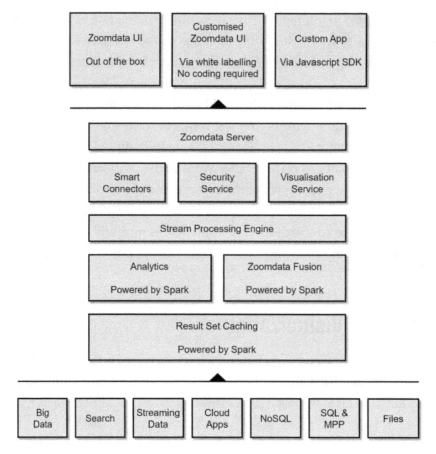

Figure 9-1. *High-level architecture of Zoomdata*

Deep Integration with Apache Spark

One of the most impressive features of Zoomdata is its deep integration with Apache Spark. As previously discussed, Zoomdata uses Spark to implement its streaming feature. Zoomdata also uses Spark for query optimization by caching data as Spark DataFrames, then performing the calculation, aggregation, and filtering using Spark against the cached data. Spark also powers another Zoomdata feature called SparkIt. SparkIt boosts performance of slow data sources such as S3 buckets and text files. SparkIt caches these data sets into Spark, which transforms them into queryable, high-performance Spark DataFrames.[v] Finally, Zoomdata uses Spark to implement Zoomdata Fusion, a high-performance data virtualization layer that virtualizes multiple data sources into one.

CHAPTER 9 ■ BIG DATA VISUALIZATION AND DATA WRANGLING

Zoomdata Fusion

Zoomdata uses a high-performance data virtualization layer called Zoomdata Fusion. Also powered by Apache Spark, Zoomdata makes multiple data sources appear as one source without physically moving the data sets together to a common location.[vi] Zoomdata uses Spark to cache data when optimizing cross-platform joins but pushes processing to the data source for maximum scalability. With Zoomdata Fusion, users will be able to conveniently query and join disparate data sets in different formats, significantly shortening the time to insight, increasing productivity, and reducing reliance on ETL (see Figure 9-2).

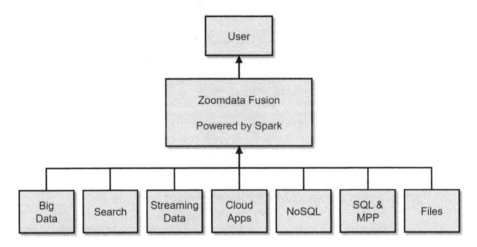

Figure 9-2. *Diagram of Zoomdata Fusion*

Data Sharpening

Zoomdata was awarded a patent on "data sharpening," an optimization technique for visualizing large data sets. Data Sharpening works by dividing a large query into a series of micro-queries. These micro-queries are then sent to the data source and executed in parallel. Results from the first micro-query is immediately returned and visualized by Zoomdata as an estimate of the final results. Zoomdata updates the visualization as new data arrives, showing incrementally better and more accurate estimates until the query completes and the full visualization is finally shown. While all of these are happening, users can still interact with the charts and launch other dashboards, without waiting for long-running queries to finish. Underneath, Zoomdata's streaming architecture makes all of these possible, breaking queries down into micro-queries and streaming results back to the user.[vii]

Support for Multiple Data Sources

Even though Zoomdata was designed for big data, it also supports the most popular MPP and SQL databases in the market including Oracle, SQL Server, Teradata, MySQL, PostgreSQL, Vertica, and Amazon Aurora.[viii] In almost all enterprise environments you will find a combination of various flavors of SQL databases and Hadoop distributions. With Zoomdata Fusion, Zoomdata allows users to easily query these disparate data sources just like it was a single data source. Other Zoomdata features that enables fast visual analytics such as data sharpening, Data DVR, and micro-queries all work on these SQL and MPP databases as well. Other popular data sources such as MongoDB, Solr, Elasticsearch, and Amazon S3 are also supported (see Figure 9-3).

Figure 9-3. *Partial list of Zoomdata's supported data sources*

Note Consult Zoomdata's website for the latest instructions on how to install Zoomdata. You also need to install Zoomdata's Kudu connector to connect to Kudu, which is available on Zoomdata's website as well.

Let's get started with an example. Start Zoomdata. You'll be presented with a login page that looks like Figure 9-4.

CHAPTER 9 BIG DATA VISUALIZATION AND DATA WRANGLING

Figure 9-4. *Zoomdata login page*

The first thing you need to do is create a new data source (see Figure 9-5).

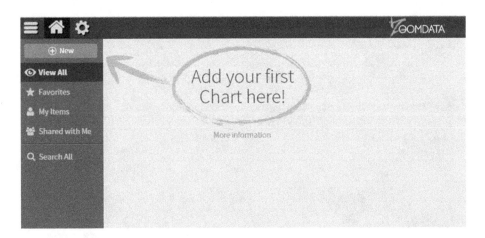

Figure 9-5. *Zoomdata homepage*

CHAPTER 9 BIG DATA VISUALIZATION AND DATA WRANGLING

Notice that the Kudu Impala data source added to the list of data sources (see Figure 9-6).

Figure 9-6. Available data sources

You will be required to input your credentials, including the JDBC URL and the location of the Kudu Master (Figure 9-7). In our example, the Kudu Master resides at 192.168.56.101:7051. Update the JDBC URL to jdbc:hive2://192.168.56.101:21050/;auth=noSasl. Note that even though we're connecting to Impala, we need to specify "hive2" in our JDBC URL. Click "Next" when done (Figure 9-8).

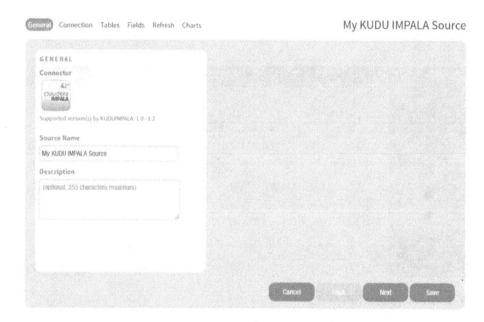

Figure 9-7. Kudu Impala Connector page

CHAPTER 9 BIG DATA VISUALIZATION AND DATA WRANGLING

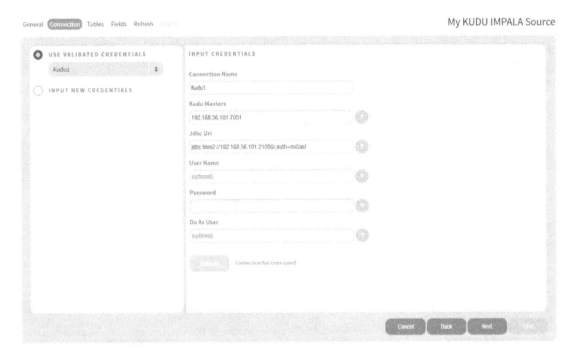

Figure 9-8. *Kudu Impala Connector Input Credentials*

Pick the table that we will be using in this example (Figure 9-9). You can also select fields (Figure 9-10). Unless you're running out of RAM, it's always a good idea to turn on caching. Since our table is from a relational data store, you don't need to enable SparkIt. If you remember, SparkIt is only for simple data sources such as text files that do not provide caching capabilities. Click Next to proceed to the next step.

CHAPTER 9   BIG DATA VISUALIZATION AND DATA WRANGLING

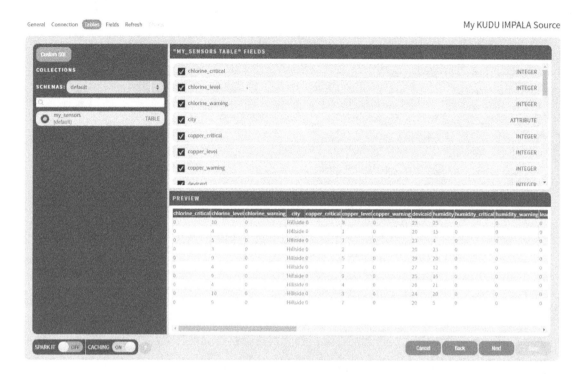

Figure 9-9. Data Source Tables

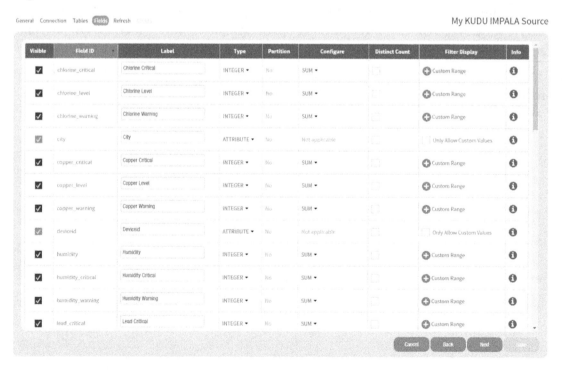

Figure 9-10. Data Source Fields

CHAPTER 9 BIG DATA VISUALIZATION AND DATA WRANGLING

Zoomdata can maintain cached result sets for certain data sources asynchronously (Figure 9-11).

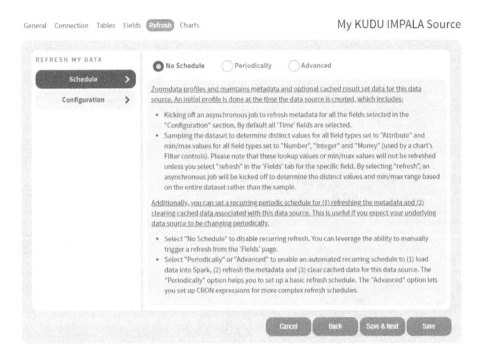

Figure 9-11. *Data Source Refresh*

You can also schedule a job to refresh the cache and metadata (Figure 9-12).

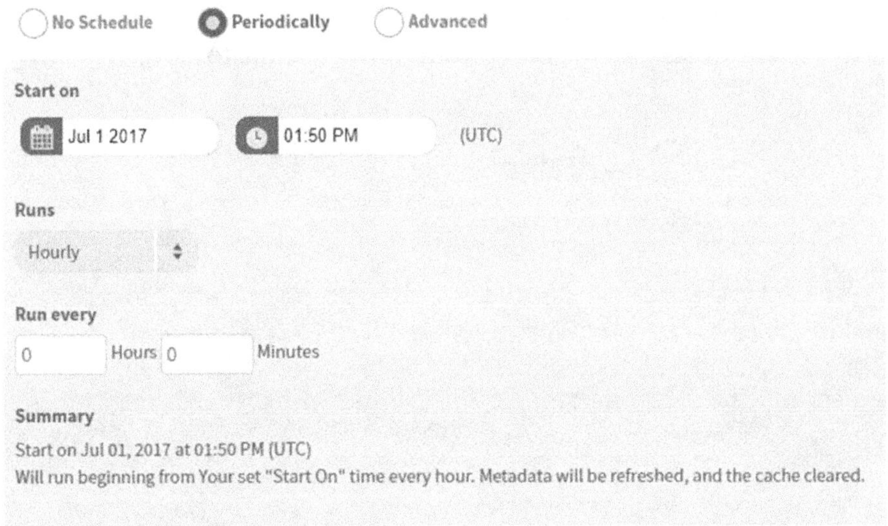

Figure 9-12. *Data Source Scheduler*

417

CHAPTER 9 BIG DATA VISUALIZATION AND DATA WRANGLING

For advanced users who are more comfortable with the cron Unix-style scheduler, a similar interface is also available (Figure 9-13).

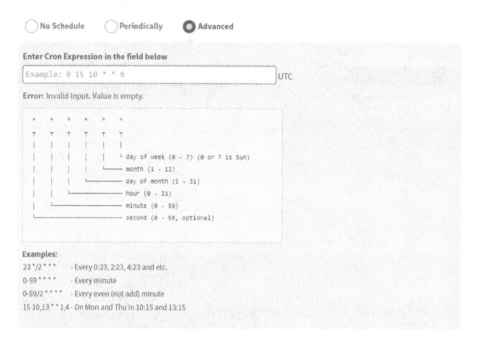

Figure 9-13. *Data Source Cron Scheduler*

To support real-time visualization, Live Mode and Playback must be enabled (Figure 9-14). As long as the data set has a time attribute, in our case the sensortimestamp, Zoomdata can visualize using the time attribute. I set the refresh rate and delay to 1 second. I also set the range from Now-1 minute to Now.

CHAPTER 9 BIG DATA VISUALIZATION AND DATA WRANGLING

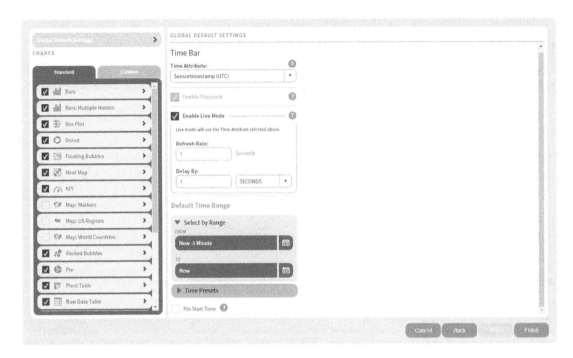

Figure 9-14. *Time Bar Global Default Settings*

You can configure the available charts (Figure 9-15).

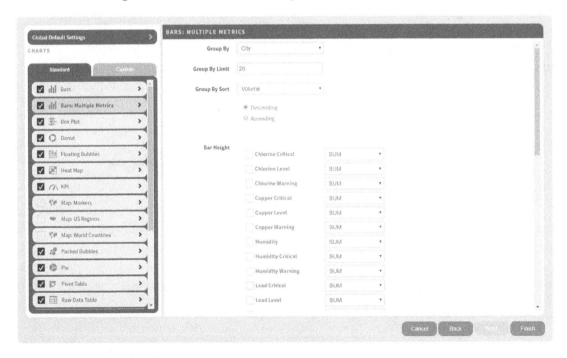

Figure 9-15. *Global Default Settings – Charts*

CHAPTER 9   BIG DATA VISUALIZATION AND DATA WRANGLING

There are different configuration options available depending on the type of chart (Figure 9-16).

Figure 9-16. *Map: World Countries – Options*

The data source will be saved after you click "Finish." You can immediately create a chart based on the source (Figure 9-17). In this case we'll pick "Bars: Multiple Metrics."

CHAPTER 9 BIG DATA VISUALIZATION AND DATA WRANGLING

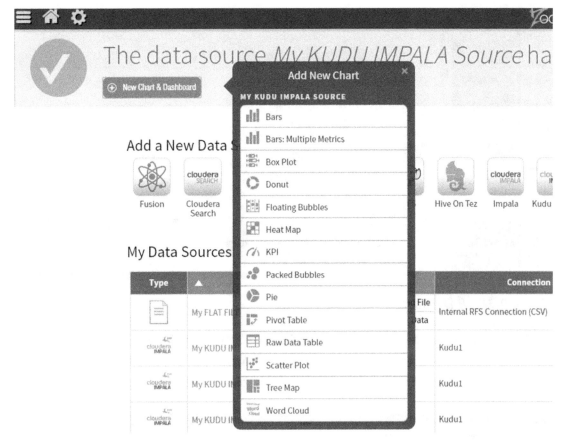

Figure 9-17. *Add New Chart*

A bar chart similar to Figure 9-18 will be displayed. You can adjust different bar chart options.

CHAPTER 9 BIG DATA VISUALIZATION AND DATA WRANGLING

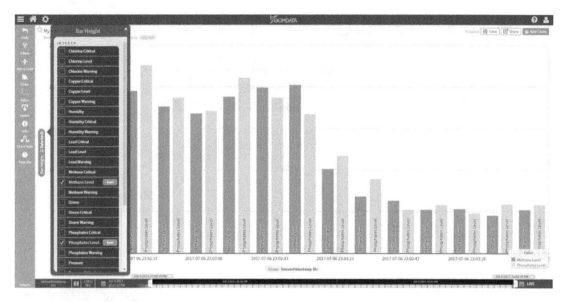

Figure 9-18. Bar Chart

When a chart is maximized, you're provided with more configuration options such as Filters, Color, and Chart Style to name a few. In the example shown in Figure 9-19, I've changed the color of the bar chart from yellow and blue to purple and green.

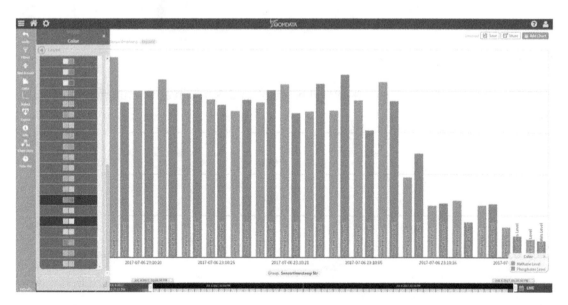

Figure 9-19. Bar Chart – Color

422

CHAPTER 9 BIG DATA VISUALIZATION AND DATA WRANGLING

To add more charts, click "Add Chart" located near the upper-right hand corner of the application (Figure 9-20).

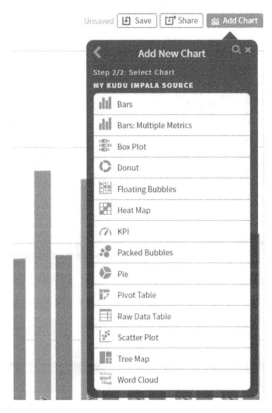

Figure 9-20. Add New Chart

CHAPTER 9   BIG DATA VISUALIZATION AND DATA WRANGLING

The pie chart is one of the most common chart types (Figure 9-21).

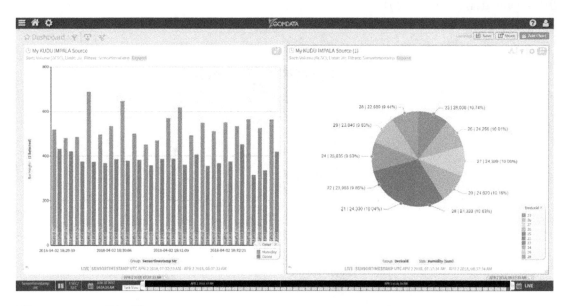

Figure 9-21. *Pie Chart*

Zoomdata has numerous chart types as shown in Figure 9-22.

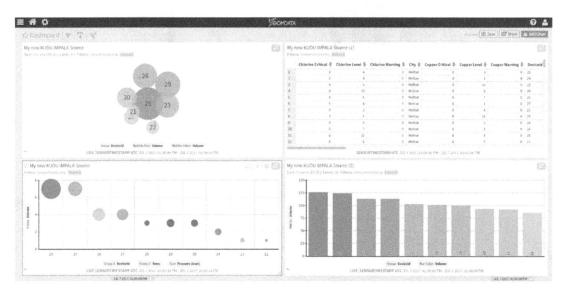

Figure 9-22. *Multiple Chart Types*

CHAPTER 9 BIG DATA VISUALIZATION AND DATA WRANGLING

Zoomdata has mapping capabilities as shown in Figure 9-23.

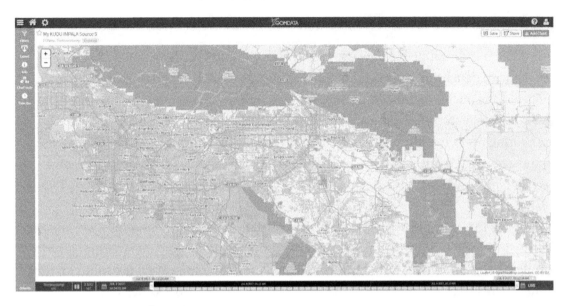

Figure 9-23. Zoomdata Map

Zoomdata can update the map with markers in real time as shown in Figure 9-24.

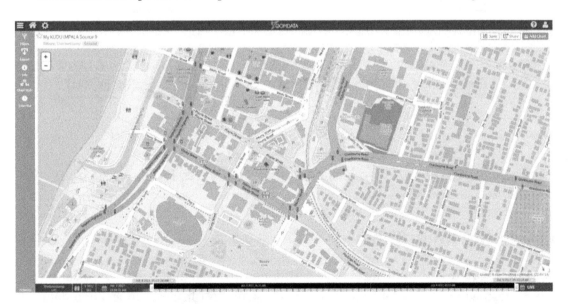

Figure 9-24. Zoomdata Map with Markers

CHAPTER 9   BIG DATA VISUALIZATION AND DATA WRANGLING

Real-Time IoT with StreamSets, Kudu, and Zoomdata

One of the most exciting big data use cases is IoT (Internet of Things). IoT enables the collection of data from sensors embedded in hardware such as smart phones, electronic gadgets, appliances, manufacturing equipment, and health care devices to name a few. Use cases range from real-time predictive health care, network pipe leakage detection, monitoring water quality, emergency alert systems, smart homes, smart cities, and connected car to name a few. IoT data can be run through rules engines for real-time decision making or machine learning models for more advanced real-time predictive analytics.

For our IoT example, we will detect network pipe leakage and monitor water quality for a fictional water utility company. We'll use a shell script to generate random data to simulate IoT data coming from sensors installed on water pipes. We'll use Kudu for data storage, StreamSets for real-time data ingestion and stream processing, and Zoomdata for real-time data visualization.

In real IoT projects, the data source will most likely be an IoT gateway that serves data over a lightweight messaging protocol such as MQTT. Kafka is also highly recommended as a buffer between the IoT gateway and the data storage to provide high availability and replayability in case old data is needed (Figure 9-25).

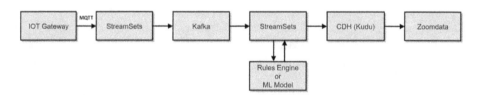

Figure 9-25. *A Typical IoT Architecture using StreamSets, Kafka, Kudu, and Zoomdata*

Create the Kudu Table

Zoomdata requires the Kudu table to be partitioned on the timestamp field. To enable live mode, the timestamp field has to be stored as a BIGINT in epoch time. Zoomdata will recognize the timestamp field as "playable" and enable live mode on the data source (Listing 9-1).

Listing 9-1. Table for the sensor data. Must be run in impala-shell

```
CREATE TABLE my_sensors (
rowid BIGINT NOT NULL,
sensortimestamp BIGINT NOT NULL,
deviceid INTEGER,
temperature INTEGER,
pressure INTEGER,
humidity INTEGER,
ozone INTEGER,
sensortimestamp_str STRING,
city STRING,
temperature_warning INTEGER,
temperature_critical INTEGER,
pressure_warning INTEGER,
pressure_critical INTEGER,
humidity_warning INTEGER,
humidity_critical INTEGER,
ozone_warning INTEGER,
ozone_critical INTEGER,
lead_level INTEGER,
copper_level INTEGER,
phosphates_level INTEGER,
methane_level INTEGER,
chlorine_level INTEGER,
lead_warning INTEGER,
lead_critical INTEGER,
copper_warning INTEGER,
copper_critical INTEGER,
phosphates_warning INTEGER,
phosphates_critical INTEGER,
methane_warning INTEGER,
methane_critical INTEGER,
chlorine_warning INTEGER,
chlorine_critical INTEGER,
PRIMARY KEY(rowid,sensortimestamp)
)
```

CHAPTER 9 BIG DATA VISUALIZATION AND DATA WRANGLING

```
PARTITION BY HASH (rowid) PARTITIONS 16,
RANGE (sensortimestamp)
(
PARTITION unix_timestamp('2017-01-01') <= VALUES <
unix_timestamp('2018-01-01'),

PARTITION unix_timestamp('2018-01-01') <= VALUES <
unix_timestamp('2019-01-01'),

PARTITION unix_timestamp('2019-01-01') <= VALUES <
unix_timestamp('2020-01-01')
)
STORED AS KUDU;
```

Test Data Source

We'll execute shell script to generate random data: generatetest.sh. I introduce a 1 second delay every 50 records so I don't overwhelm my test server (Listing 9-2).

Listing 9-2. generatetest.sh shell script to generate test data

```
#!/bin/bash

x=0
while true
do
   echo $RANDOM$RANDOM,$(( $RANDOM % 10 + 20 )),
   $(( ( RANDOM % 40 )  + 1 )),$(( ( RANDOM % 80 ) + 1 )),
   $(( ( RANDOM % 30 )  + 1 )),$(( ( RANDOM % 20 ) + 1 )),
   `date "+%Y-%m-%d %T"`,"Melton", 0,0,0,0,0,0,0,0,
   $(( $RANDOM % 10 + 1 )), $(( $RANDOM % 10 + 1 )),
   $(( $RANDOM % 10 + 1 )),$(( $RANDOM % 10 + 1 )),
   $(( $RANDOM % 10 + 1 )),0,0,0,0,0,0,0,0,0,0
        if [ "$x" = 50 ];
        then
                sleep 1
                x=0
        fi
```

```
((x++))
Done
```

The script will generate random sensor data. Listing 9-3 shows what the data looks like.

Listing 9-3. Sample test sensor data

```
891013984,23,8,4,24,11,2017-07-09 17:31:33,Melton,
0,0,0,0,0,0,0,0,2,8,3,1,4,0,0,0,0,0,0,0,0,0
1191723491,29,20,68,14,10,2017-07-09 17:31:33,Melton,
0,0,0,0,0,0,0,0,7,1,6,4,3,0,0,0,0,0,0,0,0,0
919749,24,25,67,12,10,2017-07-09 17:31:33,Melton,
0,0,0,0,0,0,0,0,4,6,10,9,4,0,0,0,0,0,0,0,0,0
2615810801,22,21,59,24,11,2017-07-09 17:31:33,Melton,
0,0,0,0,0,0,0,0,5,7,10,7,2,0,0,0,0,0,0,0,0,0
2409532223,25,6,45,21,3,2017-07-09 17:31:33,Melton,
0,0,0,0,0,0,0,0,4,1,9,4,4,0,0,0,0,0,0,0,0,0
2229524773,29,20,68,12,3,2017-07-09 17:31:33,Melton,
0,0,0,0,0,0,0,0,6,7,2,4,9,0,0,0,0,0,0,0,0,0
295358267,22,15,16,7,1,2017-07-09 17:31:33,Melton,
0,0,0,0,0,0,0,0,1,4,6,10,8,0,0,0,0,0,0,0,0,0
836218647,28,25,59,3,19,2017-07-09 17:31:33,Melton,
0,0,0,0,0,0,0,0,7,2,6,3,4,0,0,0,0,0,0,0,0,0
2379015092,24,23,23,10,14,2017-07-09 17:31:33,Melton,
0,0,0,0,0,0,0,0,2,1,10,8,7,0,0,0,0,0,0,0,0,0
189463852,20,2,10,30,16,2017-07-09 17:31:33,Melton,
0,0,0,0,0,0,0,0,8,4,7,8,7,0,0,0,0,0,0,0,0,0
1250719778,26,15,68,30,4,2017-07-09 17:31:33,Melton,
0,0,0,0,0,0,0,0,7,6,9,8,10,0,0,0,0,0,0,0,0,0
1380822028,27,32,40,11,7,2017-07-09 17:31:33,Melton,
0,0,0,0,0,0,0,0,2,6,7,6,5,0,0,0,0,0,0,0,0,0
2698312711,21,14,5,29,19,2017-07-09 17:31:33,Melton,
0,0,0,0,0,0,0,0,2,2,8,1,3,0,0,0,0,0,0,0,0,0
1300319275,23,33,52,24,4,2017-07-09 17:31:33,Melton,
0,0,0,0,0,0,0,0,2,7,1,1,3,0,0,0,0,0,0,0,0,0
2491313552,27,25,69,24,10,2017-07-09 17:31:33,Melton,
```

CHAPTER 9 BIG DATA VISUALIZATION AND DATA WRANGLING

```
0,0,0,0,0,0,0,0,8,2,4,3,8,0,0,0,0,0,0,0,0,0,0
149243062,21,24,2,15,8,2017-07-09 17:31:33,Melton,
0,0,0,0,0,0,0,0,7,3,3,5,7,0,0,0,0,0,0,0,0,0,0
```

Design the Pipeline

We can now design the StreamSets pipeline. Refer to Chapter 9 for a more in-depth discussion of StreamSets. The first thing we need to do is define an origin, or the data source. StreamSets supports different type of origins such as MQTT, JDBC, S3, Kafka, and Flume to name a few. For this example, we'll use a "File Tail" origin. Set data source to sensordata.csv. We'll run a script to populate this file later (Figure 9-26).

Figure 9-26. *StreamSets file tail origin*

The File Tail data source needs a second destination for its metadata. Let's add a Local FS destination to store metadata (Figure 9-27).

CHAPTER 9 BIG DATA VISUALIZATION AND DATA WRANGLING

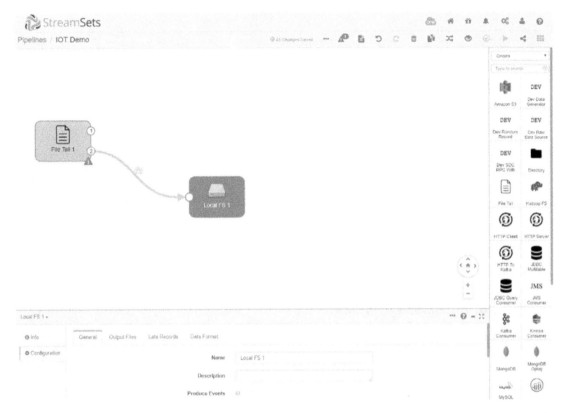

Figure 9-27. *StreamSets local file system*

We'll need a field splitter processor to split our CSV data into individual columns (Figure 9-28). We'll specify our SDC fields in the New Split Fields configuration box (Listing 9-4).

431

CHAPTER 9   BIG DATA VISUALIZATION AND DATA WRANGLING

Figure 9-28. *StreamSets field splitter*

Listing 9-4. New SDC Fields

```
[
 "/rowid",
 "/deviceid",
 "/temperature",
 "/pressure",
 "/humidity",
 "/ozone",
 "/sensortimestamp_str",
 "/city",
 "/temperature_warning",
 "/temperature_critical",
 "/pressure_warning",
 "/pressure_critical",
 "/humidity_warning",
 "/humidity_critical",
 "/ozone_warning",
 "/ozone_critical",
 "/lead_level",
 "/copper_level",
```

```
"/phosphates_level",
"/methane_level",
"/chlorine_level",
"/lead_warning",
"/lead_critical",
"/copper_warning",
"/copper_critical",
"/phosphates_warning",
"/phosphates_critical",
"/methane_warning",
"/methane_critical",
"/chlorine_warning",
"/chlorine_critical"
]
```

We'll also need a field type converter to convert the data types of the fields generated by the field splitter processor (Figure 9-29).

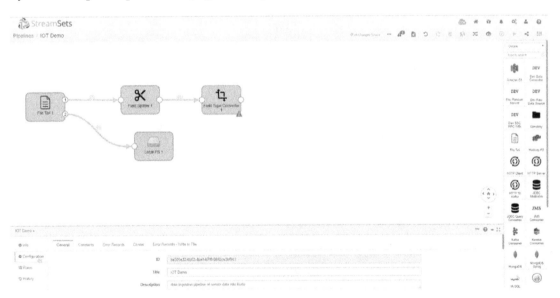

Figure 9-29. *StreamSets field converter*

When configuring the field type converter processor, we have to specify the fields that we need to convert and the data type to convert the data into. All fields are set to STRING by default by SDC. You need to convert the fields to their correct data types (Figure 9-30).

CHAPTER 9   BIG DATA VISUALIZATION AND DATA WRANGLING

Figure 9-30. StreamSets field converter

Kudu expects the timestamp to be in Unix time format. We'll use a JavaScript Evaluator (Figure 9-31) and write code to convert the timestamp from STRING to Unix time format as BIGINT^{tx} (Listing 9-5). Note that this limitation has been addressed in recent versions of Kudu. However, Zoomdata still expects the date to be in Unix time format. Consult Chapter 7 to learn more about StreamSets evaluators.

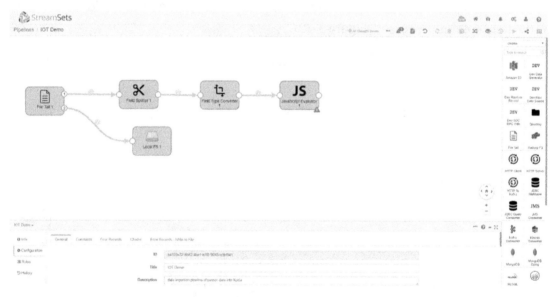

Figure 9-31. StreamSets Javascript Evaluator

CHAPTER 9 BIG DATA VISUALIZATION AND DATA WRANGLING

Listing 9-5. Javascript Evaluator code

```
Date.prototype.getUnixTime = function() { return this.getTime()/1000|0 };
if(!Date.now) Date.now = function() { return new Date(); }
Date.time = function() { return Date.now().getUnixTime(); }

for(var i = 0; i < records.length; i++) {
  try {

    var someDate = new Date(records[i].value['sensortimestamp_str']);
    var theUnixTime = someDate.getUnixTime();

    records[i].value['sensortimestamp'] = theUnixTime;

    output.write(records[i]);

  } catch (e) {
    error.write(records[i], e);
  }
```

Finally, we use a Kudu destination as the final destination of our data. Make sure you map the SDC fields to the correct Kudu column names (Figure 9-32).

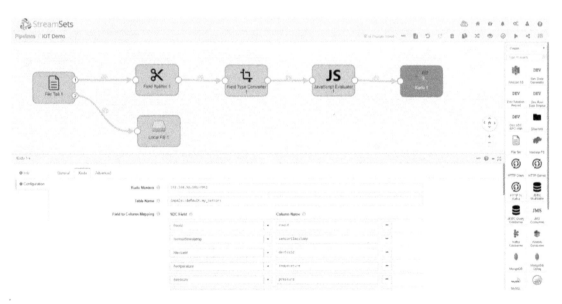

Figure 9-32. StreamSets Kudu destination

435

CHAPTER 9 BIG DATA VISUALIZATION AND DATA WRANGLING

That's it on the StreamSets end.

Configure Zoomdata

Let's configure a data source for Zoomdata. We'll use the Kudu Impala connector (Figure 9-33).

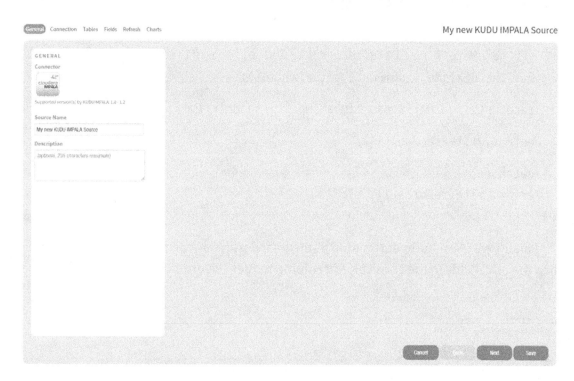

Figure 9-33. *Zoomdata data source*

CHAPTER 9 BIG DATA VISUALIZATION AND DATA WRANGLING

Enter the connection information such as the IP address and port number for the Kudu Masters, the JDBC URL, and so on. Make sure to validate your connection (Figure 9-34).

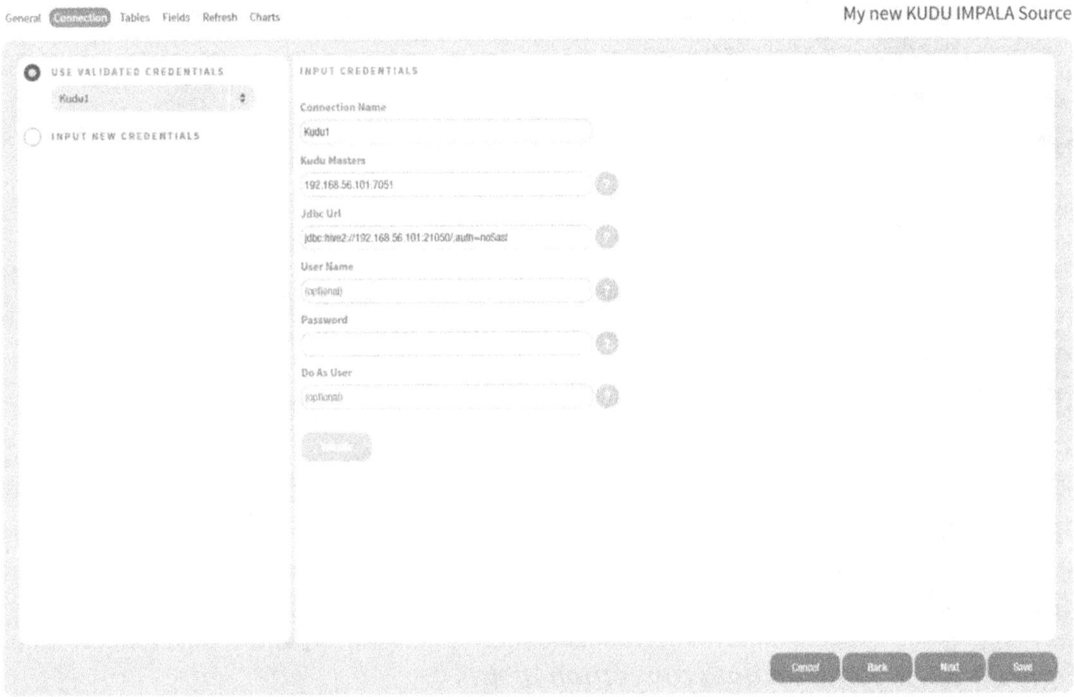

Figure 9-34. Zoomdata data connection input credentials

CHAPTER 9   BIG DATA VISUALIZATION AND DATA WRANGLING

Choose the table that you just created earlier. You can exclude fields if you wish (Figure 9-35 and Figure 9-36).

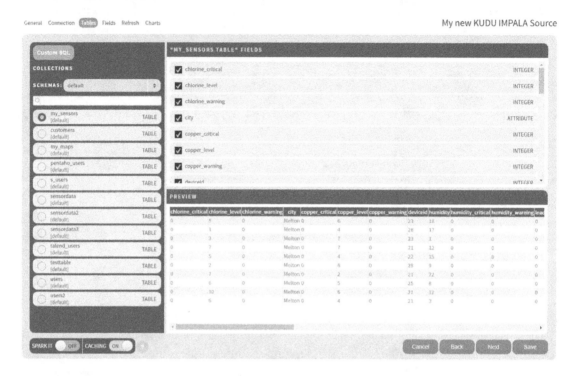

Figure 9-35. *Zoomdata data connection tables*

CHAPTER 9   BIG DATA VISUALIZATION AND DATA WRANGLING

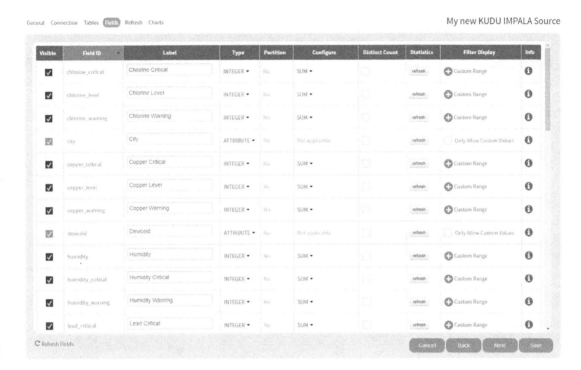

Figure 9-36. *Zoomdata data connection fields*

You can schedule data caching if you wish. If you have a large data set, caching can help improve performance and scalability (Figure 9-37).

CHAPTER 9   BIG DATA VISUALIZATION AND DATA WRANGLING

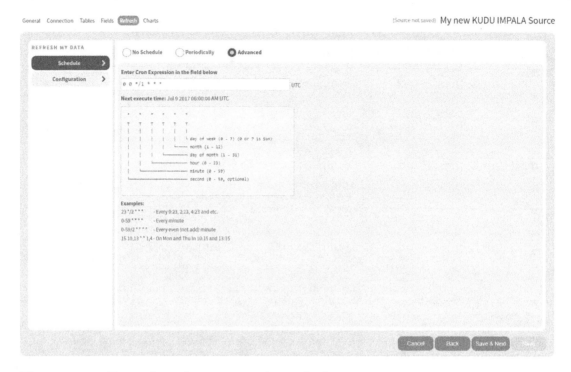

Figure 9-37.* Zoomdata data connection refresh*

Configure the time bar. Make sure to enable live mode and playback (Figure 9-38).

CHAPTER 9 BIG DATA VISUALIZATION AND DATA WRANGLING

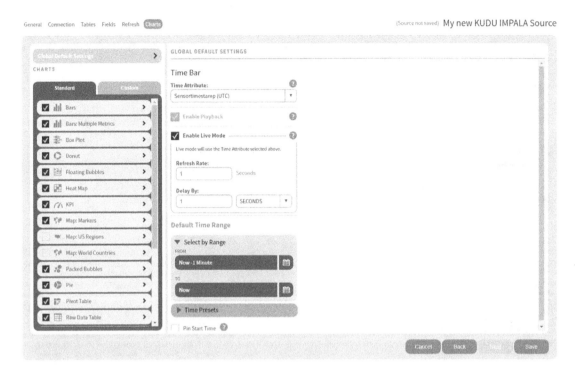

Figure 9-38. *Zoomdata time bar*

You can also set default configuration options for charts (Figure 9-39).

441

CHAPTER 9 BIG DATA VISUALIZATION AND DATA WRANGLING

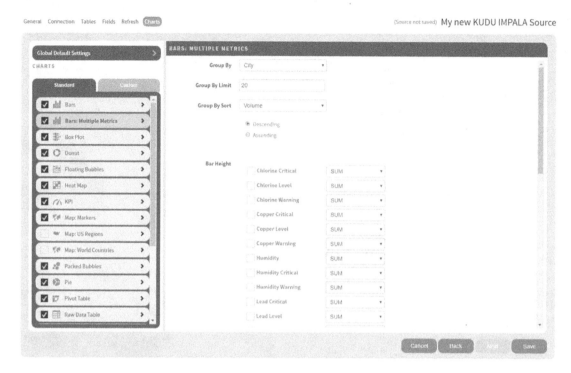

Figure 9-39. Zoomdata time bar

Save the data source. You can now start designing the user interface (Figure 9-40).

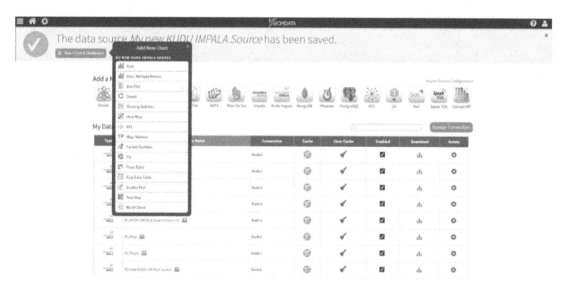

Figure 9-40. Zoomdata add new chart

442

CHAPTER 9   BIG DATA VISUALIZATION AND DATA WRANGLING

I choose Bars: Multiple Metrics as my first chart. You won't see any data yet because we haven't started the StreamSets pipeline (Figure 9-41).

Figure 9-41. *An empty chart*

Let's start generating some data. On the server where we installed StreamSets, open a terminal window and run the script. Redirect the output to the File tail destination: ./generatetest.csv >> sensordata.csv.

Start the StreamSets pipeline. In a second, you'll see data coming in StreamSets via the different statistics about your pipeline such as Record Throughput, Batch Throughput, the Input and Output Record Count for each step. It will also show you if there are any errors in your pipeline (Figure 9-42).

CHAPTER 9   BIG DATA VISUALIZATION AND DATA WRANGLING

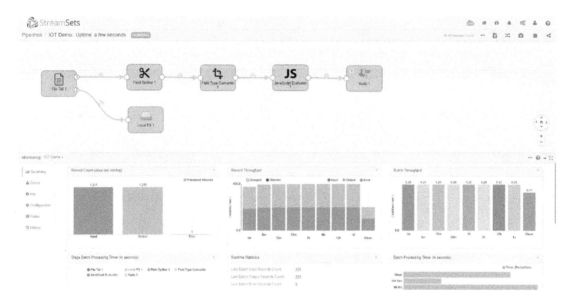

Figure 9-42. *StreamSets canvas*

As you are designing your Zoomdata dashboard, charts are immediately populated (Figure 9-43).

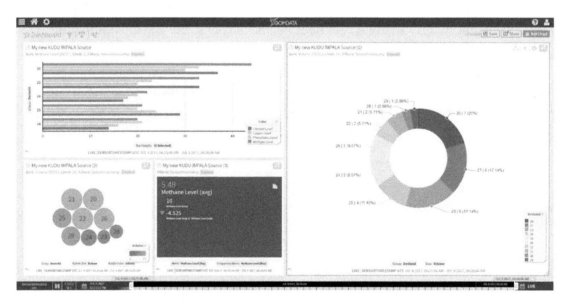

Figure 9-43. *Live Zoomdata dashboard*

Congratulations! You've just implemented a real-time IoT pipeline using StreamSets, Kudu, and ZoomData.

Data Wrangling

In most cases, data will not come in the format suitable for data analysis. Some data transformation may be needed to convert a field's data type; data cleansing may be required to handle missing or incorrect values; or data needs to be joined with other data sources such as CRM, weather data, or web logs. A survey conducted says that 80% of what data scientists do is data preparation and cleansing, and the remaining 20% is spent on the actual data analysis.[x] In traditional data warehouse and business intelligence environments, data is made fit for analysis through an extract, transform, and load (ETL) process. During off-peak hours, data is cleansed, transformed, and copied from online transaction processing (OLTP) sources to the enterprise data warehouse. Various reports, key performance indicators (KPIs), and data visualization are updated with new data and presented in corporate dashboards, ready to be consumed by users. OLAP cubes and pivot tables provide some level of interactivity. In some cases, power users are granted SQL access to a reporting database or a replicated copy of the data warehouse where they can run ad hoc SQL queries. But in general, traditional data warehousing and business intelligence are fairly static.

The advent of big data has rendered old-style ETL inadequate. Weeks or months of upfront data modeling and ETL design to accommodate new data sets is considered rigid and old-fashioned. The enterprise data warehouse, once considered an organization's central data repository, now shares that title with the enterprise "data hub" or "data lake," as more and more workloads are being moved to big data platforms. In cases where appropriate, ETL has been replaced by ELT. Extract, load, and transform (ELT) allows users to dump data with little or no transformation unto a big data platform. It's then up to the data analyst to make the data suitable for analysis based on his or her needs. ELT is in, ETL is out. EDW is rigid and structured. Big data is fast and agile.

At the same time, users have become savvier and more demanding. Massive amounts of data is available, and waiting weeks or months for a team of ETL developers to design and implement a data ingestion pipeline to make these data available for consumption is no longer an option. A new set of activities, collectively known as data wrangling, has emerged.

Data wrangling is more than just preparing data for analysis. Data wrangling is a part of the data analysis process itself. By performing data wrangling, iteratively discovering, structuring, cleaning, enriching, and validating, users get a better understanding of their data and enable them to ask better questions. As users iteratively wrangle through their data, it exposes hidden patterns, revealing new ways to analyze your data. Data wrangling is the perfect complement for big data and ELT. The pioneers of data wrangling at Stanford and Berkeley, who went on to start Trifacta, came up with six core activities involved in data wrangling (Table 9-1).[xi]

Table 9-1. Six Core Data Wrangling Activities

Activity	Description
Discovering	This is the data profiling and assessment stage.
Structuring	Set of activities to determine the structure of your data. Creating schemas, pivoting rows, adding or removing columns, etc.
Cleaning	Set of activities that includes fixing invalid or missing values, standardizing the format of specific fields (i.e., date, phone, and state), removing extra characters, etc.
Enriching	Set of activities that involves joining or blending your data with other data sources to improve the results of your data analysis.
Validating	Set of activities to check if the enrichment and data cleansing performed actually achieved its goal.
Publishing	Once you're happy with the output of your data wrangling, it's time to publish the results. The results could be input data to a data visualization tool such as Tableau or as input for a marketing campaign initiative.

Note that the data wrangling activities are iterative in nature. You will usually perform the steps multiple times and in different orders to achieve the desired results.

If you've ever used Microsoft Excel to make numeric values look like a certain format or removed certain characters using regular expressions, you've performed a simple form of data wrangling. Microsoft Excel is actually a pretty good tool for data wrangling, even though it wasn't strictly designed for such tasks. However, Microsoft Excel falls short when used on large data sets or when data is stored in a big data platform or a relational database. A new breed of interactive data wrangling tools was developed to fill the market niche. These data wrangling tools take interactive data processing to

CHAPTER 9   BIG DATA VISUALIZATION AND DATA WRANGLING

the next level, offering features that make it easy to perform tasks that would normally require coding expertise. Most of these tools automate transformation and offer relevant suggestions based on the data set provided. They have deep integration with big data platforms such as Cloudera Enterprise and can connect via Impala, Hive, Spark HDFS, or HBase. I'll show examples on how to use some of the most popular data wrangling tools.

Trifacta

Trifacta is one of the pioneers of data wrangling. Developed by Stanford PhD Sean Kandel, UC Berkeley professor Joe Hellerstein, and University of Washington and former Stanford professor Jeffrey Heer. They started a joint research project called the Stanford/Berkeley Wrangler, which eventually became Trifacta.[xii] Trifacta sells two versions of their product, Trifacta Wrangler and Trifacta Wrangler Enterprise. Trifacta Wrangler is a free desktop application aimed for individual use with small data sets. Trifacta Wrangler Enterprise is designed for teams with centralized management of security and data governance.[xiii] As mentioned earlier, Trifacta can connect to Impala to provide high-performance querying capabilities.[xiv] Please consult Trifacta's website on how to install Trifacta Wrangler.

I'll show you some of the features of Trifacta Wrangler. Trifacta organizes data wrangling activities based on "flows." Trifacta Wrangler will ask you to create a new flow when you first start the application.

The next thing you need to do is to upload a data set. Trifacta can connect to different types of data sources, but for now we'll upload a sample CSV file containing customer information. After uploading the CSV file, you can now explore the data. Click the button next to the "Details" button and select "Wrangle in new Flow" (Figure 9-44).

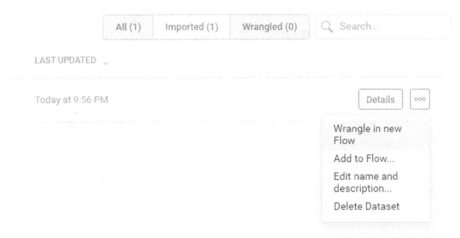

Figure 9-44. *Wrangle in new Flow*

447

CHAPTER 9   BIG DATA VISUALIZATION AND DATA WRANGLING

You will be shown the transformer page that looks like Figure 9-45. Trifacta immediately detects and shows how data in individual columns are distributed.

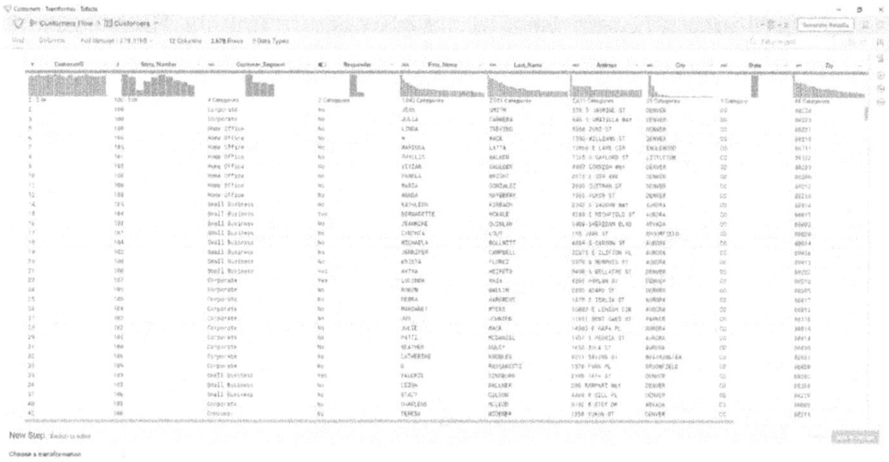

Figure 9-45.   Trifacta's transformer page

You can scroll to the left or right of the chart that shows the distribution of data to quickly view the contents of the column. This will help you determine the data that need to be processed or transformed (Figure 9-46).

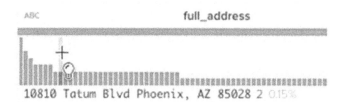

Figure 9-46.   Histogram showing data distribution

Under the column name, you'll see a data quality indicator for the specific column. Mismatched data types will be shown as red, and empty rows will be shown as gray. The indicator should be green if the data stored in that column are all valid (Figure 9-47).

448

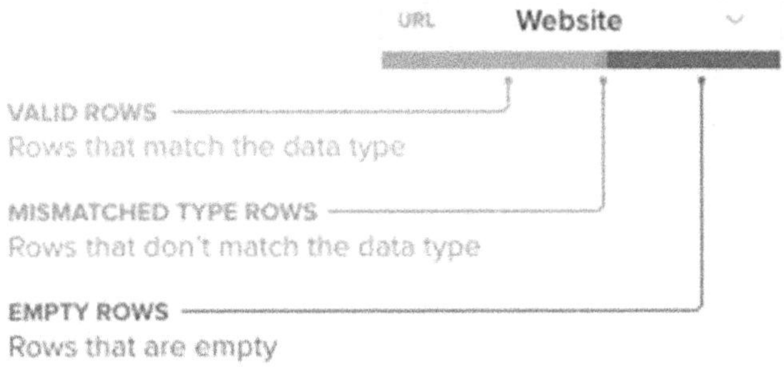

CHAPTER 9 BIG DATA VISUALIZATION AND DATA WRANGLING

Figure 9-47. Data quality bar

By clicking a specific column, you will be presented with various suggestions on how to transform data stored on that column. This feature is powered by machine learning. Trifacta learns from your past data wrangling activities to make the next suggestions (Figure 9-48).

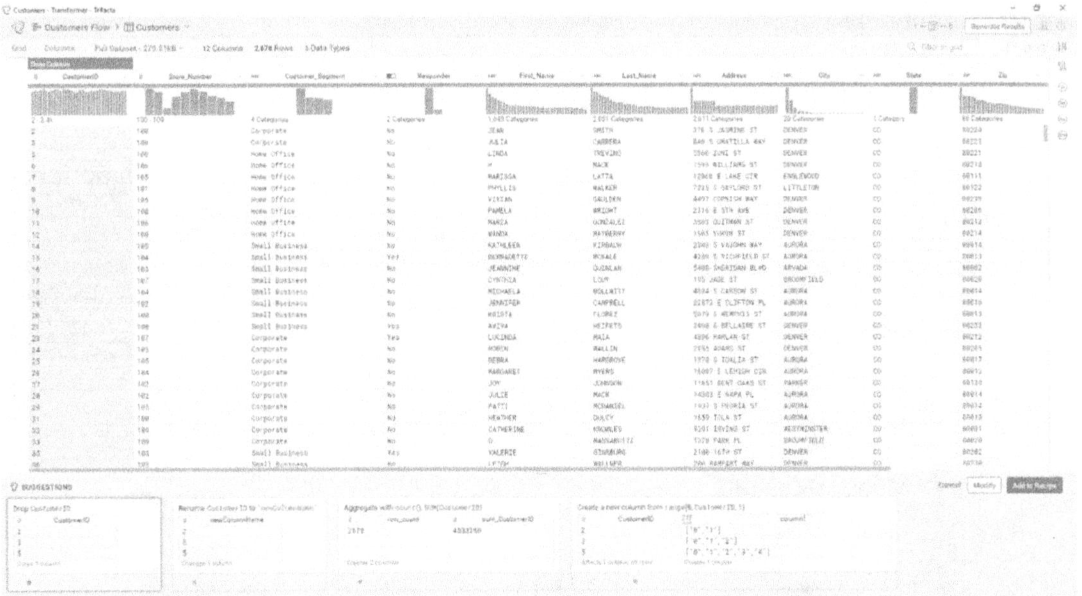

Figure 9-48. Trifacta's suggestions

CHAPTER 9 BIG DATA VISUALIZATION AND DATA WRANGLING

Different columns get different suggestions (Figure 9-49).

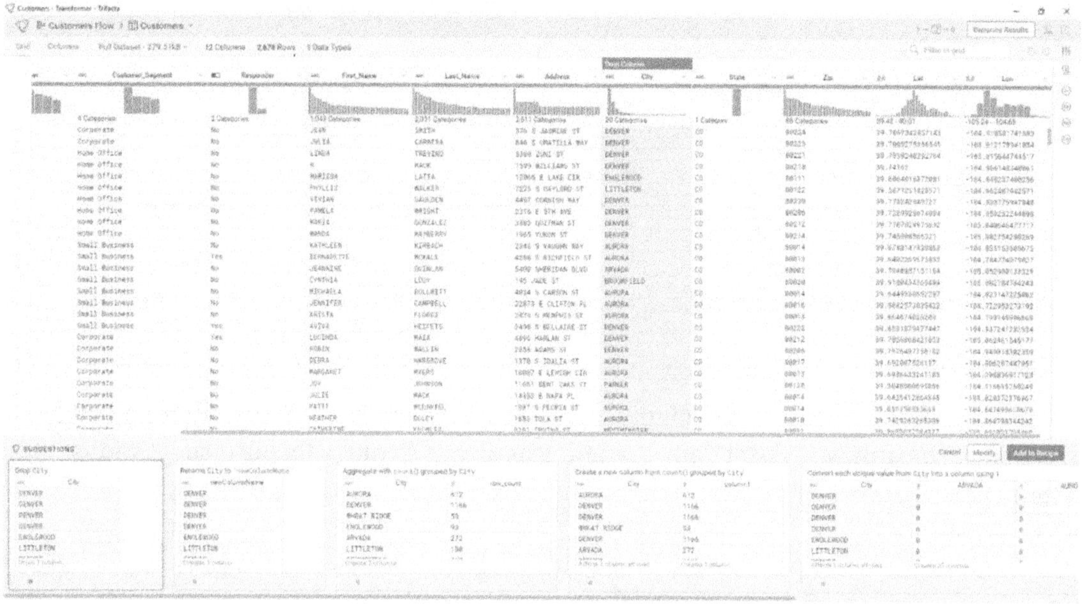

Figure 9-49. Trifacta has different suggestions for different types of data

Click the button next to the column name to get a list of data transformation options for the particular column. You can rename the column name, change data types, filter, clean, or perform aggregation on the column to name a few (Figure 9-50).

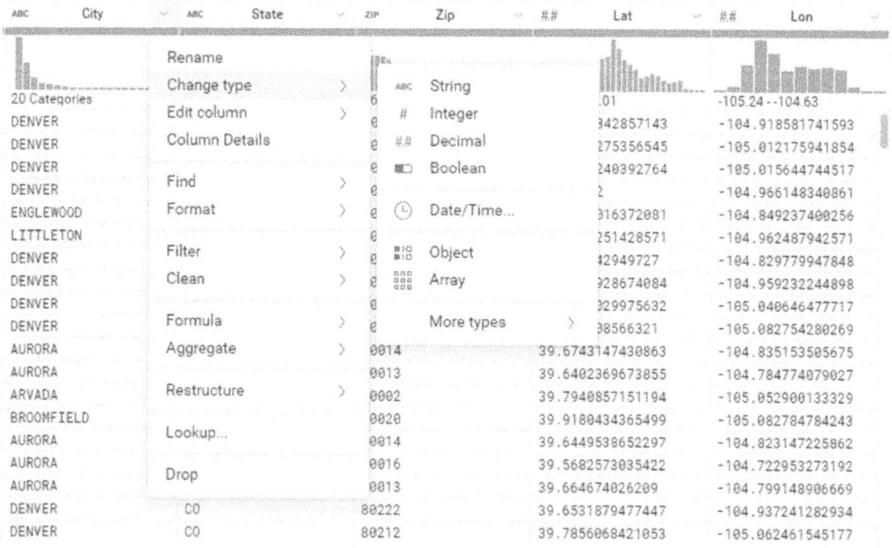

Figure 9-50. Additional transformation options

450

CHAPTER 9 BIG DATA VISUALIZATION AND DATA WRANGLING

Almost every type of data transformation is available (Figure 9-51).

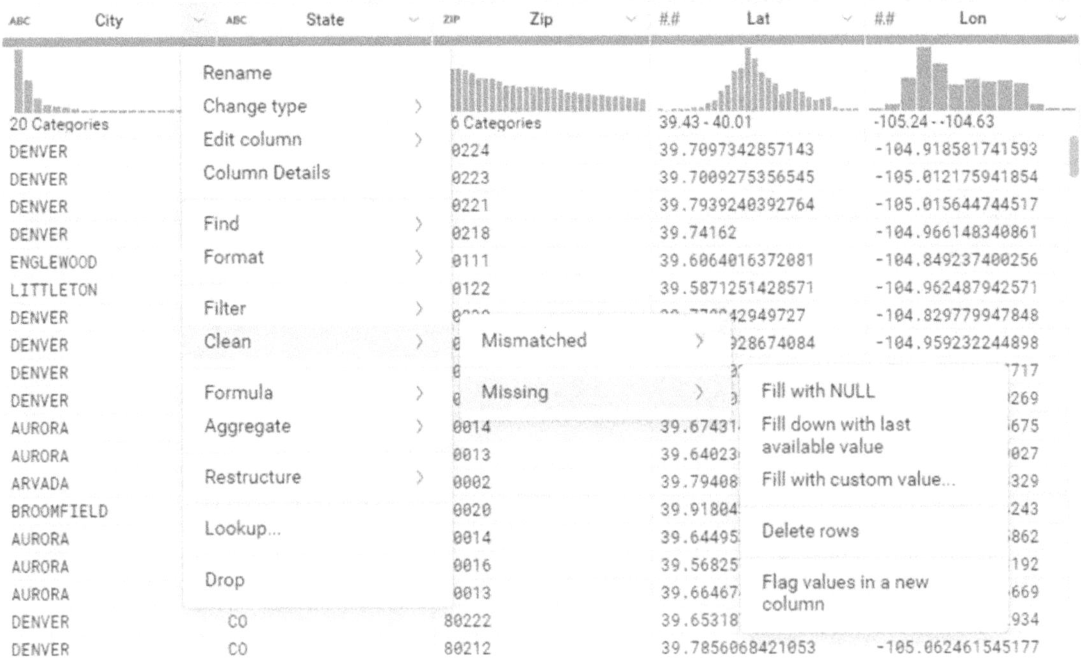

Figure 9-51. Additional transformation options

Double-clicking the column will show you another window with various statistics and useful information about the information stored in that column. Information includes top values, the most frequent values, and string length statistics to name a few (Figure 9-52).

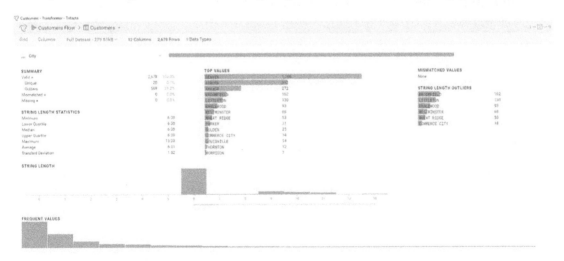

Figure 9-52. The final output from the Job Results window

451

CHAPTER 9 BIG DATA VISUALIZATION AND DATA WRANGLING

Trifacta is smart enough to know if the column includes location information such as zip code. It will try to show the information using a map (Figure 9-53).

Figure 9-53. *The final output from the Job Results window*

Let's try one of the suggested data transformations. Let's convert the last name from all uppercase to a proper last name (only the first letter is uppercase). Choose the last suggestion and click the "Modify" button to apply the changes (Figure 9-54).

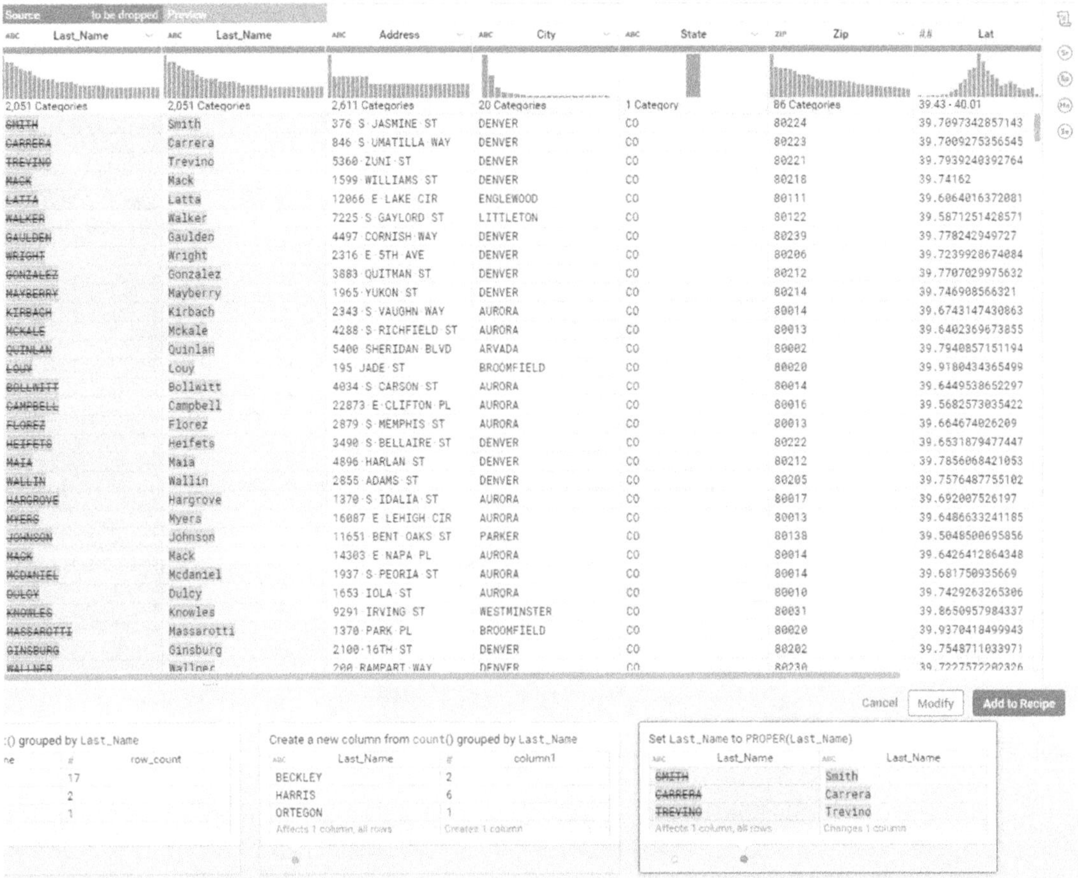

CHAPTER 9 BIG DATA VISUALIZATION AND DATA WRANGLING

Last_Name	Last_Name	Address	City	State	Zip	Zip	Lat
2,051 Categories	2,051 Categories	2,611 Categories	20 Categories	1 Category		86 Categories	39.43 - 40.01
SMITH	Smith	376 S JASMINE ST	DENVER	CO		80224	39.7097342857143
CARRERA	Carrera	846 S UMATILLA WAY	DENVER	CO		80223	39.7009275356545
TREVINO	Trevino	5360 ZUNI ST	DENVER	CO		80221	39.7939240392764
MACK	Mack	1599 WILLIAMS ST	DENVER	CO		80218	39.74162
LATTA	Latta	12066 E LAKE CIR	ENGLEWOOD	CO		80111	39.6064016372081
WALKER	Walker	7225 S GAYLORD ST	LITTLETON	CO		80122	39.5871251428571
GAULDEN	Gaulden	4497 CORNISH WAY	DENVER	CO		80239	39.778242949727
WRIGHT	Wright	2316 E 5TH AVE	DENVER	CO		80206	39.7239928674084
GONZALEZ	Gonzalez	3883 QUITMAN ST	DENVER	CO		80212	39.7707029975632
MAYBERRY	Mayberry	1965 YUKON ST	DENVER	CO		80214	39.746908566321
KIRBACH	Kirbach	2343 S VAUGHN WAY	AURORA	CO		80014	39.6743147430863
MCKALE	Mckale	4288 S RICHFIELD ST	AURORA	CO		80013	39.6402369673855
QUINLAN	Quinlan	5400 SHERIDAN BLVD	ARVADA	CO		80002	39.7940857151194
LOUY	Louy	195 JADE ST	BROOMFIELD	CO		80020	39.9180433365499
BOLLWITT	Bollwitt	4034 S CARSON ST	AURORA	CO		80014	39.6449538652297
CAMPBELL	Campbell	22873 E CLIFTON PL	AURORA	CO		80016	39.5682573035422
FLOREZ	Florez	2879 S MEMPHIS ST	AURORA	CO		80013	39.664674026209
HEIFETS	Heifets	3490 S BELLAIRE ST	DENVER	CO		80222	39.6531879477447
MAIA	Maia	4896 HARLAN ST	DENVER	CO		80212	39.7856068421053
WALLIN	Wallin	2855 ADAMS ST	DENVER	CO		80205	39.7576487755102
HARGROVE	Hargrove	1370 S IDALIA ST	AURORA	CO		80017	39.692007526197
MYERS	Myers	16087 E LEHIGH CIR	AURORA	CO		80013	39.6486633241185
JOHNSON	Johnson	11651 BENT OAKS ST	PARKER	CO		80138	39.5048500695856
MACK	Mack	14303 E NAPA PL	AURORA	CO		80014	39.6426412864348
MCDANIEL	Mcdaniel	1937 S PEORIA ST	AURORA	CO		80014	39.681750935669
DULCY	Dulcy	1653 IOLA ST	AURORA	CO		80010	39.7429263265306
KNOWLES	Knowles	9291 IRVING ST	WESTMINSTER	CO		80031	39.8650957984337
MASSAROTTI	Massarotti	1370 PARK PL	BROOMFIELD	CO		80020	39.9370418499943
GINSBURG	Ginsburg	2100 16TH ST	DENVER	CO		80202	39.7548711033971
WALLNER	Wallner	200 RAMPART WAY	DENVER	CO		80230	39.722757220326

Figure 9-54. Apply the transformation

The last name has been transformed as shown in Figure 9-55.

453

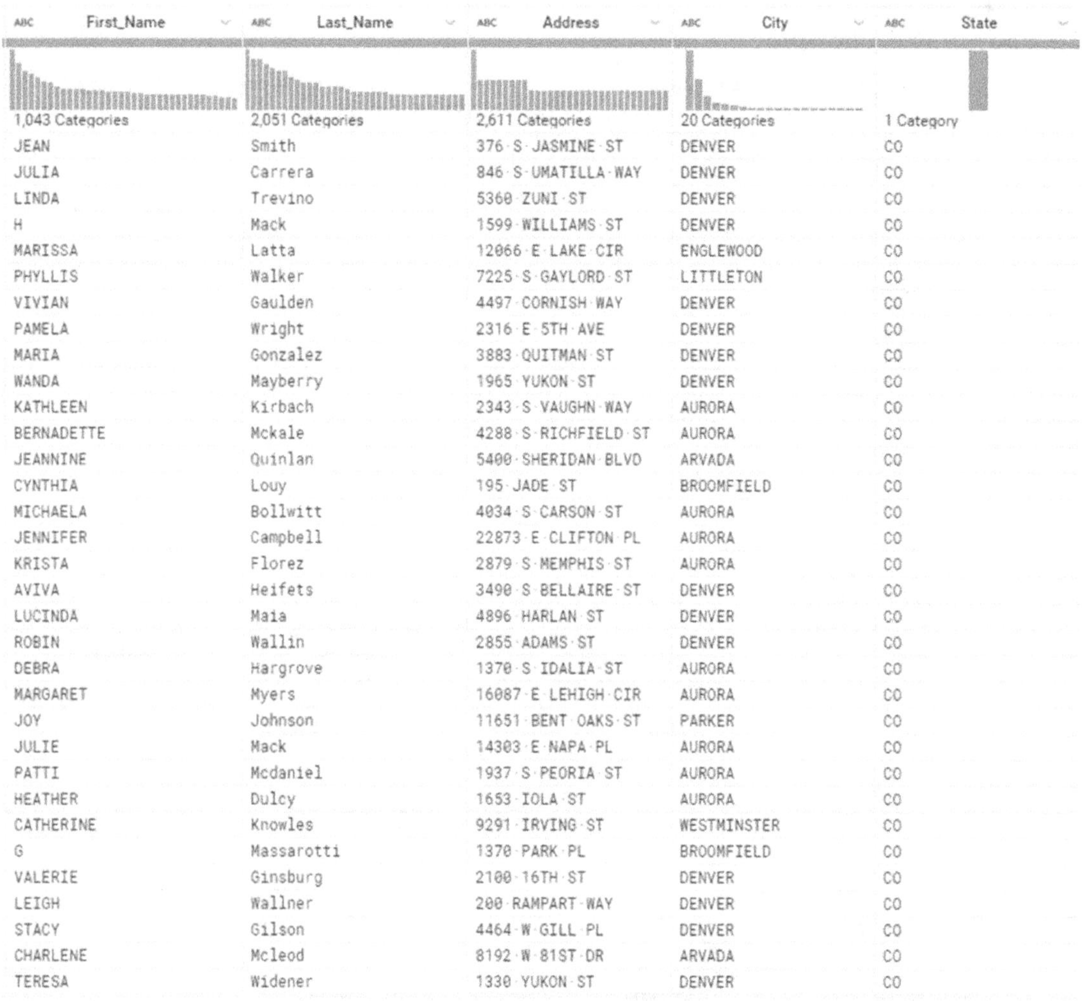

CHAPTER 9 BIG DATA VISUALIZATION AND DATA WRANGLING

First_Name	Last_Name	Address	City	State
1,043 Categories	2,051 Categories	2,611 Categories	20 Categories	1 Category
JEAN	Smith	376 S JASMINE ST	DENVER	CO
JULIA	Carrera	846 S UMATILLA WAY	DENVER	CO
LINDA	Trevino	5360 ZUNI ST	DENVER	CO
H	Mack	1599 WILLIAMS ST	DENVER	CO
MARISSA	Latta	12066 E LAKE CIR	ENGLEWOOD	CO
PHYLLIS	Walker	7225 S GAYLORD ST	LITTLETON	CO
VIVIAN	Gaulden	4497 CORNISH WAY	DENVER	CO
PAMELA	Wright	2316 E 5TH AVE	DENVER	CO
MARIA	Gonzalez	3883 QUITMAN ST	DENVER	CO
WANDA	Mayberry	1965 YUKON ST	DENVER	CO
KATHLEEN	Kirbach	2343 S VAUGHN WAY	AURORA	CO
BERNADETTE	Mckale	4288 S RICHFIELD ST	AURORA	CO
JEANNINE	Quinlan	5400 SHERIDAN BLVD	ARVADA	CO
CYNTHIA	Louy	195 JADE ST	BROOMFIELD	CO
MICHAELA	Bollwitt	4034 S CARSON ST	AURORA	CO
JENNIFER	Campbell	22873 E CLIFTON PL	AURORA	CO
KRISTA	Florez	2879 S MEMPHIS ST	AURORA	CO
AVIVA	Heifets	3490 S BELLAIRE ST	DENVER	CO
LUCINDA	Maia	4896 HARLAN ST	DENVER	CO
ROBIN	Wallin	2855 ADAMS ST	DENVER	CO
DEBRA	Hargrove	1370 S IDALIA ST	AURORA	CO
MARGARET	Myers	16087 E LEHIGH CIR	AURORA	CO
JOY	Johnson	11651 BENT OAKS ST	PARKER	CO
JULIE	Mack	14303 E NAPA PL	AURORA	CO
PATTI	Mcdaniel	1937 S PEORIA ST	AURORA	CO
HEATHER	Dulcy	1653 IOLA ST	AURORA	CO
CATHERINE	Knowles	9291 IRVING ST	WESTMINSTER	CO
G	Massarotti	1370 PARK PL	BROOMFIELD	CO
VALERIE	Ginsburg	2100 16TH ST	DENVER	CO
LEIGH	Wallner	200 RAMPART WAY	DENVER	CO
STACY	Gilson	4464 W GILL PL	DENVER	CO
CHARLENE	Mcleod	8192 W 81ST DR	ARVADA	CO
TERESA	Widener	1330 YUKON ST	DENVER	CO

Figure 9-55. *The transformation has been applied*

Click the "Generate Results" button near the upper-right corner of the page (Figure 9-56).

Figure 9-56. *Generate Results*

CHAPTER 9 BIG DATA VISUALIZATION AND DATA WRANGLING

You can save the result in a different format. Choose CSV for our data set. You can optionally compress the results if you want. Make sure the "Profile Results" option is checked to generate a profile of your results (Figure 9-57).

Figure 9-57. *Result Summary*

A "Result Summary" window will appear. It will include various statistics and information about your data. Review it to make sure the results are what you expect. Trifacta doesn't have built-in data visualization. It depends on other tools such as Tableau, Qlik, Power BI, or even Microsoft Excel to provide visualization.

Alteryx

Alteryx is another popular software development company that develops data wrangling and analytics software. The company was founded in 2010 and is based in Irvine, California. It supports several data sources such as Oracle, SQL Server, Hive, and Impala[xv] for fast ad hoc query capability.

You'll be presented with a Getting Started window the first time you start Alteryx (Figure 9-58). You can go through the tutorials, open recent workflows, or create a new workflow. For now let's create a new workflow.

CHAPTER 9   BIG DATA VISUALIZATION AND DATA WRANGLING

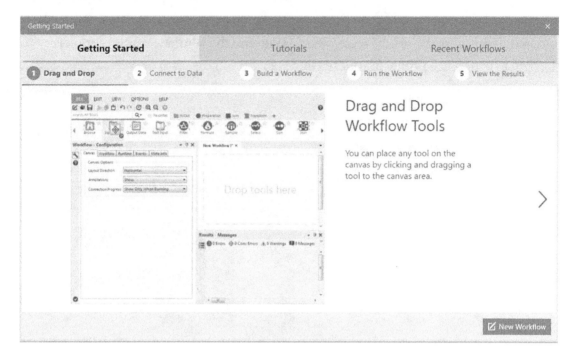

Figure 9-58. *Getting started window*

Alteryx is known for its ease of use. Most of the user's interactions with Alteryx will involve dropping, dragging, and configuring tools from the Tool Palette (Figure 9-59).

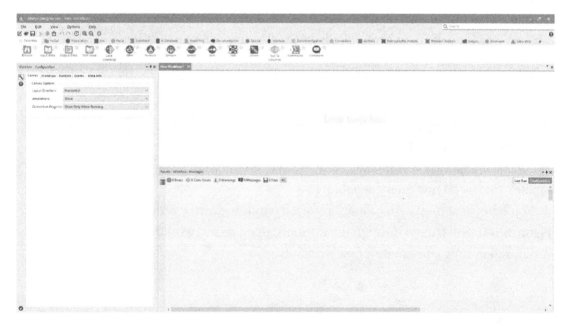

Figure 9-59. *Alteryx main window*

CHAPTER 9 BIG DATA VISUALIZATION AND DATA WRANGLING

The Tool Palette provides users with tools organized into tool categories. You drag these tools into the canvas and connect them to design your workflows. The tools let users perform different operations that users would normally perform to process and analyze data. Popular operations include select, filter, sort, join, union, summarize, and browse to name a few. These are tasks that users would normally perform by developing SQL queries. Alteryx provides a much easier and more productive method (Figure 9-60).

Figure 9-60. Tool Palette – Favorites

Alteryx also provides a myriad of tools for more complicated operations such as correlation, transposition, demographic analysis, behavior analysis, and spatial match to name a few. Explore the other tabs to familiarize yourself with the available tools (Figure 9-61).

Figure 9-61. Tool Palette – Preparation

Every workflow starts by specifying a data source using the Input Data tool. Drag and drop the Input Data tool to the workflow window. You can specify a file or connect to an external data source such as Impala, Oracle, SQL Server, or any data source that provides an ODBC interface (Figure 9-62).

CHAPTER 9 BIG DATA VISUALIZATION AND DATA WRANGLING

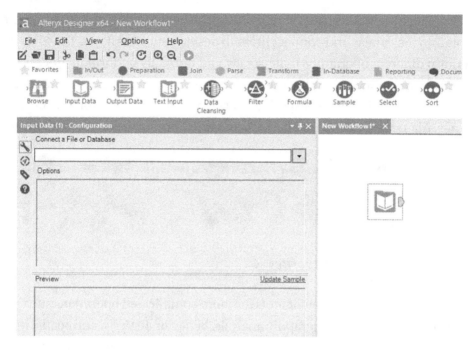

Figure 9-62. *Input Data Tool*

We'll get our data from a file in this example (Figure 9-63).

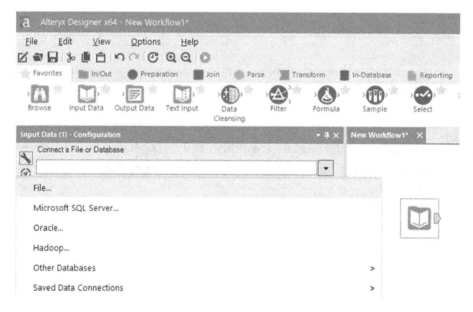

Figure 9-63. *Select file*

CHAPTER 9   BIG DATA VISUALIZATION AND DATA WRANGLING

Alteryx comes with several sample files. Let's use the Customers.csv file (Figure 9-64).

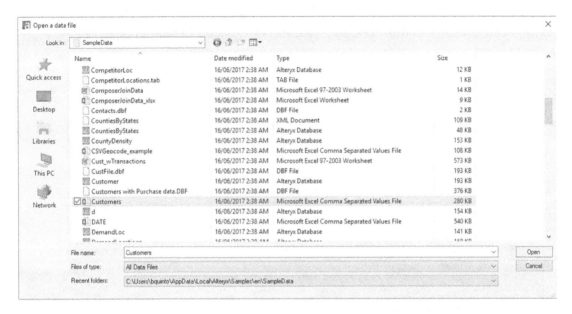

Figure 9-64. Select Customers.csv

Now that you've configured a data source, let's drag and drop the Select tool. The Select tool lets you select specific columns. The columns are all selected by default, so let's deselect some columns for this example. Make sure to connect the Input Data and Select tools as shown in Figure 9-65.

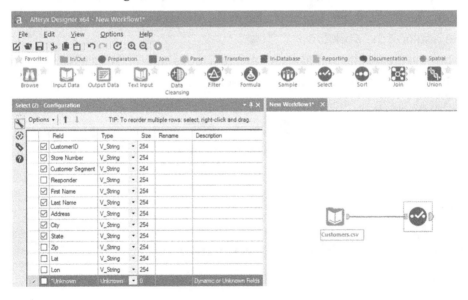

Figure 9-65. Select tool

459

CHAPTER 9 BIG DATA VISUALIZATION AND DATA WRANGLING

After we've selected a few columns, we can sort the results with the Sort tool. Connect the Select tool with the Sort tool. The Sort tool lets you specify how you would like to sort the results. In this example, we'll sort by City and Customer Segment in ascending order. You can do that via the configuration option (Figure 9-66).

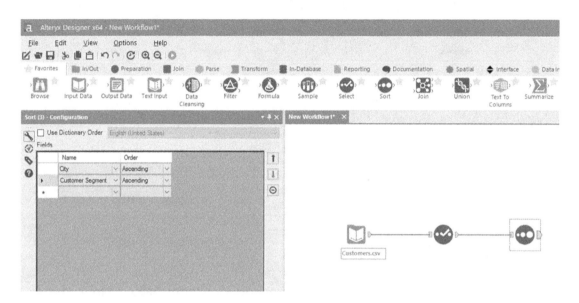

Figure 9-66. *Sort Tool*

We could go on by adding more tools and make the workflow as complex as we want, but we'll stop here for this example. Almost every Alteryx workflow ends by browsing or saving the results of the workflow.

Let's browse the results to make sure the data looks correct. Drag a Browse tool on the workflow window and make sure it connects with the Sort tool (Figure 9-67).

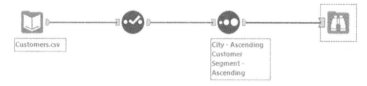

Figure 9-67. *Browse data tool*

CHAPTER 9 BIG DATA VISUALIZATION AND DATA WRANGLING

Run the workflow by clicking the green run button located in the upper corner of the window, near the Help menu item. After a few seconds, the results of your workflow will be shown in the output window. Inspect the results and see if the correct columns were selected and the sort order is as you specified earlier.

Like Trifacta, Alteryx also profiles your data and shows helpful statistics and information about it such as the distribution of data in the column, data quality information, data type of column, number of NULLs or blank values, average length, longest length, etc. In Figure 9-68, it shows statistics about the CustomerID field.

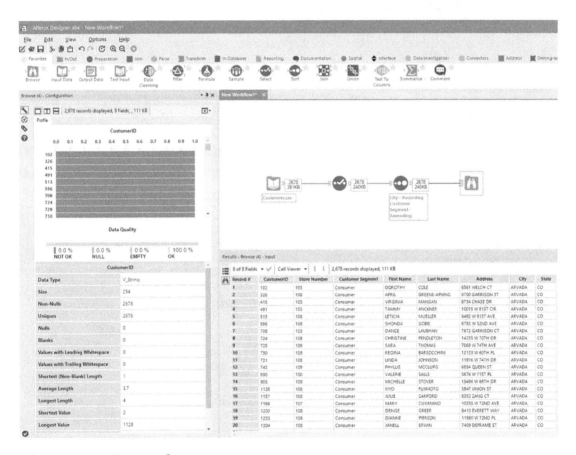

Figure 9-68. Browse data

461

CHAPTER 9 BIG DATA VISUALIZATION AND DATA WRANGLING

Click on the Customer Segment field to show statistics about the column (Figure 9-69).

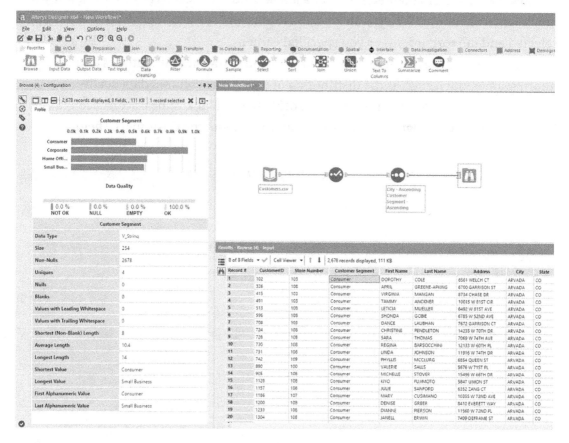

Figure 9-69. Data quality – Customer Segment field

CHAPTER 9 BIG DATA VISUALIZATION AND DATA WRANGLING

Figure 9-70 shows statistics about the City field.

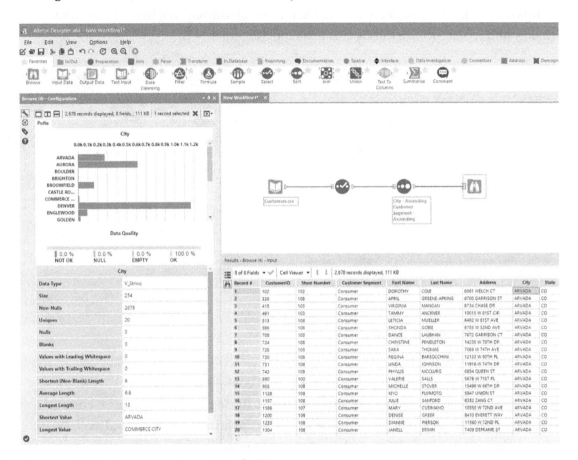

Figure 9-70. Data quality – City field

CHAPTER 9   BIG DATA VISUALIZATION AND DATA WRANGLING

As mentioned earlier, we can also save the results to a file or an external data source such as Kudu, Oracle, SQL Server, or HDFS. Delete the Browse tool from the workflow window and replace it with the Output Data tool (Figure 9-71).

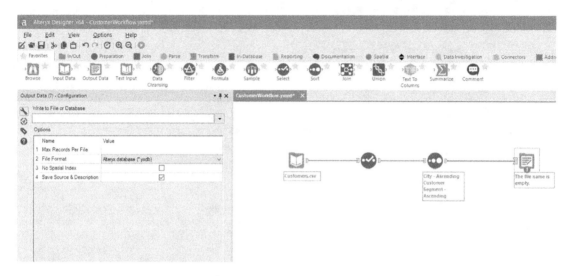

Figure 9-71. Output Data tool

Let's save the results to a file (Figure 9-71). The configuration window on the left side of the screen will let you set options on how to save the file, including the File Format. As shown, you can choose to save the file in a different format such as CSV, Excel, JSON, Tableau data extract, and Qlikview data eXchange to name a few.

Let's save the data in CSV format (Figure 9-72). After saving the results, let's inspect the file with Notepad to make sure that the data was saved correctly (Figure 9-73).

CHAPTER 9 BIG DATA VISUALIZATION AND DATA WRANGLING

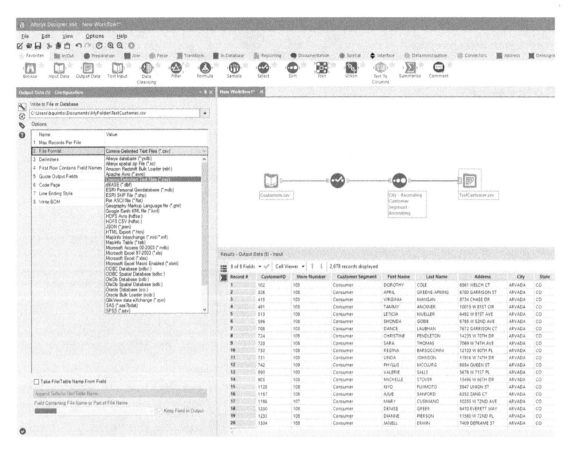

Figure 9-72. Select file format

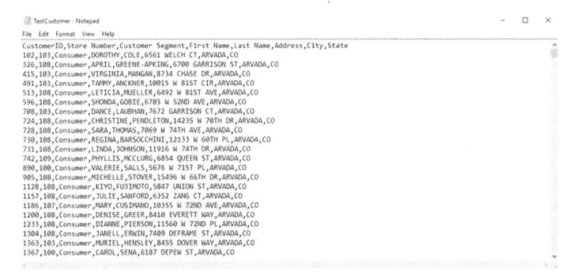

Figure 9-73. Validate saved data

Congratulations! You now have some basic understanding of and hands-on experience with Alteryx.

Datameer

Datameer is another popular data wrangling tool with built-in data visualization and machine learning capabilities. It has a spreadsheet-like interface and includes over 200 analytics functions.[xvi] Datameer is based in San Francisco, California, and was founded in 2009. Unlike Trifacta and Alteryx, Datameer has no connector for Impala, although it has connectors for Spark, HDFS, Hive, and HBase.[xvii]

Let's explore Datameer. First thing you need to do is upload a data file (Figure 9-74).

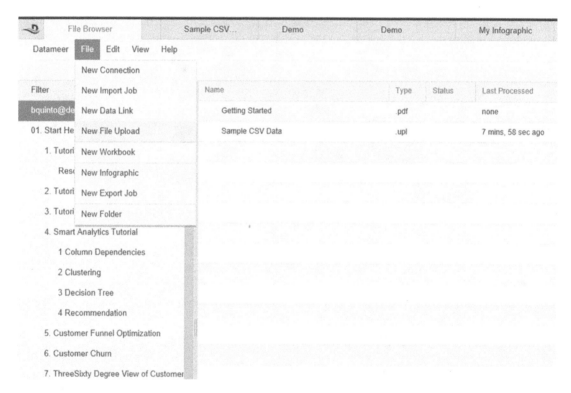

Figure 9-74. Upload data file

CHAPTER 9 BIG DATA VISUALIZATION AND DATA WRANGLING

Specify the file and file type and click Next (Figure 9-75).

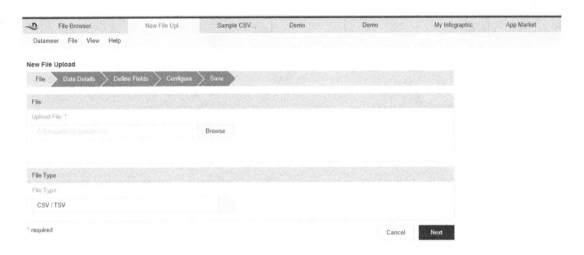

Figure 9-75. *Specify file type*

You can configure the data fields. Datameer allows you to change the data type, field name, and so on (Figure 9-76).

467

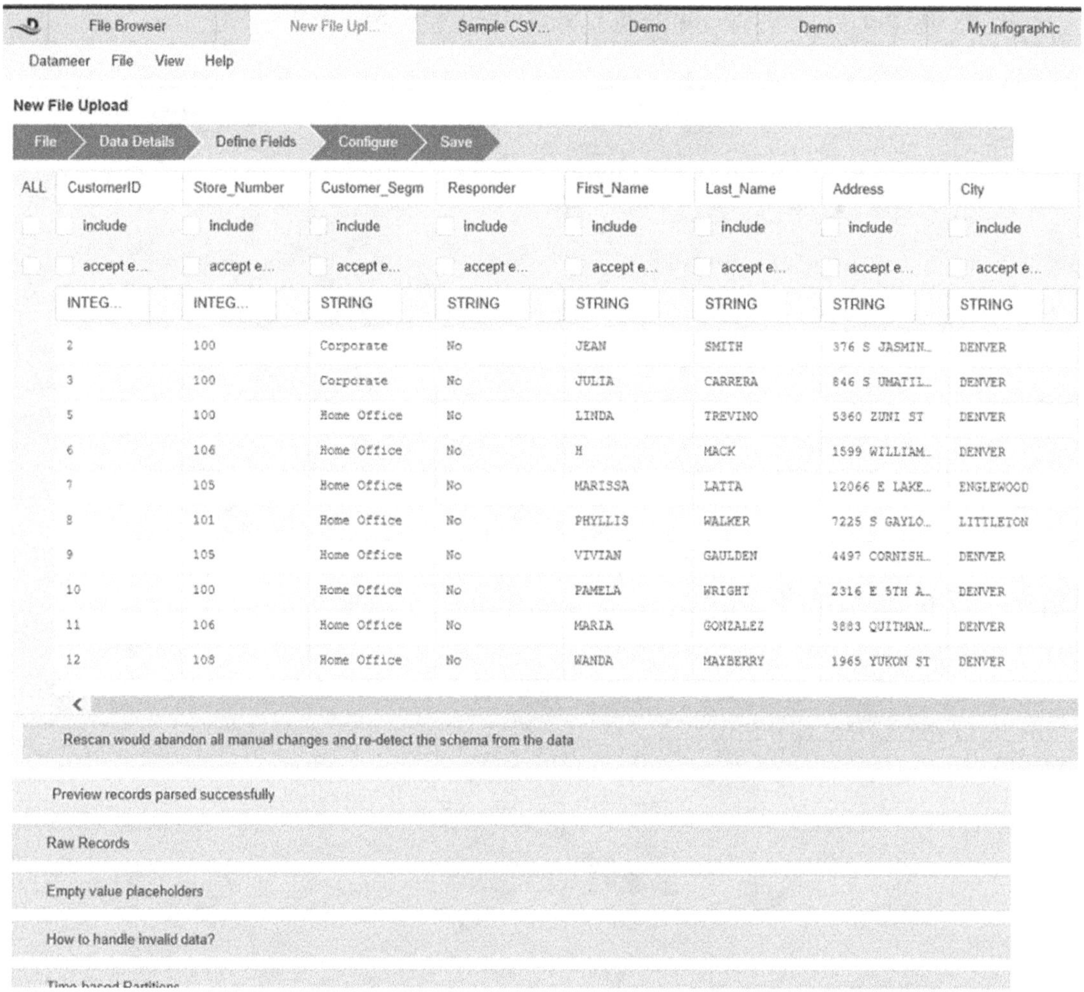

CHAPTER 9 BIG DATA VISUALIZATION AND DATA WRANGLING

ALL	CustomerID	Store_Number	Customer_Segm	Responder	First_Name	Last_Name	Address	City
	include	include	include	include	include	include	include	include
	accept e...	accept e...	accept e...	accept e...	accept e...	accept e...	accept e...	accept e...
	INTEG...	INTEG...	STRING	STRING	STRING	STRING	STRING	STRING
	2	100	Corporate	No	JEAN	SMITH	376 S JASMIN...	DENVER
	3	100	Corporate	No	JULIA	CARRERA	846 S UMATIL...	DENVER
	5	100	Home Office	No	LINDA	TREVINO	5360 ZUNI ST	DENVER
	6	106	Home Office	No	H	MACK	1599 WILLIAM...	DENVER
	7	105	Home Office	No	MARISSA	LATTA	12066 E LAKE...	ENGLEWOOD
	8	101	Home Office	No	PHYLLIS	WALKER	7225 S GAYLO...	LITTLETON
	9	105	Home Office	No	VIVIAN	GAULDEN	4497 CORNISH...	DENVER
	10	100	Home Office	No	PAMELA	WRIGHT	2316 E 5TH A...	DENVER
	11	106	Home Office	No	MARIA	GONZALEZ	3883 QUITMAN...	DENVER
	12	108	Home Office	No	WANDA	MAYBERRY	1965 YUKON ST	DENVER

Figure 9-76. *Configure the data fields*

Data is presented in spreadsheet format (Figure 9-77).

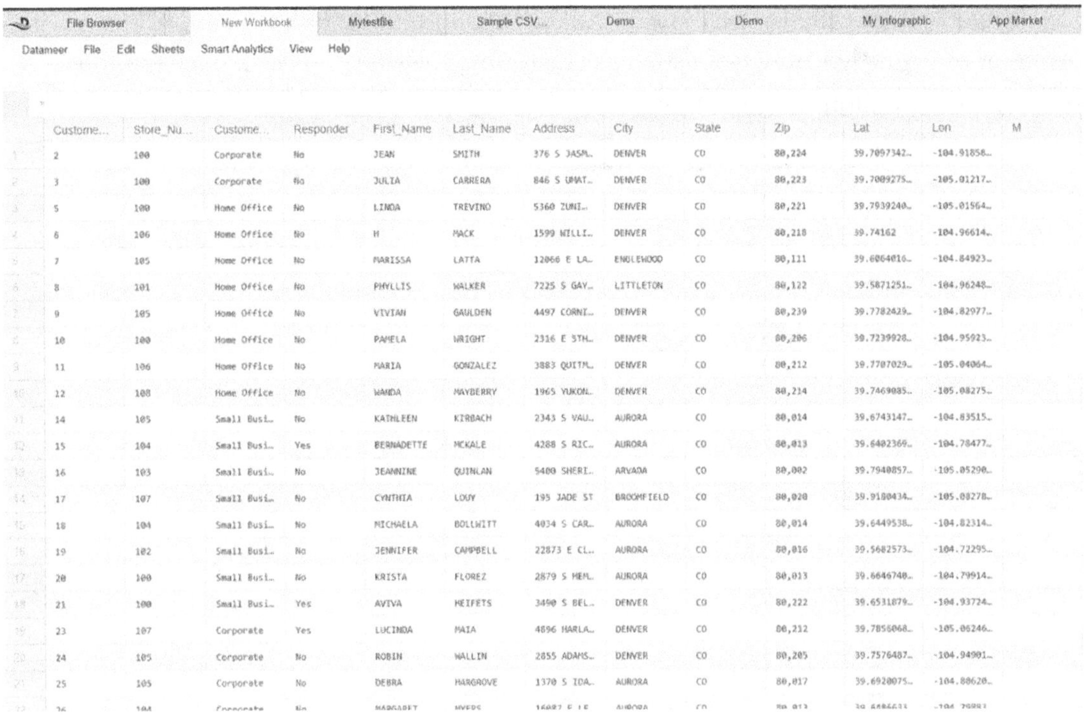

CHAPTER 9   BIG DATA VISUALIZATION AND DATA WRANGLING

Figure 9-77. Spreadsheet-like user interface

Just like Trifacta and Alteryx, Datameer profiles your data and provides statistics and information about your data fields (Figure 9-78).

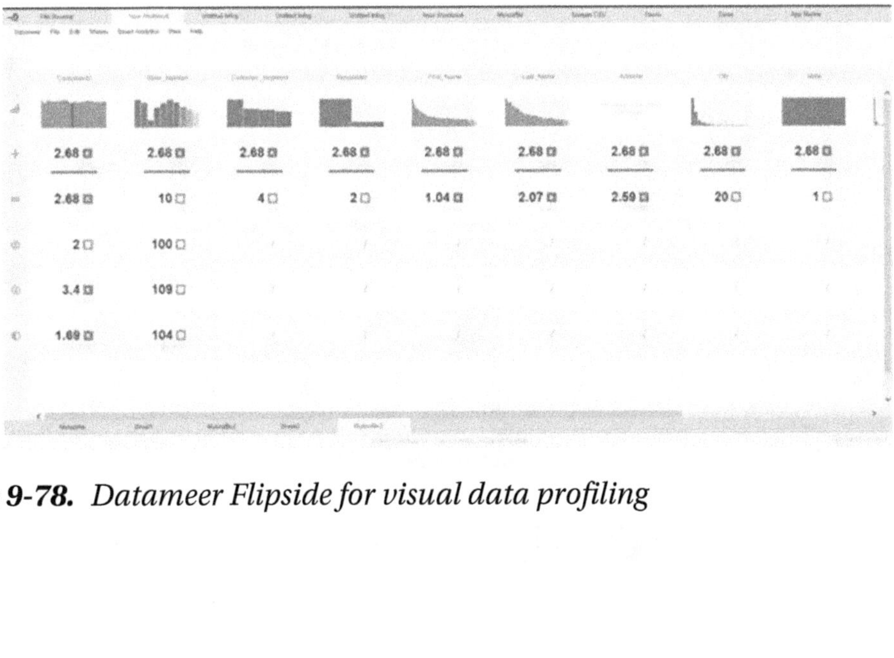

Figure 9-78. Datameer Flipside for visual data profiling

CHAPTER 9 BIG DATA VISUALIZATION AND DATA WRANGLING

Using a feature called Smart Analytics, Datameer provides built-in support for common machine learning tasks such as clustering, classification with decision trees, and recommendations (Figure 9-79).

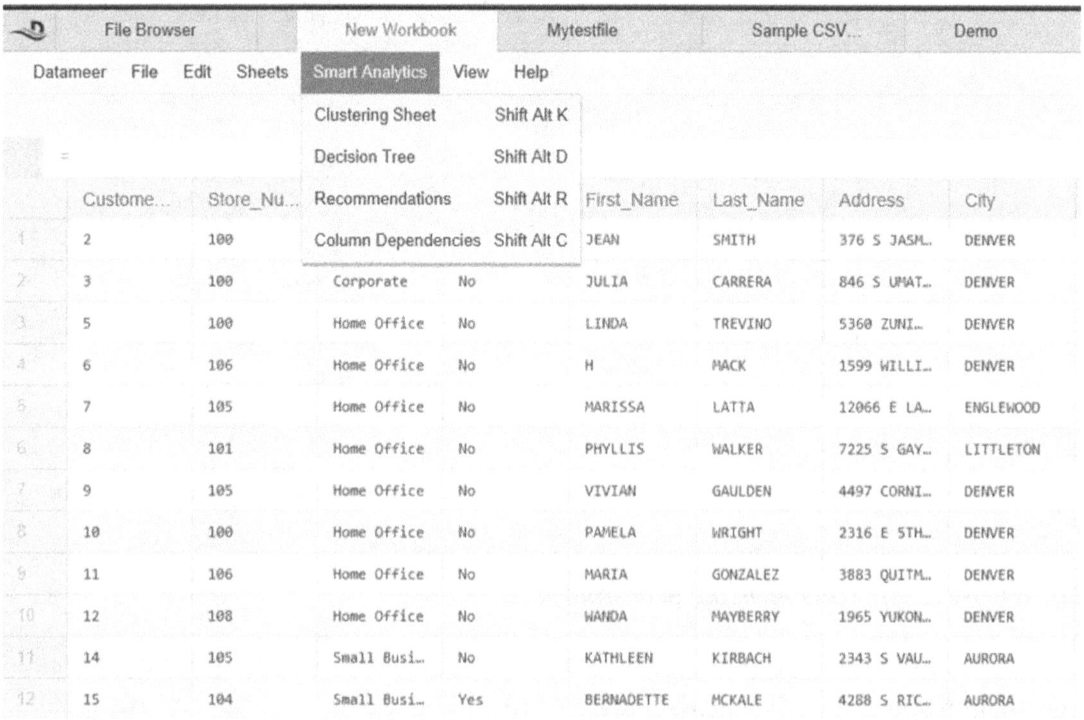

Figure 9-79. Smart Analytics

CHAPTER 9 BIG DATA VISUALIZATION AND DATA WRANGLING

Performing clustering can show some interesting patterns about our data (Figure 9-80).

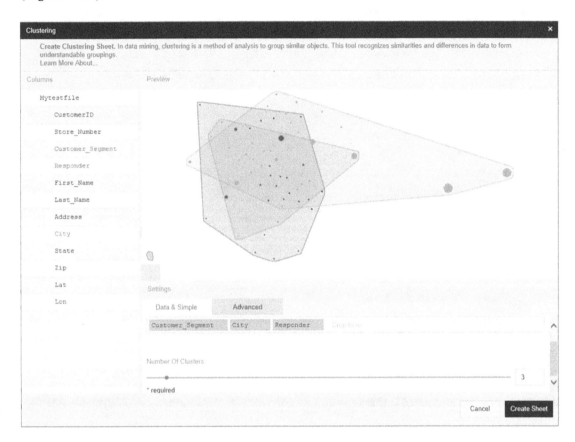

Figure 9-80. *Clustering*

The data will be grouped into clusters. The groupings will be added to your data set as another column. Each cluster will be identified by a cluster id (Figure 9-81).

CHAPTER 9 BIG DATA VISUALIZATION AND DATA WRANGLING

Address	City	State	Zip	Lat	Lon	ClusterID
4896 HARLAN ST	DENVER	CO	80,212	39.7856068...	-105.06246...	C
2855 ADAMS ST	DENVER	CO	80,205	39.7576487...	-104.94901...	B
1370 S IDALIA ST	AURORA	CO	80,017	39.6920075...	-104.80620...	B
16087 E LEHIGH CIR	AURORA	CO	80,013	39.6486633...	-104.79883...	B
11651 BENT OAKS ST	PARKER	CO	80,138	39.5048500...	-104.71669...	B
14303 E NAPA PL	AURORA	CO	80,014	39.6426412...	-104.82037...	B
1937 S PEORIA ST	AURORA	CO	80,014	39.6817509...	-104.84749...	B
1653 IOLA ST	AURORA	CO	80,010	39.7429263...	-104.86479...	B
9291 IRVING ST	WESTMINSTER	CO	80,031	39.8650957...	-105.03102...	B
1370 PARK PL	BROOMFIELD	CO	80,020	39.9370418...	-105.06904...	B
2100 16TH ST	DENVER	CO	80,202	39.7548711...	-105.00477...	C
200 RAMPART WAY	DENVER	CO	80,230	39.7227572...	-104.90062...	C
4464 W GILL PL	DENVER	CO	80,219	39.7048260...	-105.04558...	C
8192 W 81ST DR	ARVADA	CO	80,005	39.8422976...	-105.08793...	B
1330 YUKON ST	DENVER	CO	80,214	39.7378055...	-105.08243...	A
5428 S IDALIA WAY	AURORA	CO	80,015	39.6185699...	-104.80698...	A

Figure 9-81. Column ID showing the different clusters

Classification can also be performed using decision trees (Figure 9-82).

Figure 9-82. Decision trees

CHAPTER 9 BIG DATA VISUALIZATION AND DATA WRANGLING

Data will be classified via an additional prediction field as shown in Figure 9-83.

Address	City	State	Zip	Lat	Lon	ClusterID	Prediction
376 S JASMINE ST	DENVER	CO	80,224	39.7097342...	-104.91858...	B	Home Office
846 S UMATILLA WAY	DENVER	CO	80,223	39.7009275...	-105.01217...	B	Home Office
5360 ZUNI ST	DENVER	CO	80,221	39.7939240...	-105.01564...	B	Home Office
1599 WILLIAMS ST	DENVER	CO	80,218	39.74162	-104.96614...	B	Home Office
12066 E LAKE CIR	ENGLEWOOD	CO	80,111	39.6064016...	-104.84923...	B	Home Office
7225 S GAYLORD ST	LITTLETON	CO	80,122	39.5871251...	-104.96248...	B	Home Office
4497 CORNISH WAY	DENVER	CO	80,239	39.7782429...	-104.82977...	B	Home Office
2316 E 5TH AVE	DENVER	CO	80,206	39.7239928...	-104.95923...	B	Home Office
3883 QUITMAN ST	DENVER	CO	80,212	39.7707029...	-105.04064...	B	Home Office
1965 YUKON ST	DENVER	CO	80,214	39.7469085...	-105.08275...	B	Home Office
2343 S VAUGHN WAY	AURORA	CO	80,014	39.6743147...	-104.83515...	C	Small Busi...
4288 S RICHFIELD ST	AURORA	CO	80,013	39.6402369...	-104.78477...	C	Small Busi...
5400 SHERIDAN BLVD	ARVADA	CO	80,002	39.7940857...	-105.05290...	C	Small Busi...
195 JADE ST	BROOMFIELD	CO	80,020	39.9180434...	-105.08278...	C	Small Busi...
4034 S CARSON ST	AURORA	CO	80,014	39.6449538...	-104.82314...	C	Small Busi...
22873 E CLIFTON PL	AURORA	CO	80,016	39.5682573...	-104.72295...	C	Small Busi...
2879 S MEMPHIS ST	AURORA	CO	80,013	39.6646740...	-104.79914...	C	Small Busi...
3490 S BELLAIRE ST	DENVER	CO	80,222	39.6531879...	-104.93724...	C	Small Busi...

Figure 9-83. *Prediction column*

As previously mentioned, Datameer includes built-in data visualization (Figure 9-84).

CHAPTER 9   BIG DATA VISUALIZATION AND DATA WRANGLING

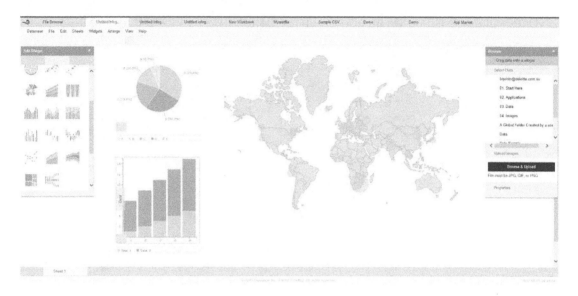

Figure 9-84. Datameer data visualization

Summary

Make sure your big data visualization tool can handle the three V's of big data (volume, variety, and velocity). If you're just in the process of selecting a tool for your organization, consider if the tool will be able to handle terabyte/petabyte-size data sets. Will your data source be purely relational or will you be analyzing semi-structured and unstructured data sets as well? Do you have requirements to ingest and process data in real time or near real time? Try leveraging your existing BI and data visualization tools and determine if they can fulfill your requirements. You may be connecting to a Hadoop cluster, but your data set may not be that large. This is a common situation nowadays, organizations using big data platforms as a cost-effective way to consolidate expensive reporting servers and data marts.

Data wrangling tools have become popular these past few years. Users have become savvier and more demanding. Who better to prepare the data than the people who understand it best? If your organization has power users who insist on preparing and transforming data themselves, then it might be worthwhile looking at some of the data wrangling tools I covered in the chapter.

References

i. Christine Vitron, James Holman; "Considerations for Adding SAS® Visual Analytics to an Existing SAS® Business Intelligence Deployment", SAS, 2018, http://support.sas.com/resources/papers/proceedings14/SAS146-2014.pdf

ii. Zoomdata; "ABOUT ZOOMDATA", Zoomdata, 2018, https://www.zoomdata.com/about-zoomdata/

iii. Zoomdata; "Real-Time & Streaming Analytics", Zoomdata, 2018, https://www.zoomdata.com/product/real-time-streaming-analytics/

iv. Zoomdata; "Real-Time & Streaming Analytics", Zoomdata, 2018, https://www.zoomdata.com/product/stream-processing-analytics/

v. https://www.zoomdata.com/product/apache-spark-data-stream-processing/

vi. Zoomdata; "Real-Time & Streaming Analytics", Zoomdata, 2018, https://www.zoomdata.com/product/stream-processing-analytics/

vii. Zoomdata; "Data Sharpening in Zoomdata", Zoomdata, 2018, https://www.zoomdata.com/docs/2.5/data-sharpening-in-zoomdata.html

viii. Zoomdata; "Real-Time & Streaming Analytics", Zoomdata, 2018, https://www.zoomdata.com/product/stream-processing-analytics/

ix. Coderwall; "How to get the correct Unix Timestamp from any Date in JavaScript", Coderwall, 2018, https://coderwall.com/p/rbfl6g/how-to-get-the-correct-unix-timestamp-from-any-date-in-javascript

x. Gil Press; "Cleaning Big Data: Most Time-Consuming, Least Enjoyable Data Science Task, Survey Says", Forbes, 2018, https://www.forbes.com/sites/gilpress/2016/03/23/data-preparation-most-time-consuming-least-enjoyable-data-science-task-survey-says/#3bfc5cdc6f63

xi. Tye Rattenbury; "Six Core Data Wrangling Activities", Datanami, 2015, https://www.datanami.com/2015/09/14/six-core-data-wrangling-activities/

xii. Stanford; "Wrangler is an interactive tool for data cleaning and transformation." Stanford, 2013, http://vis.stanford.edu/wrangler/

xiii. Trifacta; "Data Wrangling for Organizations", Trifacta, 2018, https://www.trifacta.com/products/wrangler-enterprise/

xiv. Trifacta; "The Impala Hadoop Connection: From Cloudera Impala and Apache Hadoop and beyond", 2018, https://www.trifacta.com/impala-hadoop/

xv. Alteryx; "Alteryx enables access to Cloudera in a number of ways.", Alteryx, 2018, https://www.alteryx.com/partners/cloudera

xvi. Datameer; "Cloudera and Datameer", Datameer, 2018, https://www.cloudera.com/partners/solutions/datameer.html

xvii. Datameer; "Connector", Datameer, 2018, https://www.datameer.com/product/connectors/

CHAPTER 10

Distributed In-Memory Big Data Computing

Alluxio, formerly known as Tachyon, is an open source project from UC Berkeley AMPLab. Alluxio is a distributed memory-centric storage system originally developed as a research project by Haoyuan Li in 2012, then a PhD student and a founding Apache Spark committer at AMPLab.[i] The project is the storage layer of the Berkeley Data Analytics Stack (BDAS). In 2015, Alluxio, Inc. was founded by Li to commercialize Alluxio, receiving a $7.5 million cash infusion from Andreesen Horowitz. Today, Alluxio has more than 200 contributors from 50 organizations around the world such as Intel, IBM, Yahoo, and Red Hat. Several high-profile companies are currently using Alluxio in production such as Baidu, Alibaba, Rackspace, and Barclays.[ii]

Ion Stoica, co-author of Spark, co-founder and executive chairman of DataBricks, co-director of UC Berkeley AMPLab, and PhD co-advisor to Haoyuan Li has stated, "As a layer that abstracts away the differences of existing storage systems from the cluster computing frameworks such as Apache Spark and Hadoop MapReduce, Alluxio can enable the rapid evolution of the big data storage, similarly to the way the Internet Protocol (IP) has enabled the evolution of the Internet."[iii]

Additionally, Michael Franklin, Professor of Computer Science and Director of the AMPLab at UC Berkeley said that "AMPLab has created some of the most important open source technologies in the new big data stack, including Apache Spark. Alluxio is the next project with roots in the AMPLab to have major impact. We see it playing a huge disruptive role in the evolution of the storage layer to handle the expanding range of big data use cases."[iv]

Architecture

Alluxio is a memory-centric distributed storage system and aims to be the de facto storage unification layer for big data. It provides a virtualization layer that unifies access for different storage engines such as Local FS, HDFS, S3, and NFS and computing frameworks such as Spark, MapReduce, Hive, and Presto. Figure 10-1 give you an overview of Alluxio's architecture.

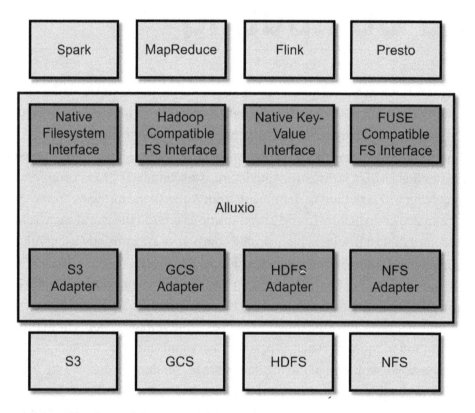

Figure 10-1. Alluxio Architecture Overview

Alluxio is the middle layer that coordinates data sharing and directs data access, while at the same time providing computing frameworks and big data applications high-performance low-latency memory speed. Alluxio integrates seamlessly with Spark and Hadoop, only requiring minor configuration changes. By taking advantage of Alluxio's unified namespace feature, applications only need to connect to Alluxio to access data stored in any of the supported storage engines. Alluxio has its own native API as well as a Hadoop-compatible file system interface. The convenience class enables users to execute code originally written for Hadoop without any code changes. A REST API provides access to other languages. We will explore the API's later in the chapter.

Alluxio's unified namespace feature does not support relational data stores such as Kudu, relational databases such as Oracle or SQL Server, or document databases such as MongoDB. Of course, writing to and from Alluxio and the storage engines mentioned are supported. Developers can use a computing framework such as Spark to create a Data Frame from a Kudu table and store it in an Alluxio file system in Parquet or CSV format, and vice versa (Figure 10-2).

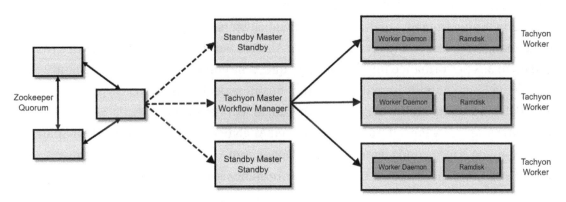

Figure 10-2. *Alluxio Technical Architecture*

Why Use Alluxio?

The typical Hadoop distribution includes more than 20 open source components. Adding another component to your technology stack is probably the furthest thing in your mind. Nevertheless, Alluxio delivers substantial benefits that will make you wonder why Alluxio is not part core Apache Spark.

CHAPTER 10 DISTRIBUTED IN-MEMORY BIG DATA COMPUTING

Significantly Improve Big Data Processing Performance and Scalability

Over the years memory has gotten cheaper, while its performance has gotten faster. Meanwhile, performance of hard drives has only gotten marginally better. There is no question that data processing in-memory is an order of magnitude faster than processing data on disk. In almost all programming paradigms, we are advised to cache data in-memory to improve performance. One of the main advantages of Apache Spark over MapReduce is its ability to cache data. Alluxio takes that to the next level, providing big data applications not just as a caching layer, but a full-blown distributed high-performance memory-centric storage system.

Baidu is operating one of the largest Alluxio clusters in the world, with 1,000 worker nodes handling more than 2PB of data. With Alluxio, Baidu is seeing an average of 10x and up to 30x performance improvement in query and processing time, significantly improving Baidu's ability to make important business decisions.[v] Barclays published an article describing their experience with Alluxio. Barclays Data Scientist Gianmario Spacagna and Harry Powell, Head of Advanced Analytics were able to tune their Spark jobs from hours to seconds using Alluxio.[vi] Qunar.com, one of China's largest travel search engines, experienced a 15x – 300x performance improvement using Alluxio.[vii]

Multiple Frameworks and Applications Can Share Data at Memory Speed

A typical Hadoop cluster has multiple sessions running different computing frameworks such as Spark and MapReduce. In case of Spark, each application gets its own executor processes, with each task within an executor running on its own JVM, isolating Spark applications from each other. This means that Spark (and MapReduce) applications have no way of sharing data, except writing to a storage system such as HDFS or S3. As shown in Figure 10-3, a Spark job and a MapReduce job are using the same data stored in HDFS or S3. In Figure 10-4, Multiple Spark jobs are using the same data with each job storing its own version of the data in its own heap space.[viii] Not only is data duplicated, but sharing data via HDFS or S3 can be slow, particularly if you're sharing large amounts of data.

CHAPTER 10 DISTRIBUTED IN-MEMORY BIG DATA COMPUTING

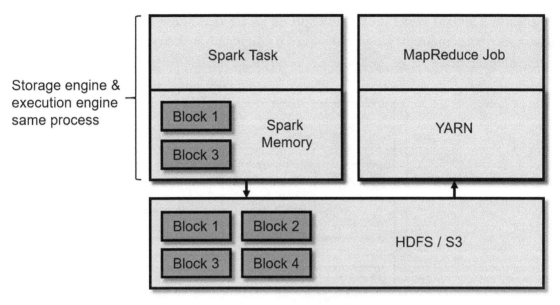

Inter-process data sharing latency & throughput slowed down by network and disk IO

Figure 10-3. Different frameworks sharing data via HDFS or S3

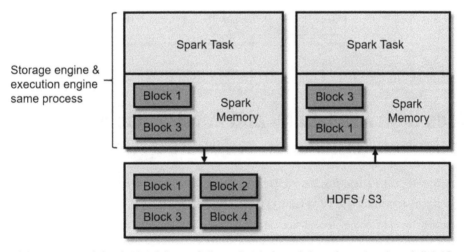

Inter-process data sharing latency & throughput slowed down by network and disk IO

Figure 10-4. Different jobs sharing data via HDFS or S3

CHAPTER 10 DISTRIBUTED IN-MEMORY BIG DATA COMPUTING

By using Alluxio as an off-heap storage (Figure 10-5), multiple frameworks and jobs can share data at memory speed, reducing data duplication, increasing throughput, and decreasing latency.

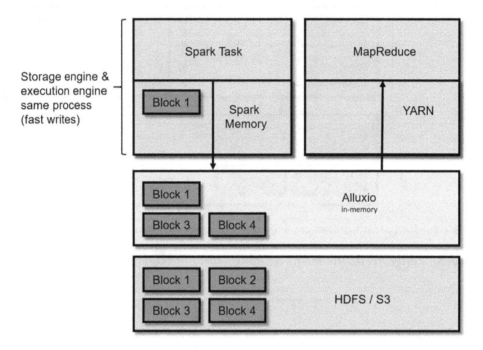

Figure 10-5. *Different jobs and frameworks sharing data at memory speed*

Provides High Availability and Persistence in Case of Application Termination or Failure

In Spark, the executor processes and the executor memory reside in the same JVM, with all cached data stored in the JVM heap space (Figure 10-6).

CHAPTER 10 DISTRIBUTED IN-MEMORY BIG DATA COMPUTING

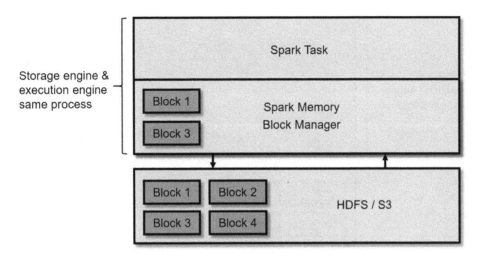

Figure 10-6. *Spark job with its own heap memory*

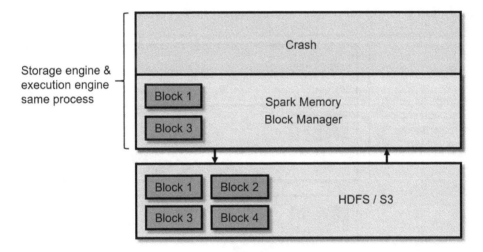

Figure 10-7. *Spark job crashes or completes*

When the job completes or for some reason the JVM crashes due to runtime exceptions, all the data cached in heap space will be lost as shown in Figures 10-7 and 10-8.

483

CHAPTER 10   DISTRIBUTED IN-MEMORY BIG DATA COMPUTING

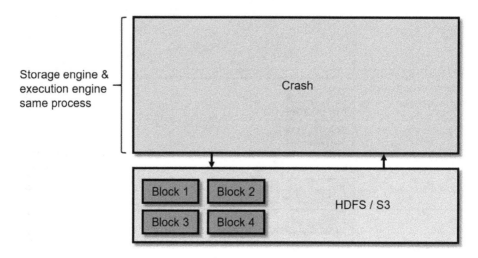

Figure 10-8. *Spark job crashes or completes. Heap space is lost*

The solution is to use Alluxio as an off-heap storage (Figure 10-9).

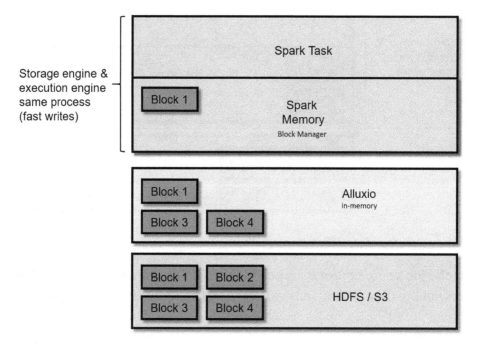

Figure 10-9. *Spark using Alluxio as off-heap storage*

In this case, even if the Spark JVM crashes, the data is still available in Alluxio (Figures 10-10 and 10-11).

CHAPTER 10 DISTRIBUTED IN-MEMORY BIG DATA COMPUTING

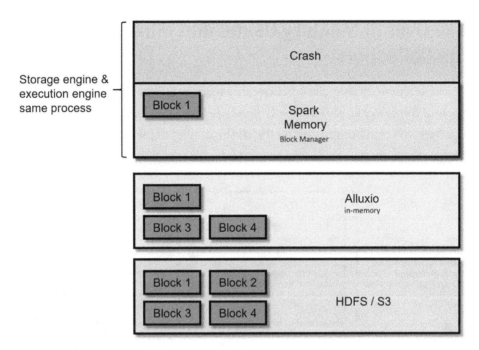

Figure 10-10. *Spark job crashes or completes*

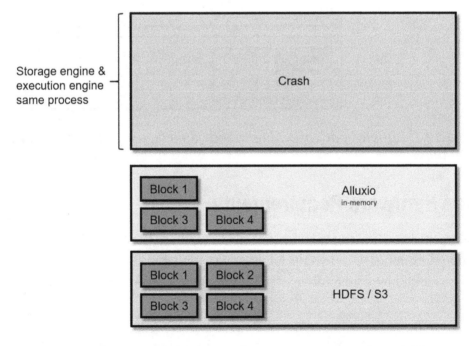

Figure 10-11. *Spark job crashes or completes. Heap space is lost. Off-heap memory is still available.*

485

CHAPTER 10 DISTRIBUTED IN-MEMORY BIG DATA COMPUTING

Optimize Overall Memory Usage and Minimize Garbage Collection

By using Alluxio, memory usage is considerably more efficient since data is shared across jobs and frameworks and because data is stored off-heap, garbage collection is minimized as well, further improving the performance of jobs and applications (Figure 10-12).

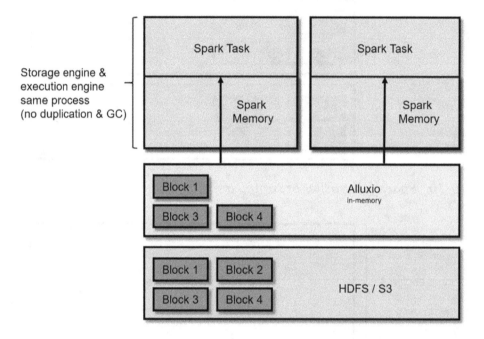

Figure 10-12. Multiple Spark and MapReduce jobs can access the same data stored in Alluxio

Reduce Hardware Requirements

Big data processing with Alluxio is significantly faster than with HDFS. IBM's tests shows Alluxio outperforming HDFS by 110x for write IO.[ix] With that kind of performance, there is less requirement for additional hardware, thus saving you in infrastructure and licensing costs.

Alluxio Components

Similar to Hadoop and other Hadoop components, Alluxio has a master/slave architecture.

Primary Master

The primary master manages the global metadata of the cluster.

Secondary Master

The secondary master manages a journal and periodically does a checkpoint.

Worker

Workers store the data and serve requests from applications to read or write data. Workers also manage local resources such as memory and disk space.

Client

The Alluxio client provides a filesystem API for users to communicate with Alluxio.

Installation

There are several ways to install Alluxio. Alluxio runs on YARN, Mesos, Docker, and EC2 to mention a few.[x] To get you started quickly, I'll install Alluxio on a single server.

Download the newest version of Alluxio from the Alluxio website.

```
wget http://alluxio.org/downloads/files/1.6.1/alluxio-1.4.0-bin.tar.gz
tar xvfz alluxio-1.4.0-bin.tar.gz
cd alluxio-1.4.0
```

Let's format the worker storage directory and Alluxio journal to prepare the worker and master.

```
./bin/alluxio format
Waiting for tasks to finish...
All tasks finished, please analyze the log at /opt/alluxio-1.4.0/bin/../
logs/task.log.
Formatting Alluxio Master @ server01
```

CHAPTER 10 DISTRIBUTED IN-MEMORY BIG DATA COMPUTING

Let's start Alluxio.

```
./bin/alluxio-start.sh local
Waiting for tasks to finish...
All tasks finished, please analyze the log at /opt/alluxio-1.4.0/bin/../
logs/task.log.
Waiting for tasks to finish...
All tasks finished, please analyze the log at /opt/alluxio-1.4.0/bin/../
logs/task.log.
Killed 0 processes on server01
Killed 0 processes on server01
Starting master @ server01. Logging to /opt/alluxio-1.4.0/logs
Formatting RamFS: /mnt/ramdisk (4000mb)
Starting worker @ server01. Logging to /opt/alluxio-1.4.0/logs
Starting proxy @ server01. Logging to /opt/alluxio-1.4.0/logs
```

I create a 100 MB file and copy it to memory. You can create a bigger file if you have more memory. List the contents of the directory.

```
./bin/alluxio fs ls /
[root@server01 alluxio-1.4.0]# ./bin/alluxio fs copyFromLocal /root/test01.
csv /

Copied /root/test01.csv to /
./bin/alluxio fs ls /
-rw-r--r--     root           root           103.39MB  05-22-2017
22:21:14:925   In Memory      /test01.csv
```

Let's persist the file from memory to the local file system.

```
./bin/alluxio fs persist /test01.csv
persisted file /test01.csv with size 108416290
```

Apache Spark and Alluxio

You access data in Alluxio similar to how you would access data stored in HDFS and S3 from Spark.

```
val dataRDD = sc.textFile("alluxio://localhost:19998/test01.csv")

val parsedRDD = dataRDD.map{_.split(",")}

case class CustomerData(userid: Long, city: String, state: String, age: Short)

val dataDF = parsedRDD.map{ a => CustomerData (a(0).toLong, a(1), a(2), a(3).toShort) }.toDF

dataDF.show()

+------+---------------+-----+---+
|userid|           city|state|age|
+------+---------------+-----+---+
|   300|       Torrance|   CA| 23|
|   302|Manhattan Beach|   CA| 21|
+------+---------------+-----+---+
```

You can also access Alluxio from MapReduce, Hive, Flink, and Presto to mention a few. Check Alluxio's online documentation for more details.

Administering Alluxio

Alluxio provides a web interface to facilitate system administration and monitoring. You get both high level and detailed information on space capacity, usage, uptime, start time, and list of files to name a few. Alluxio provides you with a web interface for the master and workers. Alluxio also provides a command-line interface for typical file system operations.

Master

You can access Alluxio's master home page by visiting `http://<Master IP Address >:19999` (Figure 10-13).

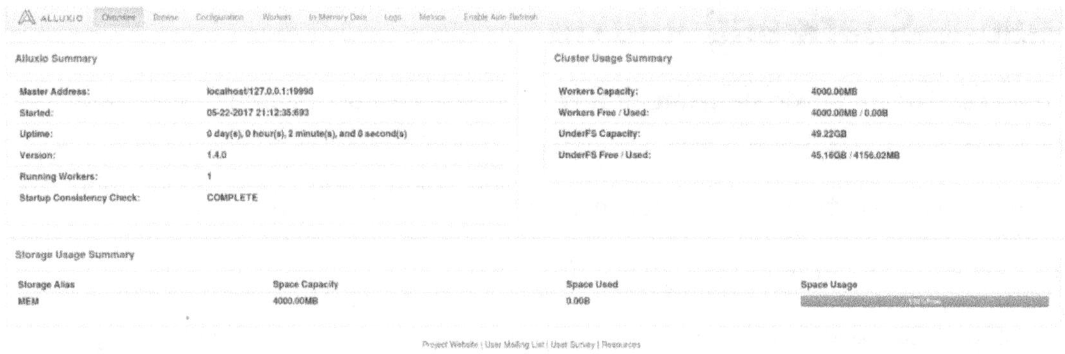

CHAPTER 10   DISTRIBUTED IN-MEMORY BIG DATA COMPUTING

Figure 10-13. Master home page

Worker

You can access each Alluxio worker's web interface by visiting http://<Worker IP Address>:30000 (Figure 10-14).

Figure 10-14. Worker home page

Apache Ignite

Ignite is another in-memory platform similar to Alluxio. GridGain Systems originally contributed Apache Ignite to the Apache Software Foundation in 2014. It was promoted to a top-level project in 2015.[xi] It is extremely versatile and can be used as an in-memory data grid, in-memory database, in-memory distributed filesystem, streaming analytics engine, and accelerator for Hadoop and Spark to mention a few.[xii]

Apache Geode

Geode is a distributed in-memory database designed for transactional application with low-latency response times and high-concurrency requirements. Pivotal submitted Geode to the Apache Incubator in 2015. It graduated from the Apache Incubator to become a top-level Apache project in November 2016. Gemfire, the commercial version of Geode, was a popular low-latency transactional system used in Wall Street trading platforms.[xiii]

Summary

Spark is a fast in-memory data processing framework. It can be made significantly faster with Alluxio by providing an off-heap storage that can be utilized to make data sharing across jobs and frameworks more efficient, minimizing garbage collection, and optimizing overall memory usage. Not only will jobs run considerably faster, but you also reduce costs due to decreased hardware requirements. Alluxio is not the only in-memory database available; Apache Ignite and Geode are viable options, as well as other commercial alternatives such as Oracle Coherence and Times Ten.

This chapter serves as an introduction to distributed in-memory computing, and Alluxio in particular. Alluxio is the default off-heap storage solution for Spark. You can learn more about Alluxio by visiting its website at Alluxio.org or Alluxio.com.

References

i. Chris A Mattman; "Apache Spark for the Incubator," Apache Spark, 2013, `http://mail-archives.apache.org/mod_mbox/incubator-general/201306.mbox/%3CCDD80F64.D5F9D%25chris.a.mattmann@jpl.nasa.gov%3E`

ii. Haoyuan Li; "Alluxio, formerly Tachyon, is Entering a New Era with 1.0 release," Alluxio, 2016, `https://www.alluxio.com/blog/alluxio-formerly-tachyon-is-entering-a-new-era-with-10-release`

CHAPTER 10   DISTRIBUTED IN-MEMORY BIG DATA COMPUTING

iii. Haoyuan Li; "Alluxio, formerly Tachyon, is Entering a New Era with 1.0 release," Alluxio, 2016, https://www.alluxio.com/blog/alluxio-formerly-tachyon-is-entering-a-new-era-with-10-release

iv. MarketWired; "Alluxio Virtualizes Distributed Storage for Petabyte Scale Computing at In-Memory Speeds," Alluxio, 2016, http://www.marketwired.com/press-release/alluxio-virtualizes-distributed-storage-petabyte-scale-computing-in-memory-speeds-2099053.htm

v. MarketWired; "Alluxio Virtualizes Distributed Storage for Petabyte Scale Computing at In-Memory Speeds," Alluxio, 2016, http://www.marketwired.com/press-release/alluxio-virtualizes-distributed-storage-petabyte-scale-computing-in-memory-speeds-2099053.htm

vi. Henry Powell, Gianmario Spacagna; "Making the Impossible Possible with Tachyon: Accelerate Spark Jobs from Hours to Seconds," DZone, 2016, https://dzone.com/articles/Accelerate-In-Memory-Processing-with-Spark-from-Hours-to-Seconds-With-Tachyon

vii. Haoyuan Li; "Alluxio Keynote at Strata+Hadoop World Beijing 2016," Alluxio, 2016, https://www.slideshare.net/Alluxio/alluxio-keynote-at-stratahadoop-world-beijing-2016-65172341

viii. Mingfei S.; "Getting Started with Tachyon by Use Cases," Intel, 2016, https://software.intel.com/en-us/blogs/2016/02/04/getting-started-with-tachyon-by-use-cases

ix. Gil Vernik; "Tachyon for ultra-fast Big Data processing," IBM, 2015, https://www.ibm.com/blogs/research/2015/08/tachyon-for-ultra-fast-big-data-processing/

x. Alluxio; "Quick Start Guide," Alluxio, 2018, https://www.alluxio.org/docs/1.6/en/Getting-Started.html

xi. Nikita Ivanov; "Fire up big data processing with Apache Ignite," InfoWorld, 2016, https://www.infoworld.com/article/3135070/data-center/fire-up-big-data-processing-with-apache-ignite.html

xii. GridGain; "The Foundation of the GridGain In-Memory Computing Platform," GridGain, 2018, https://www.gridgain.com/technology/apache-ignite

xiii. Apache Software Foundation; "The Apache Software Foundation Announces Apache® Geode™ as a Top-Level Project," GlobeNewsWire, 2018, https://globenewswire.com/news-release/2016/11/21/891611/0/en/The-Apache-Software-Foundation-Announces-Apache-Geode-as-a-Top-Level-Project.html

CHAPTER 11

Big Data Governance and Management

Data governance is the set of policies and processes used to formally manage an organizations' data. Data governance is a wide topic and encompasses several functions[i] such as data quality, metadata management, master data management, and data security to mention a few. The main goal of data governance is to ensure that an organization's data is secure and trustworthy.

Data governance is perhaps one of the most important parts of an organization's information management strategy. Data governance issues such as lack of data quality or compromised data security have the ability to sink data-driven projects or cause massive revenue lost. For organizations such as banks and government agencies, data governance is a must. There are several data governance and management frameworks to choose from such as the DAMA framework. The DAMA framework codifies a collection of knowledge areas and processes in a guide known as "The DAMA Guide to the Data Management Body of Knowledge or the DAMA-DMBoK Guide."[ii]

It's not the goal of this chapter to provide an exhaustive coverage of data governance. Entire books have been written about data governance. Instead, my aim is to introduce you to data governance from a big data perspective. I use Cloudera Navigator to show common data governance tasks on big data platforms. To learn more about data governance, I recommend *Data Governance: How to Design, Deploy and Sustain an Effective Data Governance Program* by John Ladley (Morgan Kaufmann, 2012). To learn more about Cloudera Navigator, visit Cloudera's website.

CHAPTER 11   BIG DATA GOVERNANCE AND MANAGEMENT

Data Governance for Big Data

Because data is consolidated in a big data platform (also known as a "data lake" or "data hub"), data governance becomes even more important. Storing millions of files and tables in a single platform can become a data management nightmare. Imagine having multiple versions of the same data ingested from different data sources or having to wade through millions of files and tables to look for a particular piece of information. Data governance is critical in data warehouse environments, but the importance is magnified when it comes to big data. You get a real appreciation of data governance once you start working with hundreds of terabytes or petabytes of data stored in different types and formats. A successful big data strategy relies on a well-executed data governance plan.

Cloudera Navigator

Cloudera includes a data governance tool called Cloudera Navigator to help manage and organize the data stored in the data lake. In some organizations, there is a usually a designated person or a group, depending on the size of the organization, responsible for data governance. They are referred to as data stewards. Some teams combine the role of data stewards with that of administrators, architects, or a data quality engineer. However your team is structured, there should be a formal data governance framework enforced and followed in your organization.

Data stewards and administrators can use Cloudera Navigator to proactively manage and monitor data stored in the data lake. Cloudera Navigator allows you to perform tasks such as metadata management, data classification, data auditing and access control, explore data lineage, perform impact analysis, enforce policies, and automate data life cycle.

You can access Cloudera Navigator from Cloudera Manager. Navigate to Clusters, then click Cloudera Navigator as shown in Figure 11-1. You can also access Cloudera Navigator directly by pointing your browser to the hostname where Cloudera Metadata Server is installed on port 7187.

CHAPTER 11   BIG DATA GOVERNANCE AND MANAGEMENT

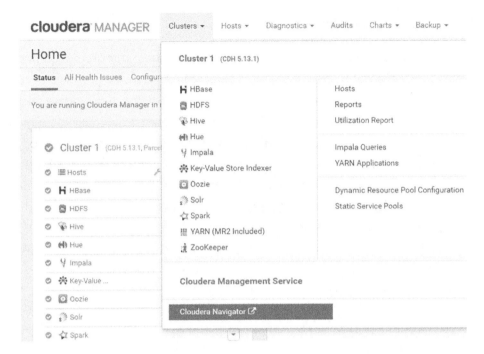

Figure 11-1. *How to access Cloudera Navigator from Cloudera Manager*

When you log in, you'll see a search box where you can enter entity names, metadata tags, or any keyword that can help you find the entity that you want to search for. On the left side of the window are filters that can help you perform more refined faceted searches as shown in Figure 11-2.

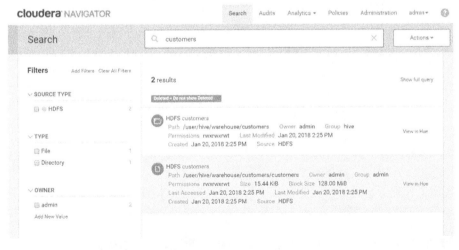

Figure 11-2. *Cloudera Navigator User Interface*

497

Chapter 11 Big Data Governance and Management

Metadata Management

Metadata management involves managing data that describes your organization's data.[iii] For example, metadata can be a file's creation date, table size, or file permissions to name a few. Cloudera Navigator allows users to search for, define, and update properties and tag entities such as files, tables, and directories. There are three types of metadata[iv]: Technical Metadata, which includes the file or table name; date and time creation; size permissions and owner, to mention a few. Users are not allowed to modify technical metadata. Custom Metadata are key-value pairs that can be added to entities *before and after* entities are created. Managed Metadata includes tags, descriptions, and key-value pairs that can be added or updated only after entities are created. Metadata is the basis for most data governance tasks such as data classification, searches, and policy enforcement. I add a description and three metadata tags to the customer file stored in HDFS (annie, berkeley, and marketing) as shown in Figure 11-3.

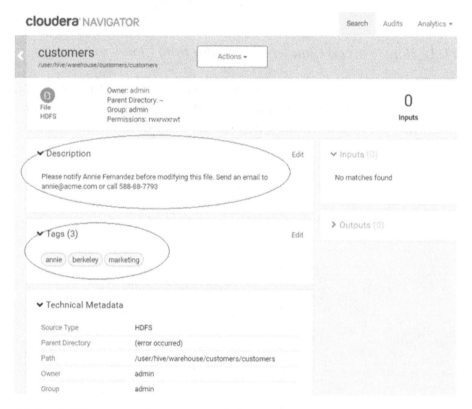

Figure 11-3. Adding metadata tags and description to entities

CHAPTER 11 BIG DATA GOVERNANCE AND MANAGEMENT

Searching for "marketing" returns the customer file that I tagged earlier as shown in Figure 11-4.

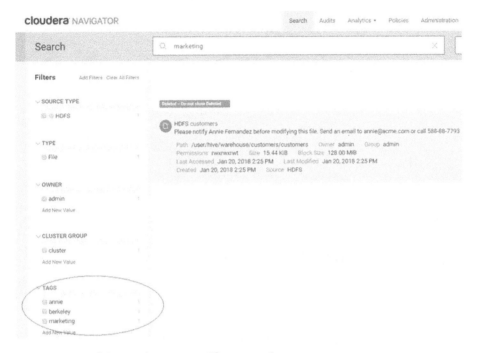

Figure 11-4. Searching using a specific metadata tag

Data Classification

Metadata tags and descriptions can be very powerful. Tags can be used to classify data. For example, you can tag files and tables as "marketing" to identify the entities that belong to the marketing department, or you can tag entities as "finance" to let everyone know that the particular entity is owned by the finance department and so on. You can use a combination of tags and key-value pairs to classify your data any way you want. You can use the tags to quickly search for entities that are classified together as shown in Figure 11-5.

499

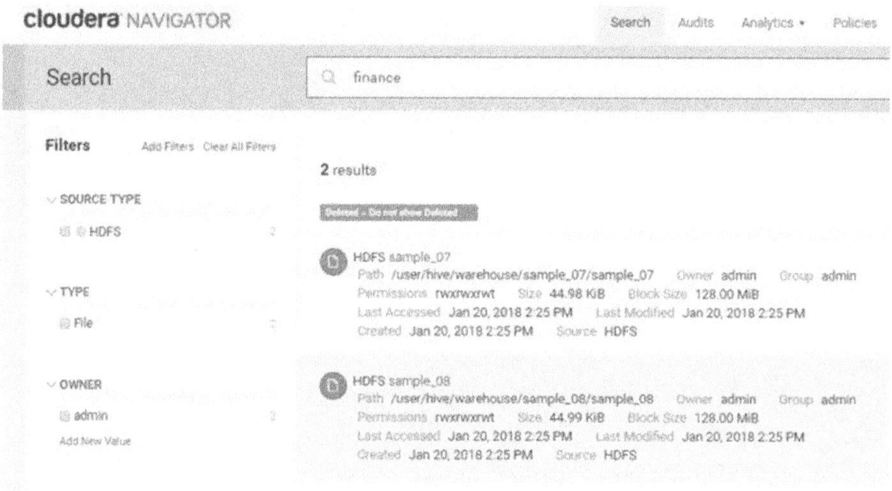

CHAPTER 11 BIG DATA GOVERNANCE AND MANAGEMENT

Figure 11-5. Classifying data using metadata

Data Lineage and Impact Analysis

Cloudera Navigator allows users to view data lineage. A data lineage shows an entity's relationship with other entities. It shows all the data transformation that was performed against a particular entity (Figure 11-6). It shows how a particular entity was created, the original data source, and who performed the data transformation. Inspecting an entity's data lineage is an effective way to determine its trustworthiness. It is also useful in application development, helping architects determine the impact of schema changes to other entities.

Figure 11-6. Data lineage showing the actual SQL statement used to create the table

Auditing and Access Control

Cloudera Navigator provides auditing capability allowing data stewards to monitor audit events on services such as HBase, HDFS, Hive, HUE, Impala, Navigator Metadata Server, Sentry, and Solr.[v] Cloudera Navigator monitors events when users successfully and unsuccessfully log in; create or delete columns, tables, files, or folders or sets; and grants

CHAPTER 11 BIG DATA GOVERNANCE AND MANAGEMENT

or revokes object permissions to mention a few available audit events. Auditing security events ensures that data hasn't been altered by someone who doesn't have the necessary permissions. This is especially useful in multitenant environments where multiple users have access to the data lake. Keeping track of access attempts and incorrect object permissions in a big data environment can be difficult. Cloudera Navigator records the user ID, object name, IP address, and the exact command or query that was executed.[vi] Figure 11-7 shows an audit event with the timestamp, username, IP Address, resource, and the exact command that was executed.

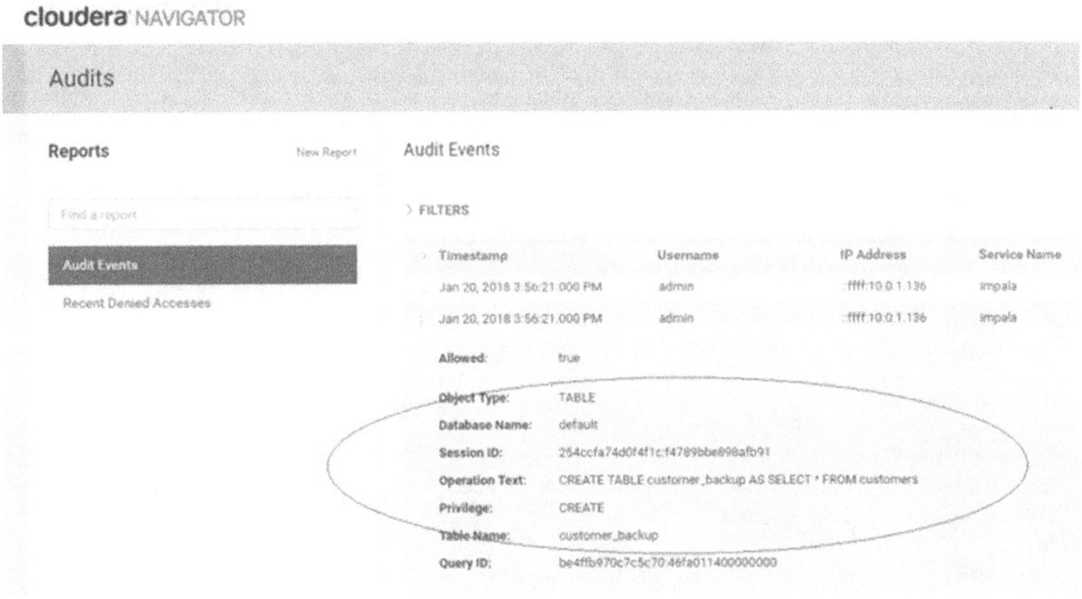

Figure 11-7. Searching using a specific metadata tag

Policy Enforcement and Data Lifecycle Automation

Cloudera Navigator allows data stewards to define policies that can be used to automate certain actions like executing commands, adding custom metadata tags, and sending a message to a JMS message queue.[vii] For example, you can create policies to automatically archive data after a specified period of time, or move data to a secure location based on an entity's metadata.

Cloudera Navigator REST API

Cloudera Navigator has REST API that developers can use to provide additional functionality or automate tasks (Figure 11-8). The API uses HTTP basic authentication and uses the same username and password that you use to log in to Cloudera Navigator.

Figure 11-8. Cloudera Navigator REST API

Cloudera Navigator Encrypt

Cloudera Navigator includes Cloudera Navigator Encrypt, Cloudera Navigator Key Trustee Server, and Cloudera Navigator Key HSM to provide an enterprise encryption and key management solution. Discussion of Navigator Encrypt is beyond the scope of this chapter. Please consult Cloudera's website for more information about Cloudera Navigator Encrypt.

Other Data Governance Tools

This is by no means an exhaustive list, but it should give you an idea of data governance tools that are available in the market today.

Apache Atlas

Apache Atlas is a data governance and metadata framework that is used mainly by Hortonworks. It provides similar features and functionalities provided by Cloudera Navigator, such as data classification, auditing, data lineage, and a policy engine.[viii] It also provides a REST API and integrates with third-party commercial data governance tools.

Informatica Metadata Manager and Enterprise Data Catalog

Informatica Metadata Manager is a metadata management tool from Informatica that integrates with Cloudera Navigator. Metadata Manager provides end-to-end enterprise data lineage. Cloudera Navigator only manages data stored in the data lake, and anything outside of the data lake is out of reach of Cloudera Navigator. Metadata Manager provides a complete view of data flow and transformation from data warehouses into data lakes and business intelligence tools across the entire ETL pipeline, providing data stewards with greater visibility into the chain of data transformation and business rules that have been applied to data.[ix] Informatica Enterprise Data Catalog is Informatica's next-generation data governance tool and will soon replace Metadata Manager. Enterprise Data Catalog uses machine learning to classify and organize data stored in your data lake. Informatica also has a master data management and data quality solution.

Collibra

Collibra is another enterprise data governance tool that integrates with Cloudera Navigator. Collibra Data Governance features a business glossary, data dictionary, reference data management, data stewardship automation, ingestion control, analytical model governance, and oversight of Hadoop jobs.

Waterline Data

Waterline Data is another data governance tool that uses machine learning to automatically recommend metadata tags and match it with business glossary terms. Waterline Data claims that it can reduce manual tagging of data by over 80%. The machine learning algorithm automatically improves based on the data steward's acceptance or rejection of the tags.[x]

Smartlogic

Smartlogic is slightly different than the other data governance tools I discussed. Smartlogic provides a content intelligence platform that combines rules engines and natural language processing to automatically classify data via application of metadata tags.[xi] It integrates with Solr, Sharepoint, and Marklogic and has partnerships with Oracle, Microsoft, Lucidworks, and SAP.

Summary

Cloudera Navigator is an enterprise data governance tool that enables organizations to properly manage and organize their data. The goal is to prevent the data lake from becoming a "data swamp" by making sure that your data is secure and trustworthy. Cloudera Navigator is extremely useful and should be part of your toolset; however it's not a comprehensive data governance and management tool. For example, it doesn't handle master data management and data quality management. As discussed earlier, you may need to integrate Cloudera Navigator with third-party data governance tools such as Informatica Master Data Management or Data Quality if you require such functionalities. Other tools such as Informatica Enterprise Information Catalog, Waterline Data, and Smartlogic use machine learning to build taxonomies and ontologies to automatically organize data, providing more advanced data governance capabilities.

References

i. A.R. Guess; "The Difference Between Data Governance & Data Management," Dataversity, 2018, http://www.dataversity.net/the-difference-between-data-governance-data-management/

ii. Patricia Cupoli, Susan Earley, Deborah Henderson; "DAMA-DMBOK2 Framework," DAMA, 2014, https://dama.org/sites/default/files/download/DAMA-DMBOK2-Framework-V2-20140317-FINAL.pdf

iii. Margaret Rouse; "metadata management," TechTarget, 2018, http://whatis.techtarget.com/definition/metadata-management

iv. Cloudera; "Cloudera Navigator Metadata Architecture," Cloudera, 2018, https://www.cloudera.com/documentation/enterprise/5-11-x/topics/cn_iu_metadata_arch.html

v. Cloudera; "Cloudera Navigator Auditing," Cloudera, 2018, https://www.cloudera.com/documentation/enterprise/5-6-x/topics/cn_iu_audits.html#cn_topic_7

vi. Cloudera; "Data Governance in Hadoop – Part 2," Cloudera, 2018, https://vision.cloudera.com/data-governance-in-hadoop-part-2/

vii. Cloudera; "Metadata Policies," Cloudera, 2018, https://www.cloudera.com/documentation/enterprise/5-7-x/topics/navigator_policies.html#xd_583c10bfdbd326ba-7dae4aa6-147c30d0933--7c4a

viii. Apache Atlas; "Data Governance and Metadata framework for Hadoop," Apache Atlas, 2018, http://atlas.apache.org/

ix. Scott Hedricks; "End-to-end Data Lineage Now Available with Informatica Integration with Cloudera Navigator," Cloudera, 2018, https://vision.cloudera.com/end-to-end-data-lineage-now-available-with-informatica-integration-with-cloudera-navigator/

x. Waterline Data; "Manage Your Data Like an Asset," Waterline Data, 2018, https://www.waterlinedata.com/product-overview/

xi. Bloomberg; "Company Overview of Smartlogic Semaphore Ltd," Bloomberg, 2018, https://www.bloomberg.com/research/stocks/private/snapshot.asp?privcapId=37460152

CHAPTER 12

Big Data in the Cloud

Big data deployments in the cloud have increasingly become popular these past few years. The flexibility and agility of the cloud is ideal for running Hadoop clusters. The cloud significantly reduces IT cost while providing applications the ability to scale. Expanding and shrinking clusters take minutes and systems administrators are not needed for most tasks. While some organizations still prefer on-premise big data deployments, most big data environments these days are deployed on one of the three main public cloud providers.

My goal in this chapter is to give you a high-level overview of Cloudera Enterprise in the cloud. For a more complete treatment of this subject, refer to the Cloudera Reference Architecture for AWS,[i] Azure,[ii] and Google Cloud Platform[iii] Deployments.

Amazon Web Services (AWS)

AWS, launched in March 2006, is the first and largest of the three main public cloud providers.[iv] AWS offers the broadest collection of storage, application, analytics, and compute services. Because of its first mover advantage, AWS is the leading cloud provider with customers ranging from the largest corporations to innovative start-ups and small businesses.

Microsoft Azure Services

Microsoft Azure offers cloud services similar to AWS such as SQL database, storage, application, and virtual machines to name a few. Azure has started eating into Amazon's market share lately due to aggressive pricing strategy and tight integration with Microsoft's technology stack. According to a survey, overall Azure adoption grew from 20 to 34% in 2017, while AWS stayed flat at 57% among its respondents.[v] Among enterprises, Azure also reduced the AWS lead with Azure, significantly increasing implementations from 26% to 46%. AWS adoption in the enterprise increased marginally from 56% to 59%.[vi]

CHAPTER 12 BIG DATA IN THE CLOUD

Google Cloud Platform (GCP)

The Google Cloud Platform, launched in 2011, is the youngest of three main public cloud providers. Initially built to support its own services such as YouTube and Google search, Google eventually built additional cloud services including Cloud Spanner and Cloud Bigtable and opened it to the general public. GCP has gained some traction lately, but it's still not considered ready for prime time compared to AWS and Azure.

Table 12-1 provides a rough guide[vii] to help you map equivalent services[viii] across the three providers.

Table 12-1. *Service Comparison Between AWS, Azure, and GCP*

Amazon Web Services	Google Cloud Platform	Microsoft Azure Services
Amazon EC2	Google Compute Engine	Azure Virtual Machines
Amazon EC2 Container Service	Google Container Engine	Azure Container Service
AWS Elastic Beanstalk	Google App Engine	Azure Cloud Services and App Service
AWS Lambda	Google Cloud Functions	Azure Functions
Amazon Glacier and Amazon S3 Standard – Infrequent Access	Google Cloud Storage Nearline	
Amazon S3	Google Cloud Storage Standard	Azure Block Blobs
Amazon EC2 Container Registry	Google Container Registry	
Amazon DynamoDB	Google Cloud Datastore or Google Cloud Bigtable	Azure DocumentDB
Amazon RDS	Google Cloud SQL	Azure SQL Database
Amazon EMR and AWS Data Pipeline	Google Cloud Dataflow and Google Cloud Dataproc	Azure HD Insight
Amazon Kinesis and Amazon Simple Queue Service (SQS)	Google Cloud Pub/Sub	Azure Event Hubs and Azure Service Bus
Amazon Redshift	Google BigQuery	Azure SQL Data Warehouse and Azure Data Lake Analytics

(continued)

Table 12-1. (*continued*)

Amazon Web Services	Google Cloud Platform	Microsoft Azure Services
Amazon CloudWatch	Google Cloud Monitoring and Google Cloud Logging	Azure Application Insights and Azure Operational Insights
Amazon Elastic Load Balancing	Google Cloud Load Balancing	Azure Load Balancer
Amazon Route 53	Google Cloud DNS and Google Domains	Azure DNS
AWS Direct Connect	Google Cloud Interconnect	Azure ExpressRoute
AWS Identity and Access Management (IAM)	Google Cloud Identity and Access Management	Azure Active Directory
AWS Organizations	Google Cloud Resource Manager	Azure Resource Manager
AWS Key Management Service (KMS)	Google Cloud Key Management Service	Azure Key Vault
Amazon Inspector	Google Cloud Security Scanner	
AWS CloudFormation	Google Cloud Deployment Manager	Azure Resource Manager

Cloudera Enterprise in the Cloud

This chapter's emphasis is on running Cloudera Enterprise in the cloud. With this particular setup, the cloud providers are only responsible for providing the infrastructure. Cloudera Enterprise components and features will be utilized instead of services offered by the cloud provider. Cloudera Enterprise can integrate with some of the services provided by the cloud providers such Amazon S3 and Azure Block Blobs.

Hybrid and Multi-Cloud

You may be wondering why someone would use Cloudera Enterprise in the cloud when the three public cloud providers already offer big data services? Two reasons: vendor lock-in and data portability. Once you have your data in one of the cloud providers and completely dependent on their services, you are locked in to their platform.

It is extremely difficult to migrate your data on premise or to another cloud provider. Even if you're able to migrate your data, you still have to refactor your applications using the services provided by the destination cloud provider. Compare this with using Cloudera Enterprise in the cloud. Cloudera Enterprise is a hybrid / multi-cloud big data platform, which means that it's possible to deploy Cloudera Enterprise clusters to run on AWS, Azure, and GCP (or on premise). Migrating Cloudera Enterprise to another cloud provider is significantly easier. Because you're just basically changing infrastructure providers, migrating your applications is just a matter of updating connection details.

This flexibility is extremely important in today's ever-changing technology landscape. Companies have been migrating their on-premise CDH cluster to AWS and vice versa for years. I know of a few companies that successfully migrated their CDH clusters from AWS to Azure. Some companies have their production clusters on premise, while their development and test/QA environments are in AWS or Azure.

According to the RightScale 2017 State of the Cloud Report, Hybrid Cloud is the preferred enterprise strategy and that 85% of enterprises have a multi-cloud strategy, up from 82 percent in 2016.[ix]

Transient Clusters

With Cloudera Enterprise, you have the option to configure transient or permanent (persistent) clusters in the cloud. Transient clusters can lower costs for some use cases with usage-based pricing. The self-service model provides flexibility by allowing users to deploy and administer their own clusters. It is appropriate for data science or data engineering jobs that are executed intermittently. You start a cluster, run your jobs, store your data in an object store such as S3 (so your data is still available for future use), then you shut down your cluster after you're done. These steps are sometimes scripted so everything is automated. In this type of deployment model, storage and compute are decoupled.[x] Examples of transient workloads include ETL jobs, training machine learning models, ad hoc data analysis, and development and testing/QA workflows[xi] to mention a few.

Persistent Clusters

For clusters that need to be available at all times, persistent clusters are more appropriate. These persistent clusters are perpetually "on" clusters in the cloud and have similar requirements to on-premise clusters. These are also known as "lift-and-shift" clusters.

These clusters are typically part of an enterprise environment where the use cases are more strategic. These clusters usually have high availability and disaster recovery, resource management, security and data governance. Examples of persistent clusters are large, multi-user clusters, BI/Analytic clusters, HBase and Kafka clusters to mention a few.[xii] Pricing for a persistent cluster is usually traditional node-based licensing.

Cloudera Director

Cloudera Enterprise provides a powerful cloud administration tool called Cloudera Director (see Figure 12-1). Cloudera Director enables Cloudera's hybrid and multi-cloud capabilities by providing flexible and easy-to-use self-service features for deploying and administering Cloudera Enterprise in cloud environments such as AWS, Azure, and GCP. Cloudera Director includes a client component and a REST API that can be used to automate and script cloud deployments.

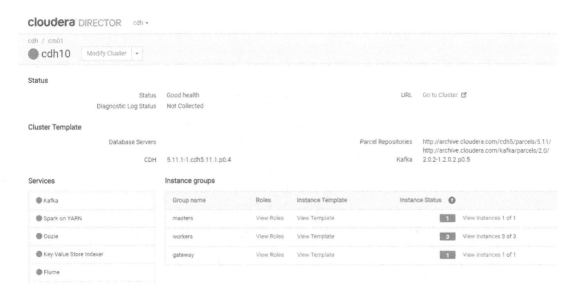

Figure 12-1. Cloudera Director

Cloudera Director Architecture

Cloudera Director makes it extremely easy for users to deploy and maintain Cloudera environments on all three public cloud providers. As you can see in Figure 12-2, the actual administration and monitoring of the CDH cluster is still performed by Cloudera

Manager, while the data governance, security, auditing, and metadata management are performed by Cloudera Navigator. You start to appreciate the value of Cloudera Director once your organization starts deploying multiple clusters in the cloud.

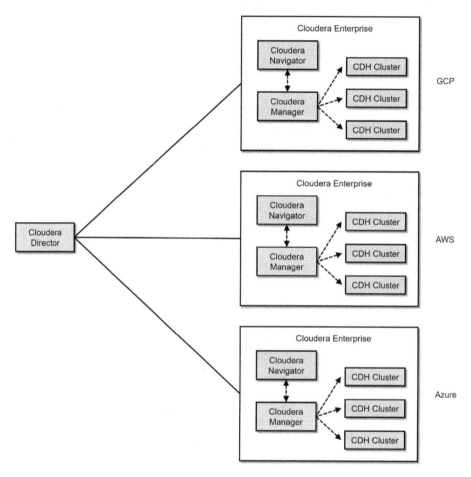

Figure 12-2. Cloudera Director high-level architecture

Cloudera Director Client

Cloudera Director includes client software that you can use to deploy and manage clusters from the command line. It is usually used to quickly deploy test and development environments. The client uses a configuration file to automate the cluster deployment.

Cloudera Director REST API

Every functionality of Cloudera Director can be accessed via a REST API. You can use the REST API to automate the task of installing and managing Cloudera cloud deployments. You can call the REST API from a script or application written in Python and Java. To learn more about Cloudera Director's REST API, refer to the Cloudera Director SDK Github page.[xiii]

Cloudera on AWS

We'll focus on running Cloudera Enterprise on AWS in this chapter since AWS is the most popular public cloud provider. Most of the concepts I talk about in this chapter also apply to Cloudera Enterprise on Azure and GCP. I'll present a high-level overview of the components and requirements of deploying Cloudera Enterprise on AWS. Providing a comprehensive guide to deploying Cloudera Enterprise on AWS is beyond the scope of this chapter. For more details, refer to the Cloudera Enterprise Reference Architecture for AWS Deployments.[xiv]

Regions and Availability Zones

In AWS, services are deployed in self-contained geographical locations known as Regions. Regions are further subdivided into availability zones. Availability zones are separate locations within a region where you can deploy your clusters. Cloudera Enterprise deployments can only reside in a single availability zone.[xv]

Virtual Private Cloud (VPC)

AWS has a concept of a virtual private cloud or VPC.[xvi] A VPC presents a virtual network that acts like a traditional network.

Security Groups

Security groups are like firewalls in AWS. Security groups enable users to open or close ports, restrict IP addresses, and allow or disallow certain network traffic to EC2 instances. Cloudera Enterprise clusters require three security groups for your cluster, flume or ingest nodes, and edge or gateway nodes.[xvii] Figure 12-3 shows a high-level view of how security groups are implemented in AWS.

CHAPTER 12 BIG DATA IN THE CLOUD

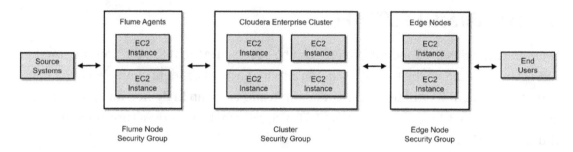

Figure 12-3. Cloudera Enterprise on AWS Security Groups

EC2 Instance

An EC2 instance is sort of like a virtual server with CPU, RAM, network, and storage associated with it. Amazon provides several EC2 instance types optimized for various use cases. The instance types are classified into several categories to help you choose the correct EC2 instance types for your application. Instance types categories include General Purpose, Compute Optimized, Memory Optimized, Accelerated Computing, and Storage Optimized. Cloudera recommends appropriate EC2 instance types for each CDH cluster node type.[xviii]

Master Nodes

EBS-optimized instances are required when using EBS storage for your master nodes.

- c4.2xlarge, c4.4xlarge, c4.8xlarge
- m4.xlarge, m4.2xlarge, m4.4xlarge, m4.8xlarge, m4.10xlarge, m4.16xlarge
- r4.xlarge, r4.2xlarge, r4.4xlarge, r4.8xlarge, r4.16xlarge

The following instance types are recommended if deploying master nodes using ephemeral disk.

- c3.8xlarge
- d2.8xlarge
- i2.8xlarge
- i3.16xlarge
- r3.8xlarge

Worker Nodes

EBS-optimized instances are required when using EBS storage for your worker nodes.

- c4.2xlarge, c4.4xlarge, c4.8xlarge
- m4.xlarge, m4.2xlarge, m4.4xlarge, m4.10xlarge
- r4.xlarge, r4.2xlarge, r4.4xlarge, r4.8xlarge, r4.16xlarge
- EBS-optimized D2, I2, or R3 instance types

The following instance types are recommended if deploying master nodes using ephemeral disk.

- c3.8xlarge
- d2.8xlarge
- i2.8xlarge, i2.16xlarge
- i3.8xlarge, i3.16xlarge
- r3.8xlarge

Note Cloudera recommends d2.8xlarge, i2.8xlarge, or i3.8xlarge instances for CDH clusters that heavily utilize Impala.

Edge Nodes

There are no specific recommendations for edge nodes. You can choose any instance type as long as it has enough resources for your workload.

Relational Databases

Cloudera Enterprise requires a relational database such as MySQL, PostgreSQL, or Oracle to store databases used by Cloudera Manager, Hive and Impala metastore, Hue and Oozie. You can use an EC2 instance or RDS to house your relational database.

Cloudera Director

Cloudera recommends using c3.large or c4.large instances for Cloudera Director.[xix]

Storage

I describe the different types of storage supported by Cloudera Enterprise.

Ephemeral

Ephemeral or instance storage is lost if you stop or terminate the EC2 instance attached to it (it will survive an instance restart). Ephemeral storage is appropriate for HDFS data directories for performance reasons since reading from ephemeral storage doesn't require reading remotely over the network.

You should have a backup or disaster recovery plan when using ephemeral storage. Copying data to S3 using distcp, backing up your data using Cloudera Manager's Backup and Data Recovery (BDR) feature, or using a data transfer or replication tool such as StreamSets Data Collector can help prevent data loss.

Elastic Block Storage (EBS)

Compared to ephemeral storage, Elastic Block Storage (EBS) provides permanent block level storage to EC2 instances. Data is not lost even if you stop or terminate your EC2 instance. There are several types of EBS storage such as ST1, SC1, and GP2 volumes. Cloudera recommends GP2 for all EBS-backed instances. Cloudera also recommends that GP2 volumes be at least 100 GB in size to maintain sufficient IOPS.[xx]

S3

S3 is an object store offered as a service by Amazon. S3 is much cheaper than traditional storage engines, which makes it ideal for storing backups and "cold" data such as old historical data. For transient workloads, S3 is also the ideal data store for persisting results from your ETL or BI/Analytic workflows.

Some of S3's IO characteristics make it unsuitable for some use cases. First is latency. Reading data from S3 is several times slower than reading data from HDFS or Kudu. If performance is important, the best practice is to store "hot" data in HDFS or Kudu and "cold" data in S3. Second, S3 is an eventually consistent data store, which means that data written to it might not be available for reading for a period of time, usually a few milliseconds or seconds. There are documented cases where the lag was several hours.[xxi] This can cause issues in applications that require read consistency.

Cloudera Enterprise and S3

Several Cloudera Enterprise services can interact with S3. The Cloudera S3 Connector enables secure S3 access in Hue.[xxii] Hive, Hive on Spark, Spark, and Impala can read and write data to S3. For more details on how to access S3 data using Impala, refer to Chapter 3. For details on how to access S3 using Spark, refer to Chapter 5.

S3Guard

As mentioned earlier, S3 is an eventually consistent data store. However, most Hadoop applications require read consistency to work correctly. Most storage engines shipped with Cloudera Enterprise (HDFS, HBase, and Kudu) provide read consistency (Solr is eventually consistent).

S3Guard was developed by the open source community to deal with issues caused by S3's eventual consistency.[xxiii] S3Guard doesn't "fix" S3's consistency model, it works by recording all metadata changes written to S3 to an external "metadata store"; since this metadata store is supposed to be consistent, applications can use the information to complete missing metadata that may not yet be available in S3. S3Guard ships with CDH.

Cloudera Enterprise on AWS Architecture

Figure 12-4 shows what a typical Cloudera Enterprise environment looks like on AWS.[xxiv]

CHAPTER 12   BIG DATA IN THE CLOUD

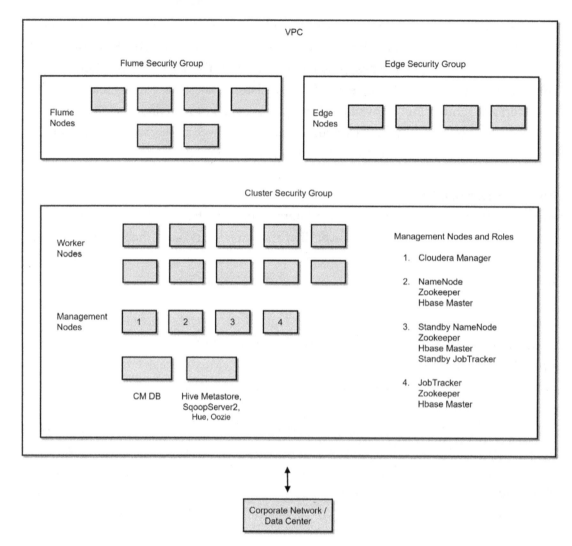

Figure 12-4.   Cloudera Enterprise on AWS Architecture

Deploying Cloudera Enterprise on AWS Using Cloudera Director

There is no better way to learn than actually deploying a cluster on AWS using Cloudera Director. Note that this installation is only for test or development environments.

You need to download and install Cloudera Director on an EC2 instance. Setting up the AWS environment is beyond the scope of this book. If you need guidance on how to set up your AWS environment for Cloudera Director such as setting up your VPC, security groups, and EC2 instances, refer to the Cloudera Director User Guide.[xxv]

I assume you've started your EC2 instance. We'll use Red Hat/CentOS 7.x for this example (Ubuntu and Red Hat/CentOS 6.x are also supported).

Log in to your EC2 instance using your pem file.

```
ssh -i mypemfile.pem ec2-user@cloudera_director_private_ip
```

Note You may have to use a different ssh login depending on the AMI you're using. In your AWS console, check the usage instructions in the Instances page. You may have to use centos or root to log in instead of ec2-user.

You need to install a supported version of the Oracle JDK. JDK 7 and 8 are currently supported. You can download the rpm from Oracle's website.

```
sudo yum localinstall jdk-version-linux-x64.rpm
```

Add the Cloudera Director repository

```
cd /etc/yum.repos.d/
sudo wget http://archive.cloudera.com/director/redhat/7/x86_64/director/cloudera-director.repo
```

Once you've updated your repository, you can now install the Cloudera Director server and client software.

```
sudo yum install cloudera-director-server cloudera-director-client
```

Start Cloudera Director server.

```
sudo service cloudera-director-server start
```

Sometimes the firewall in Red Hat/CentoOS 7 is enabled. This will prevent you from connecting to Cloudera Director. Disable and stop the firewall.

```
sudo systemctl disable firewalld
sudo systemctl stop firewalld
```

CHAPTER 12 ■ BIG DATA IN THE CLOUD

Note Cloudera does not recommend configuring your security groups to allow Internet access to your EC2 instance's public IP address. Cloudera recommends you use a SOCKS proxy server to connect to your cluster nodes and Cloudera Manager. Refer to the Cloudera Director User Guide for more details on how to configure a SOCKS proxy server. For testing purposes, you can open the security groups to allow access to the public IP address. Just don't forget to close it!

Open the web browser and go to the public or private IP address on port 7189 of the EC2 instance running Cloudera Director. Log in using "admin" on both username and password fields (see Figure 12-5).

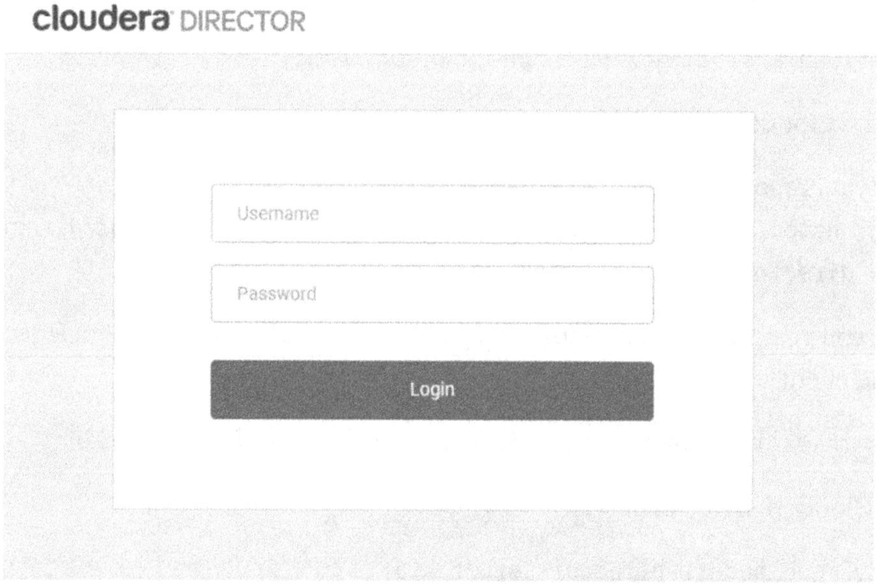

Figure 12-5. Cloudera Director Login Page

CHAPTER 12 BIG DATA IN THE CLOUD

You'll be presented with a dashboard to assist you in adding a Cloudera Manager or a cluster. Let's add a Cloudera Manager (see Figure 12-6).

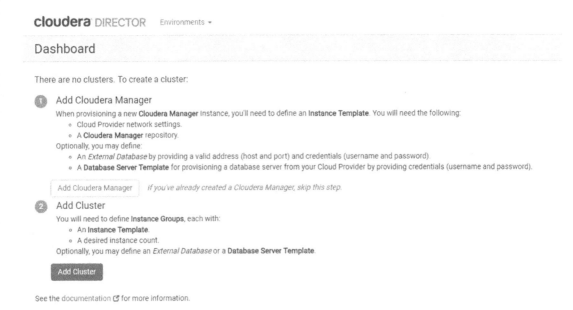

Figure 12-6. Cloudera Director Dashboard

Fill in the necessary information to create a Cloudera Manager instance such as Instance type, AMI ID, Security group ID, VPC subnet ID, and so on as shown in Figure 12-7.

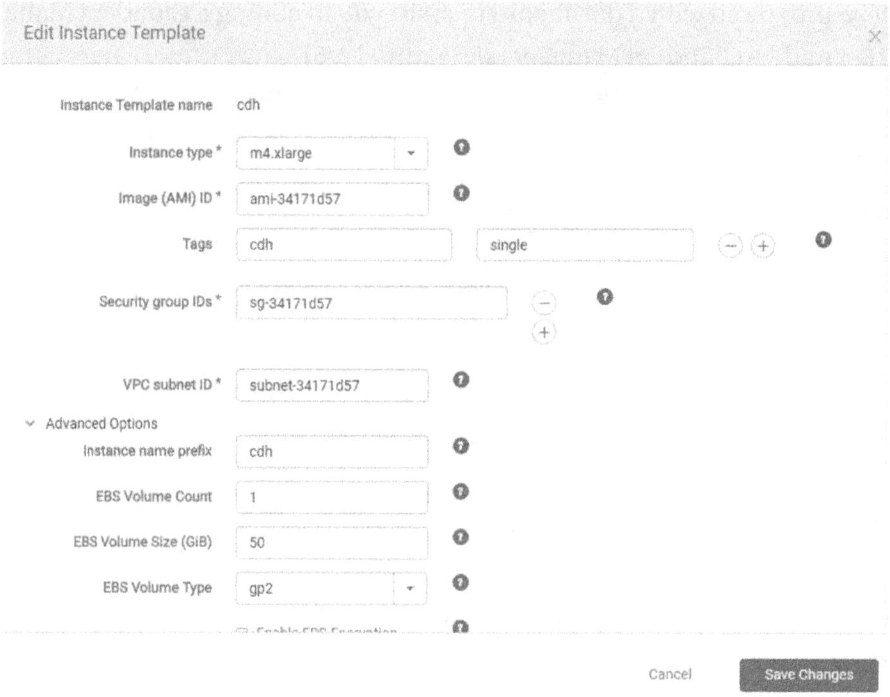

Figure 12-7. *Edit Instance Template*

Scroll down to enter more details about your instance (see Figure 12-8). Consult the AWS documentation for more information on the different options. Save the changes when you're done.

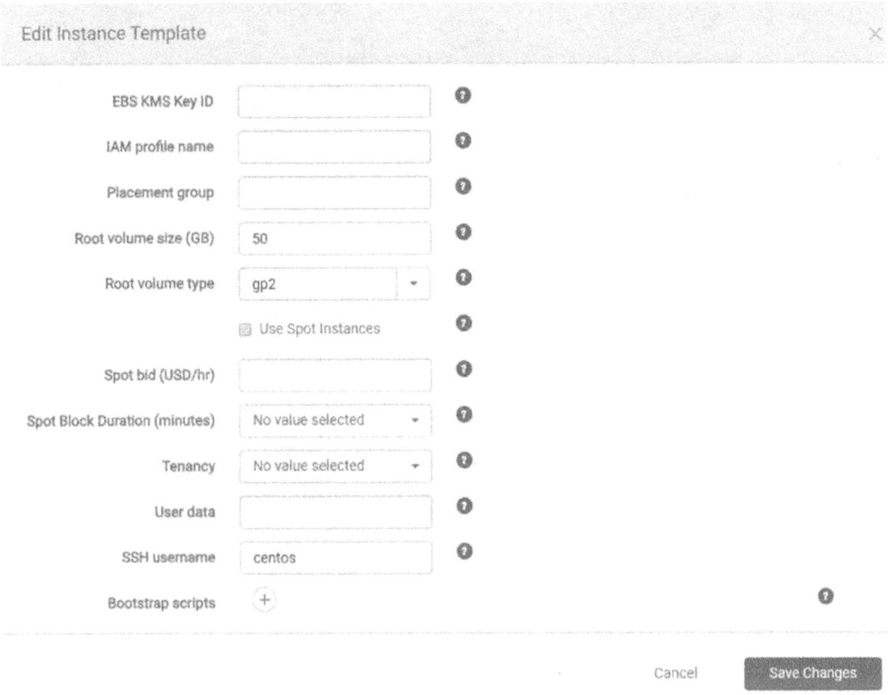

CHAPTER 12 BIG DATA IN THE CLOUD

Figure 12-8. Edit Instance Template Page 2

Enter details about your Cloudera Manager instance such as the name and password (see Figure 12-9).

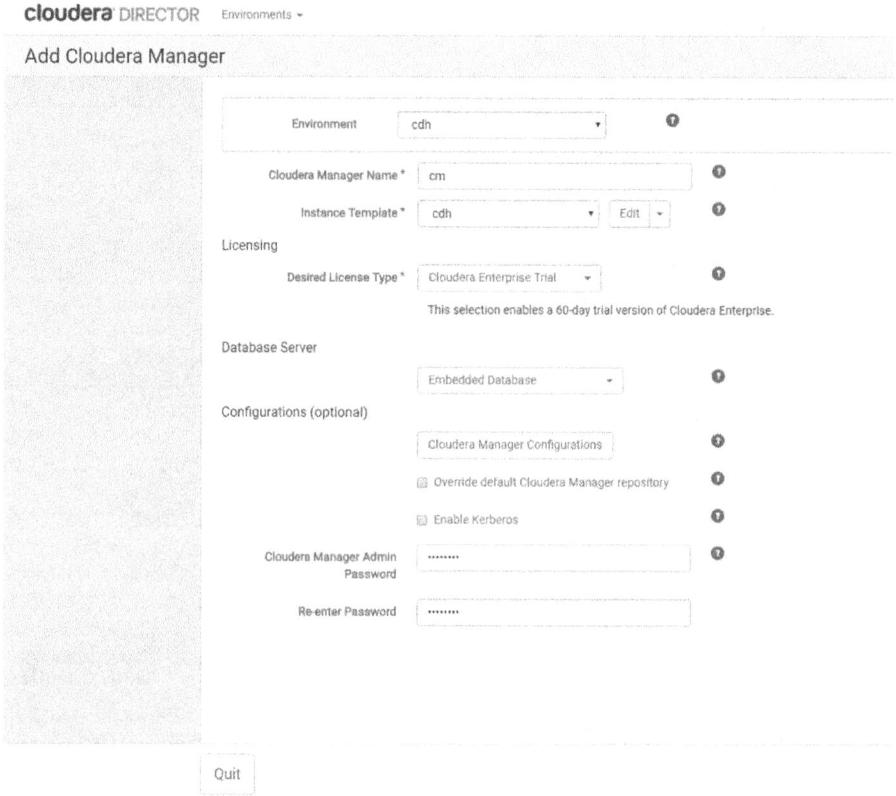

CHAPTER 12 BIG DATA IN THE CLOUD

Figure 12-9. Add Cloudera Manager

While the Cloudera Manager instance is being created, you can start creating the cluster. Cloudera Director gives you the option of creating a cluster from scratch or to clone from an existing cluster. You also have to enter the cluster name as well as choose the services you want to run on your cluster (see Figure 12-10).

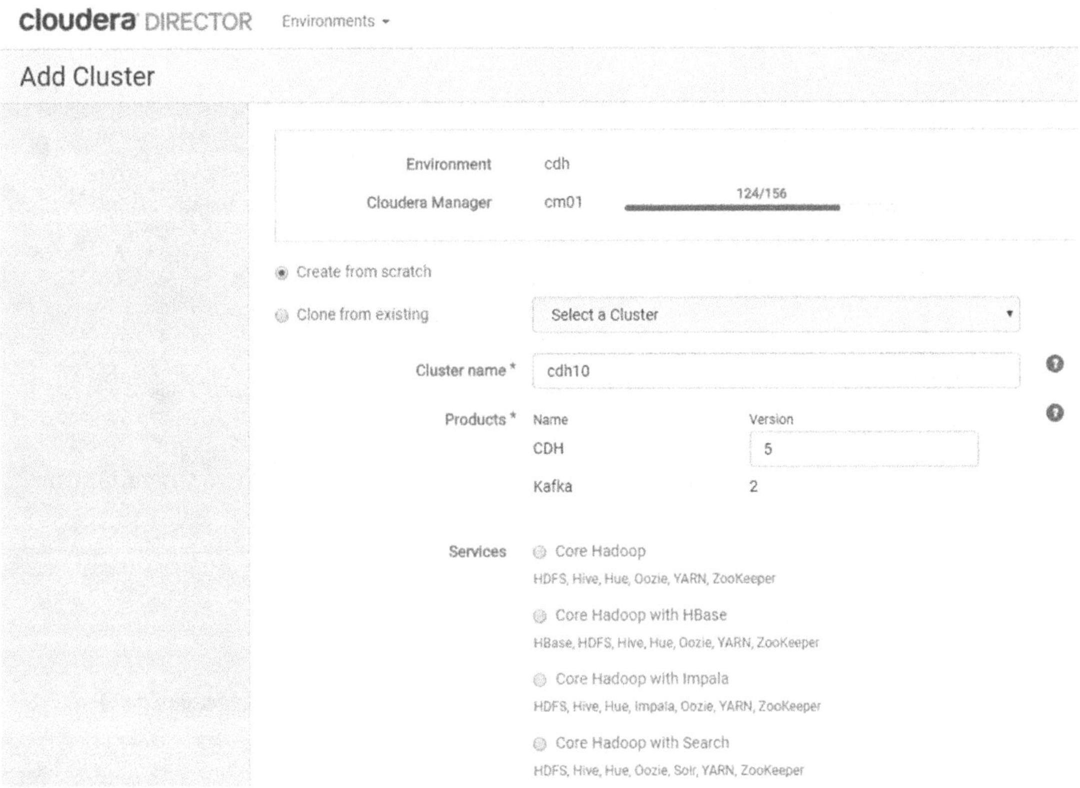

CHAPTER 12 ■ BIG DATA IN THE CLOUD

Figure 12-10. Add Cluster

Cloudera Director lets you easily configure your cluster nodes as shown in Figure 12-11. Note that there is a required minimum of three nodes for your workers.

Figure 12-11. Cluster nodes

525

CHAPTER 12   BIG DATA IN THE CLOUD

Confirm the number of nodes in your cluster. Click OK when you're ready (see Figure 12-12).

Figure 12-12. *Confirmation to create compute instance*

You'll see a progress indicator for both Cloudera Manager and Cluster installations (see Figure 12-13).

Figure 12-13. *Cloudera Manager and Cluster installation*

Click "Next" when the installation is done (see Figure 12-14).

Figure 12-14. *Cloudera Manager and Cluster installation completed*

CHAPTER 12 BIG DATA IN THE CLOUD

The Cloudera Director dashboard will be shown (see Figure 12-15).

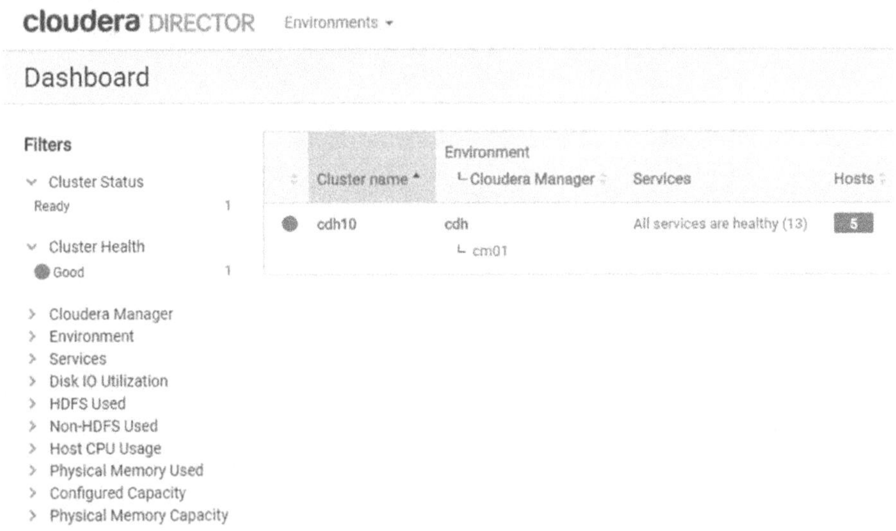

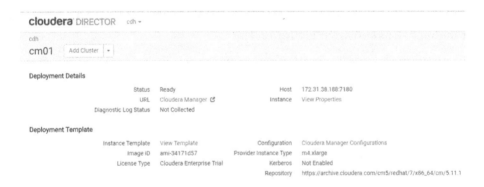

Figure 12-15. Cloudera Director Dashboard

Click on the name of the Cloudera Manager installation to get more information about Cloudera Manager (see Figure 12-16).

Figure 12-16. Cloudera Director Dashboard – Cloudera Manager Details

CHAPTER 12 ■ BIG DATA IN THE CLOUD

Or you can click on the name of the Cloudera Enterprise cluster to get more information about the cluster such as the services installed and the instance group status (see Figure 12-17).

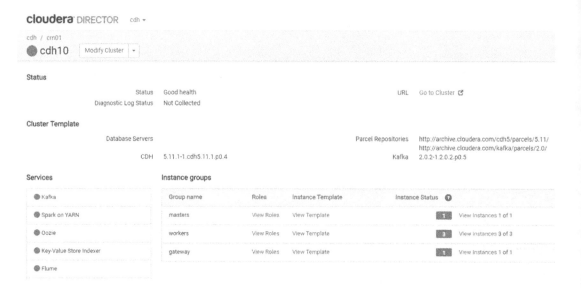

Figure 12-17. Cloudera Director Dashboard – Cluster Details

Click the link "Go to Cluster" to open Cloudera Manager. You'll see the Cloudera Manager main page as shown in Figure 12-18.

CHAPTER 12 ■ BIG DATA IN THE CLOUD

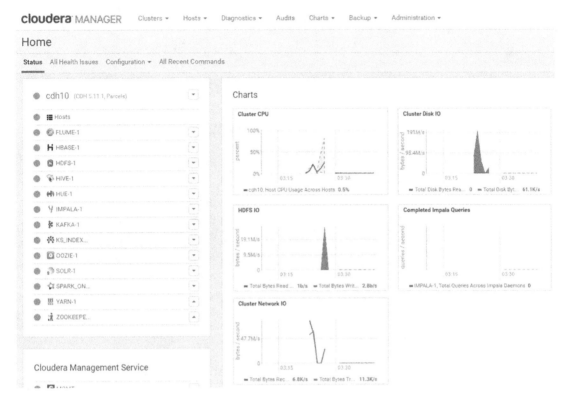

***Figure 12-18.** Cloudera Manager main page*

Navigate to the "Host" tab to get a list of cluster nodes (see Figure 12-19).

CHAPTER 12 BIG DATA IN THE CLOUD

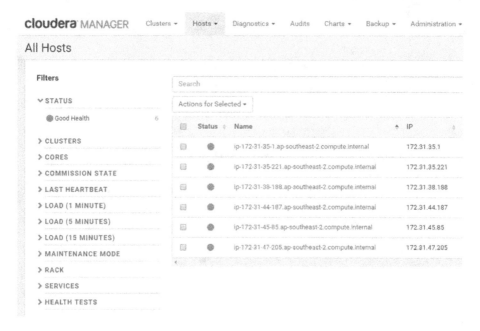

Figure 12-19. All Hosts

Expand "Roles" to get a list of services running on each node (see Figure 12-20).

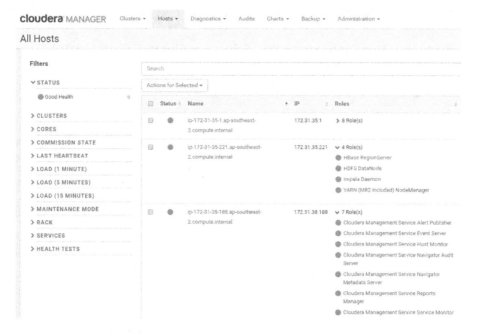

Figure 12-20. All Host – Roles

Congratulations! You've just installed a test Cloudera Enterprise cluster in AWS using Cloudera Director.

Cloudera Enterprise on Azure and GCP

For details on how to deploy Cloudera on Azure and GCP, refer to the Cloudera Director User Guide.[xxvi] The technology may be different, but most of the concepts discussed in deploying Cloudera Enterprise on AWS also apply to Azure and GCP.

Note You may run into issues with ntp[xxvii] when running Kudu on Microsoft Azure. Disabling time synchronization should fix the problem.

Cloudera Altus

According to Cloudera, Cloudera Altus[xxviii] is a "platform-as-a-service that makes it easy and cost-effective to process large-scale data sets in the cloud." Geared toward data engineers, it enables submission and execution of data pipelines created in Spark, Hive on Spark, Hive on MapReduce, and MapReduce. The data engineer doesn't need to worry about creating a cluster since the cluster is automatically created on demand. You pay by the hour and the number of worker nodes, as well as the type of AWS EC2 instance type you choose for your cluster.[xxix] For now, Cloudera Altus only runs on AWS but support for Azure is coming soon.

If you've ever created a transient cluster using Cloudera Director to run your ETL pipelines, consider Cloudera Altus as a more streamlined and automated version. Cloudera Altus is in many ways similar to Amazon EMR.

Amazon Elastic MapReduce (EMR)

Amazon EMR was the first service that let users dynamically spin up transient clusters in the cloud using EC2 instances. EMR lets users access other Amazon services and store data on Amazon data stores such as DynamoDB and S3.[xxx] EMR enjoys significant market share, targeting companies with transient big data workloads.

Chapter 12   Big Data in the Cloud

Databricks

Databricks is another cloud-based big data platform. Founded in 2013 by the same team who created Apache Spark at UC Berkeley's AMPLab. While Amazon and Cloudera provide other data processing framework in addition to Spark (Hive and MapReduce), Databricks is focused mainly on Spark. Databricks recently raised another $140 million in venture funding, bringing the total funding raised by the company to $247 million,[xxxi] giving it enough war chest to compete with other big data platforms such as Cloudera, Amazon, MapR, and Hortonworks.

Summary

Big data deployments in the cloud can help organizations get value out of their data faster, cheaper, and more effectively. However, not all platforms are created equal. Some have more features than others. Some provide more flexibility. Some are easier to use. There are plenty of options when it comes to picking the right big data and cloud platform for your organization. I hope the information I provided in this chapter will help you make the right decision.

References

i. Cloudera; "Cloudera Enterprise Reference Architecture for AWS Deployments," Cloudera, 2017, http://www.cloudera.com/documentation/other/reference-architecture/PDF/cloudera_ref_arch_aws.pdf

ii. Cloudera; "Cloudera Enterprise Reference Architecture for Azure Deployments," Cloudera, 2017, http://www.cloudera.com/documentation/other/reference-architecture/PDF/cloudera_ref_arch_azure.pdf

iii. Cloudera; "Cloudera Enterprise Reference Architecture for Google Cloud Platform Deployments," Cloudera, 2017, http://www.cloudera.com/documentation/other/reference-architecture/PDF/cloudera_ref_arch_gcp.pdf

CHAPTER 12 BIG DATA IN THE CLOUD

iv. Wolf, Damien; "Who wins the three-way cloud battle? Google vs. Azure vs. AWS," ReadWrite, 2017, https://readwrite.com/2017/02/20/wins-three-way-cloud-battle-google-vs-azure-vs-aws-dl1/

v. RightScale; "RIGHTSCALE 2017 STATE OF THE CLOUD REPORT UNCOVERS CLOUD ADOPTION TRENDS," RightScale, 2017, https://www.rightscale.com/press-releases/rightscale-2017-state-of-the-cloud-report-uncovers-cloud-adoption-trends

vi. RightScale; "RIGHTSCALE 2017 STATE OF THE CLOUD REPORT UNCOVERS CLOUD ADOPTION TRENDS," RightScale, 2017, https://www.rightscale.com/press-releases/rightscale-2017-state-of-the-cloud-report-uncovers-cloud-adoption-trends

vii. Google; "Map AWS services to Google Cloud Platform products," Google, 2017, https://cloud.google.com/free/docs/map-aws-google-cloud-platform

viii. Google; "Map Azure services to Google Cloud Platform products," Google, 2017, https://cloud.google.com/free/docs/map-azure-google-cloud-platform

ix. RightScale; "RIGHTSCALE 2017 STATE OF THE CLOUD REPORT UNCOVERS CLOUD ADOPTION TRENDS," RightScale, 2017, https://www.rightscale.com/press-releases/rightscale-2017-state-of-the-cloud-report-uncovers-cloud-adoption-trends

x. Cloudera; "Cloudera Enterprise in the Cloud," Cloudera, 2016, http://www.clouderaworldtokyo.com/session-download/B5-ClouderaUpdateNov2016.pdf

xi. Cloudera; "Cloudera Enterprise in the Cloud," Cloudera, 2016, http://www.clouderaworldtokyo.com/session-download/B5-ClouderaUpdateNov2016.pdf

533

CHAPTER 12 BIG DATA IN THE CLOUD

xii. Cloudera; "Cloudera Enterprise in the Cloud," Cloudera, 2016, http://www.clouderaworldtokyo.com/session-download/B5-ClouderaUpdateNov2016.pdf

xiii. Cloudera; "Cloudera Director API clients," Cloudera, 2018, https://github.com/cloudera/director-sdk

xiv. Cloudera; "Cloudera Enterprise Reference Architecture for AWS Deployments," Cloudera, 2018, http://www.cloudera.com/documentation/other/reference-architecture/PDF/cloudera_ref_arch_aws.pdf

xv. Cloudera; "Cloudera Enterprise Reference Architecture for AWS Deployments," Cloudera, 2018, http://www.cloudera.com/documentation/other/reference-architecture/PDF/cloudera_ref_arch_aws.pdf

xvi. Amazon; "Amazon Virtual Private Cloud Overview," Amazon, 2018, http://docs.aws.amazon.com/AmazonVPC/latest/GettingStartedGuide/ExerciseOverview.html

xvii. Cloudera; "Cloudera Enterprise Reference Architecture for AWS Deployments," Cloudera, 2018, http://www.cloudera.com/documentation/other/reference-architecture/PDF/cloudera_ref_arch_aws.pdf

xviii. Cloudera; "Cloudera Enterprise Reference Architecture for AWS Deployments," Cloudera, 2018, http://www.cloudera.com/documentation/other/reference-architecture/PDF/cloudera_ref_arch_aws.pdf

xix. Cloudera; "Creating an EC2 Instance for Cloudera Director," Cloudera, 2018, https://www.cloudera.com/documentation/director/1-5-x/topics/director_deployment_start_launcher.html

xx. Cloudera; "Cloudera Enterprise Reference Architecture for AWS Deployments," Cloudera, 2018, http://www.cloudera.com/documentation/other/reference-architecture/PDF/cloudera_ref_arch_aws.pdf

CHAPTER 12 BIG DATA IN THE CLOUD

xxi. GitHub; "S3BinaryCacheStore is eventually consistent," GitHub, 2018, https://github.com/NixOS/nix/issues/1420

xxii. Cloudera; "How to Enable S3 Cloud Storage in Hue," Cloudera, 2018, https://www.cloudera.com/documentation/enterprise/5-11-x/topics/hue_use_s3_enable.html

xxiii. Cloudera; "Introducing S3Guard: S3 Consistency for Apache Hadoop," Cloudera, 2018, https://blog.cloudera.com/blog/2017/08/introducing-s3guard-s3-consistency-for-apache-hadoop/

xxiv. Cloudera; "Cloudera Enterprise Reference Architecture for AWS Deployments," Cloudera, 2018, http://www.cloudera.com/documentation/other/reference-architecture/PDF/cloudera_ref_arch_aws.pdf

xxv. Cloudera; "Cloudera Director User Guide," Cloudera, 2018, https://www.cloudera.com/documentation/director/latest/PDF/cloudera-director.pdf

xxvi. Cloudera; "Cloudera Director User Guide," Cloudera, 2018, https://www.cloudera.com/documentation/director/latest/PDF/cloudera-director.pdf

xxvii. Cloudera; "kudu service are getting down frequently," Cloudera, 2018, https://community.cloudera.com/t5/Interactive-Short-cycle-SQL/kudu-service-are-getting-down-frequently/m-p/56122

xxviii. Cloudera; "Cloudera Altus Cloud PaaS," Cloudera, 2018, https://www.cloudera.com/products/altus.html

xxix. ZDNet; "Cloudera introduces Altus, offering Hadoop jobs as a Service," Cloudera, 2018, http://www.zdnet.com/article/cloudera-introduces-altus-offering-hadoop-jobs-as-a-service/

xxx. Amazon; "Amazon EMR," Cloudera, 2018, https://aws.amazon.com/emr/

CHAPTER 12   BIG DATA IN THE CLOUD

xxxi. BusinessInsider; "Four-year-old startup Databricks raised another $140 million to solve the hardest problem in AI," BusinessInsider, 2017, https://www.businessinsider.com.au/databricks-ai-funding-valuation-2017-8?r=US&IR=T

CHAPTER 13

Big Data Case Studies

Big data has disrupted entire industries. Innovative use case in the fields of financial services, telecommunications, transportation, health care, retail, insurance, utilities, energy, and technology (to mention a few) have revolutionized the way organizations manage, process, and analyze data. In this chapter, I present real big data case studies from six innovative companies: Navistar, Cerner, British Telecom, Shopzilla, Thomson Reuters, and Mastercard. Information and details about the case studies are referenced from Cloudera's website: www.Cloudera.com.

Navistar

Navistar is one of the leading manufacturers of commercial buses, trucks, defense vehicles, and engines.

Use Cases

Navistar's use cases include predictive maintenance, remote diagnostics and route optimization. Unscheduled vehicle repairs and breakdowns are costly and inefficient. When service interruption occurs, the impact can be significant. Vehicle owners usually lose US$1,000 in revenue per vehicle daily. Scheduled vehicle maintenance based on mileage is primitive and doesn't address Navistar's mounting vehicle maintenance problems. A more modern approach that involves real-time data monitoring and predictive analytics is needed. Furthermore, Navistar's traditional data warehouse was unable to support the increasing amount of real-time, high-volume telematics and sensor data that they were ingesting.[i]

> *As we collected more data, the analytic process slowed to a near halt on our legacy systems.*
>
> —Ashish Bayas, CTO at Navistar

Solution

Navistar built an IoT-enabled remote diagnostic platform on Cloudera Enterprise that ingests over 70 telematics and sensor data feeds from more than 300,000 connected vehicles. The data is further enriched with third-party data such as meteorological, geolocation, traffic, vehicle usage, historical warranty, and parts inventory information. The platform uses machine learning to proactively detect vehicle issues and predict vehicle maintenance requirements. Navistar also uses the platform to help prevent accidents and promote road safety. After building a prototype in September 2014, it took Navistar just six months to put the platform into production.

Using IoT devices, machine learning and predictive analytics powered by Cloudera, Navistar has completely overhauled the way we sell, maintain and service our customers' vehicle fleets.

—Ashish Bayas, CTO at Navistar

Technology and Applications

- Data Platform: Cloudera Enterprise
- Workloads: Analytic Database, Data Science & Engineering
- Components: Apache Spark, Apache Impala (incubating), Apache Kafka
- BI & Analytics Tools: Information Builders WebFOCUS In-Document Analytics, Microsoft Power BI, Microsoft SQL Server Analytic Services Models, Microsoft SQL Server Reporting Services, SAS Enterprise Guide, Tableau Desktop, Tableau Server
- Data Science Tools: Python, R, Scala
- ETL Tool: IBM InfoSphere DataStage

With Cloudera, we can analyze data in ways and speeds that were not previously possible. We can evaluate billions of rows of data from connected vehicles in hours, not weeks, to enable predictive maintenance.

—Terry Kline, CIO, Navistar

Outcome

Navistar is now able to provide proactive vehicle diagnostics and real-time monitoring services for its customers. Cloudera Enterprise enabled Navistar to build a highly scalable real-time IoT platform to derive valuable insights from multiple data sources. The platform helped Navistar customers reduce maintenance costs by up to 40% while early detection of vehicle issues also reduces vehicle downtime by up to 40%.

> *The results are overwhelmingly positive. Using real-time big data to frame business decisions and deploy proactive maintenance has opened new revenue streams and delivered additional customer value.*
>
> —Troy Clarke, CEO, Navistar

Cerner

Cerner is a leader in the health care IT space, providing solutions to thousands of facilities, such as hospitals, ambulatory offices, and physicians' offices.

Use Cases

Cerner's goal is to consolidate the world's health care data into a common platform in order to reduce cost, increase efficiency of delivering health care, and improve patient outcomes. The project requires several challenges: The data must be secure, auditable and easy to explore.

> *Our vision is to bring all of this information into a common platform and then make sense of it – and it turns out, this is actually a very challenging problem.*[ii]
>
> —David Edwards, Vice President and Fellow, Cerner

Solution

Cerner accomplished its goal by implementing a comprehensive view of population health powered by Cloudera Enterprise. The big data platform currently stores two petabytes of data, ingesting data from multiple sources such as electronic medical records (EMR), HL7 feeds, Health Information Exchange information, claims data, and custom extracts from different several proprietary and client-owned data sources. Cerner uses Apache Kafka to ingest real-time data into HBase or HDFS using Apache Storm. Cerner is exploring augmenting its platform with other real-time components such as Apache Flume, Apache Samza, and Apache Spark.[iii]

Data is transferred from Cloudera platform to Cerner's data mart running HP Vertica, providing access to SAS and SAP Business Objects users for reporting and analysis. Cerner utilizes the data to help them determine risks and opportunities for improvement across a population of people. Cerner also leverages SAS on Cloudera Enterprise for data science initiatives such as building prediction models for avoiding hospital readmissions. The Cerner team is evaluating Cloudera Search (Solr) and Impala to allow hundreds of users across the organization direct access to data stored in Cloudera Enterprise. Security is extremely important for Cerner and choosing Cloudera, one of the most secure Hadoop distributions in the market as their big data platform, it gave them confidence that patient data will be secure and completely protected.

> We're able to achieve much better outcomes, both patient-related and financial, than we ever could by just looking at pieces of the puzzle individually. It all comes down to bringing everything together and being able to extract value for any requirement. The enterprise data hub topology allows us to do exactly that.
>
> —Ryan Brush, Senior Director and Distinguished Engineer, Cerner

Technology and Applications

- Hadoop Platform: Cloudera Enterprise, Data Hub Edition
- Components in Use: Apache Crunch, Apache HBase, Apache Hive, Apache Kafka, Apache Oozie, Apache Storm, Cloudera Manager, MapReduce
- Servers: HP
- Data Mart: HP Vertica
- BI and Analytics Tools: SAP Business Objects, SAS

Outcome

By consolidating health care data from multiple sources, Cerner was able to get a far more complete view of any patient, trend, or condition, helping them achieve much better patient-related and financial outcomes. For example, the big data platform gave them the ability to predict the probability of patient re-admission. Utilizing the same platform, Cerner was also able to accurately predict early onset of sepsis in patients.

> *Our clients are reporting that the new system has actually saved hundreds of lives by being able to predict if a patient is septic more effectively than they could before.*
>
> —Ryan Brush, Senior Director and Distinguished Engineer, Cerner

British Telecom

BT is one of the leading telecommunications companies in the United Kingdom with over 18 million customers and operations in 180 countries.

Use Cases

Like every organization, British Telecom is required to provide its business units with the most relevant and up-to-date information. Their legacy ETL system, built on traditional relational databases, could not scale and can barely process close to one billion rows of data in a timely manner. Their ETL jobs were taking more than 24 hours to process 24 hours of data. Its business units had to content with working with day-old data.[iv]

> *We had a proposal to re-platform the system to a new relational database. But as we sat down, our discussion turned to Hadoop. We realized we basically had a data velocity problem. We had to process the data faster and increase the volume that we could ingest—both of which Hadoop excels at.*
>
> —Phillip Radley, Chief Data Architect, BT

Solution

BT implemented a Cloudera Enterprise cluster and replaced their batch ETL jobs with MapReduce code. The platform not only solved BT's ETL problem, but also addressed other data management challenges to help BT accelerate the delivery of new product offerings.

Because data is consolidated in a single, cost-effective infrastructure, it enabled BT to gain a unified 360-view of its data across its multiple business units. The platform will also enable BT to archive data longer from 1 year to more than 10 years and implement mission-critical data management and analytic use cases.

Soon, the company plans to use Apache Spark to combine batch, streaming, and interactive analytics, and Impala enables the business intelligence (BI) teams to perform SQL queries on the data.

Technology and Applications

- Hadoop Platform: Cloudera Enterprise, Data Hub Edition
- Hadoop Components: Apache Hive, Apache Pig, Apache Sentry, Apache Spark, Cloudera Manager, Cloudera Navigator, Impala

Outcome

Moving ETL and data processing to Hadoop enabled BT to increase data velocity, providing business users with the information they need when they need it.

- Processes 5x more customer data
- Increased data velocity by 15x
- Delivered ROI of 200–250% in one year
- The move also delivered substantial cost savings for BT

> *We were able to increase data velocity by a factor of 15. We're processing five times the data in a third of the time. The business sponsors don't know that we moved to Hadoop and they don't care. All they know is that they're now working with today's data instead of yesterdays.*
>
> —Phillip Radley, Chief Data Architect, BT

CHAPTER 13 BIG DATA CASE STUDIES

Shopzilla (Connexity)

Shopzilla is a leading e-commerce company headquartered in Los Angeles, California, with 100 million unique visitors connected to 100 million products from tens of thousands of retailers.[v]

Use Cases

Shopzilla has an existing 500-terabyte Oracle Enterprise Data Warehouse that's growing 5 terabytes a day. With the amount of data and processing required to crunch through 100 million products per day, Shopzilla's legacy data warehouse has exceeded its capacity and was unable to scale further, taking hours to process data per day.

Solution

Shopzilla implemented a hybrid environment by complementing its Oracle Enterprise Data Warehouse with a Cloudera Enterprise cluster. Low-value ETL and data processing is handled by the CDH cluster. Using Apache Sqoop, aggregated data is then transferred to the Oracle EDW, freeing it to do what it was designed to do, serving analytics and reports to business users. Shopzilla plans to utilize Apache Impala and Apache Spark in the near future.[vi] The CDH cluster is utilized to support online price comparison services, SEO, SEM, merchandising, audience scoring, and data science workloads.

> *Data scientists don't typically need to consume data warehouse resources now because all of the most recent data is available in Cloudera via R or Mahout. We needed enormous processing capabilities, scalability, full redundancy, and extensive storage – all at a cost-effective price. Our Cloudera platform provides all that and more.*
>
> —Rony Sawdayi, Vice President, Engineering, Connexity

> *We are able to answer complex questions, such as how a user is behaving on a particular site and what ads would be most effective, as well as execute other sophisticated data mining queries. It improves Connexity's ability to provide relevant results to users, and this is a core tenet of our business.*
>
> —Paramjit Singh, Director of Data, Connexity

Technology and Applications

- Data Platform: Cloudera Enterprise

- Hadoop Components: Apache HBase, Apache Hive, Apache Mahout, Apache Pig, Apache Spark, Apache Sqoop, Cloudera Impala, Cloudera Manager

- Servers: Dell

- EDW: Oracle

- BI & Analytic Tools: Oracle BI Enterprise Edition (OBIEE); R

Outcome

With Cloudera Enterprise, Connexity can now process data from 15,000 feeds and 100 million products from retailers in a matter of hours instead of several days. A new architecture is being tested and will further decreases processing time to minutes. The faster performance also enables Connexity to score and bid on 10 million keywords every day,[vii] enabling its search engine marketing activities to scale and reach 100 million unique visitors and collect billions of data points that can be utilized for highly targeted marketing and innovative data analytics.

> *Our legacy system delivers great performance for analytics and reporting, but didn't have the bandwidth for the intensive data transformations we needed – it would take hours to process 100 million products per day. We needed enormous processing capabilities, scalability, full redundancy, and extensive storage – at a cost-effective price. Our Cloudera platform provides all that and more, while complementing our current data warehouse system. We were able to reduce latency from days to hours and soon minutes.*
>
> —Paramjit Singh, Director of Data, Connexity

Thomson Reuters

Thomson Reuters is a leading mass media and information corporation that provides professionals with trusted information.

Use Cases

Thomson Reuters aims to classify tweets and distinguish fake news and opinions from real news in 40 milliseconds.[viii]

Solution

Thomson Reuters turned to machine learning and advanced analytics to build Reuters Tracer, a "bot journalist in training," Reuters Tracer analyzes 13 million tweets every day, processing events to determine if the tweet is real news or an opinion or fake news.[ix] Thomson Reuters uses Cloudera Enterprise and Apache Spark to provide machine learning capabilities needed to implement Reuters Tracer. Spark's fast in-memory features enables Reuter Tracer to process and derive meaning from millions of tweets in just 40 milliseconds.

> *To assist in evaluating the veracity of an event, we rely on hundreds of features and have trained the platform to look at the history and diversity of sources, the language used in tweets, propagation patterns, and much more, just as an investigative journalist would do.*
>
> —Sameena Shah, Director of Research and Lead Scientist on Reuters Tracer

> *Cloudera provides us with state-of-the-art technology to help us analyze data, synthesize text, and extract value and meaning from data to deliver the insights that our customers are looking for. The whole application is very fast. It takes less than 40 milliseconds to capture and detect events.*
>
> —Khalid Al-Kofahi, Head, Corporate Research & Development, Thomson Reuters

Technology and Applications

- Data Platform: Cloudera Enterprise
- Workloads: Data Science & Engineering
- Hadoop Components: Apache Spark

CHAPTER 13   BIG DATA CASE STUDIES

Outcome

- Revealed news worthy events ahead of major news outlets
- Distinguishes newsworthy tweets from rumors and fake news across 13 million tweets in 40 milliseconds

We are in the business of building information-based solutions for our professional customers in the financial, legal, tax, and accounting industries, and for Reuters, one of the leading news organizations. With Reuters Tracer, we can alert our customers when market-moving events happen as they are reported, without delays. We have dozens and dozens of examples where Reuters Tracer discovered ground-breaking events ahead of major news organizations. Additionally, because we help journalists discover events, they can focus on higher value-add work as opposed to just reporting on events.

—Khalid Al-Kofahi, Head, Corporate Research & Development, Thomson Reuters

Mastercard

Mastercard is a leader in global payments that connects billions of consumers and millions of organizations around the world.

Use Cases

Mastercard built an anti-fraud system called MATCH (Mastercard Alert to Control High-risk Merchants) that allows users to search Mastercard's proprietary database containing hundreds of millions of fraudulent businesses. As time went by, it became evident that MATCH's phonetic-based lookup feature could not provide the versatility to satisfy the growing needs of MATCH users. Additionally, the relational database management system (RDBMS) that is powering MATCH could not keep up with the growing volume of data.[x]

Solution

Mastercard implemented a new anti-fraud solution based on Cloudera Search (powered by Apache Solr), an integrated part of CDH that provides full-text search and faceted navigation. Cloudera Search provided increased scalability, richer search functionality, and better search accuracy. The new solution can use several search algorithms and new scoring capabilities that were previously hard to implement on their legacy RDBMS. The new platform will also allow Mastercard to add more data sets as opportunities arise.

Technology and Applications

- Apache Hadoop Platform: Cloudera Enterprise, Data Hub Edition
- Apache Hadoop Components: Apache Solr, Cloudera Search, Hue

Outcome

The new Cloudera-based solution is helping Mastercard easily identify fraudulent merchants to reduce risk. Mastercard users experienced dramatically improved search accuracy, increasing the number of supported search annually 5X, with 25X increase in searches per customer per day. This has allowed Mastercard to expand to new markets resulting to increase in revenue.

Summary

My goal is to provide inspiration to encourage you to start your own big data use cases using effective and proven methodologies. I hope you found this chapter useful.

References

i. Cloudera; "Navistar: Reducing Maintenance Costs more than 30 percent for Connected Vehicles," Cloudera, 2018, https://www.cloudera.com/more/customers/navistar.html

ii. Cloudera; "Cerner: Saving Lives with Big Data Analytics that Predict Patient Conditions," Cloudera, 2018, https://www.cloudera.com/more/customers/cerner.html

iii. Cloudera; "Cloudera Cerner Case Study: Saving Lives with Big Data Analytics that Predict Patient Conditions," Cloudera, 2018, https://www.cloudera.com/content/dam/www/marketing/resources/case-studies/cloudera-cerner-casestudy.pdf.landing.html

iv. https://www.cloudera.com/more/customers/bt.html

v. https://www.cloudera.com/more/customers/connexity.html

vi. https://globenewswire.com/news-release/2014/08/05/656022/10092934/en/Shopzilla-Implements-a-Cloudera-Enterprise-Data-Hub-to-Enhance-its-EDW-and-Capture-Unparalleled-Retail-Insights.html

vii. https://www.cloudera.com/content/dam/www/marketing/resources/case-studies/connexity-complements-the-edw-with-cloudera-to-improve-retail-insights.pdf.landing.html

viii. https://www.cloudera.com/content/dam/www/marketing/resources/case-studies/Cloudera_Thomson_Reuters_Case_Study.pdf.landing.html

ix. https://www.cloudera.com/more/customers/thomson-reuters.html

x. https://www.cloudera.com/more/customers/mastercard.html

Index

A

Active-active dual ingest, Kafka
    MirrorMaker, 45–46
    Spark streaming, 45
    StreamSets, 46
Alluxio, 477
    administering
        master, 489
        worker, 490
    Apache Spark and, 489
    architecture, 478–479
    components
        client, 487
        primary master, 487
        secondary master, 487
        worker, 487
    installation, 487
    use
        big data processing performance and scalability, 480
        high availability and persistence, 482–485
        memory usage and minimize garbage collection, 486
        multiple frameworks and applications, 480, 482
        reduce hardware requirements, 486
Alteryx, 455
    Browse data tool, 461
    City field, 464
    CSV format, 464
    Customer Segment field, 463
    Input Data tool, 457
    Output Data tool, 464
    selecting files, 459
    Select tool, 459
    Sort tool, 460
    Tool Palette, 457
Amazon Elastic MapReduce (EMR), 531
Amazon Web Services (AWS), 507
    Cloudera on
        Amazon EMR, 531
        architecture, 517–518
        on Azure and GCP, 531
        Cloudera Altus, 531
        databricks, 532
        EBS, 516
        EC2 instance, 514–516
        ephemeral or instance storage, 516
        regions and availability zones, 513
        S3, 516–517
        security groups, 513
        using Cloudera Director, 518–531
        VPC, 513
Apache Geode, *see* Geode
Apache Hadoop platform, 7
Apache Ignite, *see* Ignite
Apache Impala, *see* Impala
Apache Kudu, *see* Kudu
Apache, *see* Spark

INDEX

B

Backup and Disaster Recovery (BDR), 36
Berkeley Data Analytics Stack (BDAS), 477
Big data, 1
Big data integration players
    Apache NIFI, 361
    IBM InfoSphere DataStage, 361
    Informatica, 360
    Oracle Data Integrator, 360
    SSIS, 360
    Syncsort, 361
Big data visualization
    architecture, 409–410
    deep integration with
        Apache Spark, 410
    real-time data visualization, 409
    SAS Visual Analytics, 408
    self-service BI and analytics, 408–409
    Zoomdata, 408
    Zoomdata Fusion, 411
Big Data warehousing 101
    dimensional modeling, 381
        dimension tables, 382
        facts, 381
        slowly changing dimensions, 384
        snowflake schema, 383
        star schema, 382
    with Impala and Kudu, 384–386, 404
        DimCustomer, 387
        DimDate, 389
        dimensions tables, 400
        DimProduct, 388
        example, 402–405
        function uuid(), 399
        SQL Server, 390–392
        structure of Kudu tables, 400–402
        tables, 392–393, 395–396, 398

BIGINT data type, 17
British Telecom (BT), 541
    outcome, 542
    solution, 542
    technology and applications, 542
    use cases, 541

C

Cask Data Application
        Platform (CDAP), 41, 290
Cerner, 539
    outcome, 541
    solution, 539
    technology and applications, 540
    use cases, 539
Cloudera Enterprise, 447, 509
Cloudera Enterprise Backup and Disaster
        Recovery (BDR), 36
Cloudera Navigator, 496–497
    auditing and access control, 500
    data classification, 499–500
    data lineage and impact analysis, 500
    Encrypt, 502
    metadata management, 498–499
    policy enforcement and data lifecycle
        automation, 501
    REST API, 502
    user interface, 497
CREATE TABLE AS (CTAS), 34–35

D

Dataflow Performance
        Manager (DPM), 289–290
Data governance, 495
    for big data, 496
    Cloudera Navigator, 496–497

INDEX

auditing and access control, 500
data classification, 499–500
data lineage and impact
   analysis, 500
Encrypt, 502
metadata management, 498–499
policy enforcement and data
   lifecycle automation, 501
REST API, 502
user interface, 497
tools
   Apache Atlas, 503
   Collibra, 503
   Informatica Metadata Manager and
      enterprise data catalog, 503
   Smartlogic, 504
   Waterline Data, 504
Data ingestion, 231
Data ingestion with native tools, 362
   Kudu and Spark, 362–365
      Flafka, 368
      Kafka, 367–368
      Spark Streaming, 369
   Sqoop, 369–370
Datameer, 466
   clustering, 471–472
   data fields, 467
   data file, 467
   data visualization, 474
   prediction field, 473
   Smart Analytics, 470
   spreadsheet, 469
Data sharpening, 411
   support for multiple data
      sources, 412
      charts, 419–421, 423–425
      data sources, 414

fields, 417
Kudu Impala Connector
   page, 414
login page, 413
to refresh, 417
scheduler, 418
tables, 415
Zoomdata, 412
Zoomdata map, 425
Data warehouse platforms, 1
Data wrangling, 290, 446
   activities, 446
   Alteryx, 455
      Browse data tool, 461
      City field, 464
      CSV format, 464
      Customer Segment field, 463
      Input Data tool, 457
      Output Data tool, 464
      selecting files, 459
      Select tool, 459
      Sort tool, 460
      Tool Palette, 457
   Datameer, 466
      clustering, 471–472
      data fields, 467
      data file, 467
      data visualization, 474
      prediction field, 473
      Smart Analytics, 470
      spreadsheet, 469
   Trifacta, 447
      data distribution, 448
      data transformation, 451–454
      results, 455
      suggestions, 450
      transformer page, 448

551

INDEX

E, F

Elastic Block Storage (EBS), 516
Enterprise Data Warehouse (EDW), 375
    era of big data, 376
    modernization, 376
        analytics offloading and active archiving, 379
        data consolidation, 379–380
        ETL offloading, 378
        Impala and Kudu *vs.* traditional data warehouse platform, 377
    replatforming, 380
ETL offloading, 11
ETL tools
    CDAP, 41
    Pentaho, 42
    Talend, 42

G

Geode, 491
Google Cloud Platform (GCP), 508–509
GraphX, 152
GridGain Systems, 490

H

Hadoop platforms, 3
Hadoop User Experience (HUE), 96–97
Hash partitioning, 17
Hash-range partitioning, 18–19
HBase, 9, 12
HDFS, 10
Hive, 384
Hybrid and multi-cloud, 509–510

I, J

Ignite, 490
Impala, 2–3, 11, 57
    architecture, 57–58
        Amazon S3, 61
        catalog service, 59
        daemon, 58
        file format, 62
        Hadoop Ecosystem, 59
        HBase, 60–61
        HDFS, 59–60
        Hive, 59
        Kudu, 62
        Statestore, 59
    complex types, 76
        querying deeply nested collections, 78
        querying struct fields, 77
        querying using ANSI-92 SQL, 79
    in enterprise, 98
    external tables, 104
    HUE, 96–97
    internal tables, 103
    JDBC with Apache, 109
    and Kudu
        data types, 16
        hash partitioning, 17
        hash-range partitioning, 18–19
        integration works, 15
        range partitioning, 18
        table partitioning, 17
        TIMESTAMP, 17
        uuid() function, 15
    performance recommendations
        creating aggregate or summary tables, 95
        denormalization, 94

Parquet, 94
small files problem, 94
statistics, 95
tables partitioning, 94
performance tuning and monitoring, 84
Cloudera Manager, 87–88, 90–93
explaining, 85
profile, 86
summary, 85
shell, 79–82, 84
SQL
data types, 63
Server and Oracle, 110–111
statements (*see* Statements, SQL)
UDFs, 76
workload and resource management, 95
Admission Control, 95–96
Impyla, 43
Internet of Things (IoT), 2–3, 11, 231, 236, 426
Internet Protocol (IP), 477

K

Kafka, active-active dual ingest with
Spark streaming, 45
StreamSets, 46
using MirrorMaker, 45–46
Kudu, 2–3, 101, 159
active-active dual ingest, 45
backup via CTAS, 34–35
C++ client API, 29
changing data
deleting rows, 105
inserting rows, 104
updating rows, 105
upserting rows, 105

changing schema, 106
client API, 36–37, 370
Cloudera Manager, 47
cluster, 35
concepts, 12–13
data types, 102
ETL tools (*see* ETL tools)
file system, 49
high availability tools, 44
Java Client API, 24
JDBC with Apache, 109
Lambda architecture, 8–9
limitations, 51
loadgen, 49
MapReduce, 370
master, 49
master-slave architecture, 13–14
Master Web UI, 47
metrics, 48
NFS/SAN volume, 36
partitioning, 106
hash, 106
hash-hash, 108
hash-range, 107
list, 108
range, 106
primary key, 101
Python client API, 27
relational database (*see* Relational data management)
security, 52
StreamSets, 40–41
table, 50
tablets, 50
Tablet Server, 51
Tablet Server Web UI, 48
validate cluster health, 48

INDEX

Kudu context, 160
    deleting data, 164
    feature store for Spark
        MLlib, 222–228
    into HBase, 205–208
    inserting
        Amazon S3 into, 195–196
        data, 161–162
        HBase into, 188–193
        JSON into, 171–172
        MySQL into, 173–177
        Solr into, 194
        SQL Server into, 178–181, 183–185, 187
        XML into spark-xml package, 168–170
    inserting CSV into, 166
        programmatically specifying schema, 167
        using spark-csv package, 166
    Kudu table, 165
    into MySQL, 196–198
    into Oracle, 201–205
    rows to Parquet, 208–209
    selecting data, 165
    Spark Streaming and Kudu, 218–221
    SQL and Oracle dataframes, 210, 212
    into SQL Server, 198–201
    and SQL Server dataframes into Oracle, 214–217
    upserting data, 163
Kudu Master Web UI, 47
Kudu Tablet Server Web UI, 48

L

Lambda architecture, 8–9

M

Machine learning platforms, 12
MapReduce, 370
Massively Parallel Processing (MPP), 375
Mastercard, 546
    outcome, 547
    solution, 547
    technology and applications, 547
    use cases, 546
Microsoft Azure services, 507
MirrorMaker, 45–46
Multi-version concurrency control (MVCC), 14–15

N

Navistar, 537
    outcome, 539
    solution, 538
    technology and applications, 538
    use cases, 537
Next-generation big data integration tools
    data ingestion to Kudu with transformation, 328–331
    data ingestion with Kudu, 290–295, 297, 299–300, 302–306
    data transformation, 355–357, 359
    ingest CSV into HDFS and Kudu, 306–307, 309–310, 312, 314–320, 322–325, 327
    ingesting CSV files to Kudu, 342–347, 349
    PDI, 306
    SQL Server to Kudu, 331, 333–341, 349–352, 354–355
    Talend Open Studio, 341

INDEX

O

Online analytic processing (OLAP), 10
Online transaction processing (OLTP), 445

P, Q

Parquet, 16
Pentaho Data Integration (PDI), 306
Pentaho PDI, 41–42
Persistent clusters, 510
    Cloudera Director, 511
        architecture, 511–512
        on AWS (*see* Amazon Web Services (AWS), Cloudera on)
        client, 512
        REST API, 513
pyodbc, 44
Python package
    Impyla, 43
    pyodbc, 44
    SQLAlchemy, 44

R

Raft Consensus algorithm, 13
Range partitioning, 18
RDD, 119
    caching, 127
    creating, 120
        actions, 126
        coalesce, 125
        collect, 126
        count, 126
        distinct values, 122
        filter, 121
        foreach, 127
        inner join, 123
        keys, 122
        map, 120–121
        parallelize, 120
        ReduceByKey, 122
        repartition, 125
        Right Outer Join/Left Outer Join, 124
        subtract, 124
        take, 126
        textfile, 120
        transformation, 120
        union, 124
        values, 123
    lazy evaluation, 127
Real-time data visualization, 409
Real-time IoT, 426
    architecture, 426
    Kudu table, 426–427
        StreamSets pipeline, 430–435
        test data source, 428–429
        Zoomdata, 436, 438–439, 441–445
Relational database management systems, 1
Relational data management
    data consolidation, 11
    data warehousing, 10
    ETL offloading, 11
Row-level versioning, 14

S

SAS Visual Analytics, 408
Shopzilla (Connexity), 543
    outcome, 544
    solution, 543
    technology and applications, 544
    use cases, 543

INDEX

Smart Analytics, 470
Spark, 113, 159
    Amazon S3, 142
    applications, 116
    architecture, 115–116
    cluster managers, 114
    data sources, 129
        CSV files, 129
        JSON file, 131–132
        XML, 130–131
    directed acyclic graph (DAG), 2
    HBase, 136–141
    Hive on, 152
    Microsoft Excel, 143–144
    MLlib (DataFrame-based API), 145–146
        CrossValidator, 147
        estimator, 146
        evaluator, 147
        example, 147, 149–151
        GraphX, 152
        ParamGridBuilder, 147
        pipeline, 146
        transformer, 146
    monitoring and configuration
        Cloudera Manager, 153
        Web UI, 154–156
    overview, 113–114
    Parquet, 136
    relational databases
        using JDBC, 132–135
    Secure FTP, 144
    shell, 117–118
        accumulators, 119
        broadcast variables, 119
        RDD (*see* RDD)
        SparkSession, 118
    Solr, 142–143
    SQL, dataset, and dataframes, 128
    Streaming, 152
    1.x *vs.* 2.x, 152
    YARN, 116
        client mode, 117
        cluster mode, 116
Spark 1.6.x, 159
Spark 2.x, 160
Spark and Kudu
    back up data using, 38
    context, 19–20
    DataFrame API, 19
    Data Source API, 38–39
    Flafka pipeline, 24
    flume configuration file, 23–24
    streaming, 21, 23
SQLAlchemy, 44
SQL Server Integration Services (SSIS), 360
Statements, SQL
    AND and OR, 66
    built-in functions
        abs function, 75
        abs function, 75
        fnv_hash(type v), 75
        now function, 75
        regexp_like function, 75
        uuid function, 74
    create database, 64
    create external table, 65
    create table, 64
    DESCRIBE, 70
    DISTINCT, 68
    GROUP BY and HAVING, 68
    INVALIDATE METADATA, 70
    JOIN, 69
    LIKE, 67
    LIMIT, 67
    LOAD DATA, 70
    ORDER BY, 67

refresh, 71
SELECT, 65
SET, 71
    BATCH_SIZE, 72
    LIVE_PROGRESS, 71
    MEN_LIMIT, 71
    NUM_NODES, 71
SHOW, 72
    SHOW DATABASES, 72
    SHOW FILES, 73
    SHOW TABLES, 72
    SHOW TABLE TATS, 73
subquery, 70
UNION ALL, 69
UNION and UNION DISTINCT, 69
WHERE, 66
StreamSets, 45–46
StreamSets Data Collector, 231
    console, 233
        batch-oriented data ingestion, 235–236
        IoT, 236
        real-time streaming, 234–235
    deployment options, 237
    destinations, 232
    Directory origin stage, 243–245
    DPM, 289–290
    Event Framework, 289
    executors, 233
    Expression Evaluator, 265–268, 270, 273–274
    ingesting into Kudu clusters, 281–286
    ingesting XML to Kudu, 238–242
    JavaScript evaluator, 274–281
    Origins, 232
    pipeline, 232
        configuration, 242–243
        starting, 251–254
    processors, 232
    REST API, 286–289
    stream selector, 255–257, 259, 261–265
    using, 237–238
    XML parser processor, 246, 248–251
StreamSets tool, 40–41
Structured data, 376
Symmetric Multiprocessing (SMP), 375

T

Talend Kudu, 43
Thomson Reuters, 544
    outcome, 546
    solution, 545
    technology and applications, 545
    use cases, 545
Time series applications, 11
Transient clusters, 510
Trifacta, 447
    data distribution, 448
    data transformation, 451–454
    results, 455
    suggestions, 450
    transformer page, 448

U

User-defined aggregate functions (UDAFs), 76
User-defined functions (UDFs), 76

V, W, X, Y

Virtual Private Cloud (VPC), 513

Z

Zoomdata, 218, 408
Zoomdata Fusion, 411

CPSIA information can be obtained
at www.ICGtesting.com
Printed in the USA
LVHW01s0256120718
583386LV00006B/232/P